GEOCOMPUTATION

For Melissa, Erin, Theo and Max

CHRIS BRUNSDON and **ALEX SINGLETON**

GEOCOMPUTATION

A PRACTICAL PRIMER

Los Angeles | London | New Delhi
Singapore | Washington DC

Los Angeles | London | New Delhi
Singapore | Washington DC

SAGE Publications Ltd
1 Oliver's Yard
55 City Road
London EC1Y 1SP

SAGE Publications Inc.
2455 Teller Road
Thousand Oaks, California 91320

SAGE Publications India Pvt Ltd
B 1/I 1 Mohan Cooperative Industrial Area
Mathura Road
New Delhi 110 044

SAGE Publications Asia-Pacific Pte Ltd
3 Church Street
#10-04 Samsung Hub
Singapore 049483

Editor: Robert Rojek
Assistant editors: Keri Dickens and Matt Oldfield
Production editor: Katherine Haw
Copyeditor: Richard Leigh
Proofreader: Neil Dowden
Marketing manager: Michael Ainsley
Cover design: Francis Kenney
Typeset by: C&M Digitals (P) Ltd, Chennai, India
Printed and bound by Ashford Colour Press Ltd

Editorial arrangement, Introduction, Conclusion, and Chapter 15
© Chris Brunsdon and Alex Singleton 2015

Chapter 1 © James Cheshire and Robin Lovelace 2015
Chapter 2 © Paul M. Torrens 2015
Chapter 3 © Michael Batty 2015
Chapter 4 © Andrew Crooks 2015
Chapter 5 © Mark Birkin 2015
Chapter 6 © Harvey J. Miller 2015
Chapter 7 © David Rohde and Jonathan Corcoran 2015
Chapter 8 © Alexandros Alexiou and Alex Singleton 2015
Chapter 9 © Seth Spielman and David C. Folch 2015
Chapter 10 © Daniel Lewis 2015
Chapter 11 © Melanie Tomintz, Graham Clarke and Nawaf Alfadhli 2015
Chapter 12 © Tomoki Nakaya 2015
Chapter 13 © Karyn Morrissey 2015
Chapter 14 © Sergio J. Rey 2015
Chapter 16 © Chris Brunsdon and Lex Comber 2015
Chapter 17 © Oliver O'Brien 2015
Chapter 18 © Richard Kingston 2015

First published 2015

Library of Congress Control Number: 2014940792

British Library Cataloguing in Publication data

A catalogue record for this book is available from the British Library

MIX
Paper from
responsible sources
FSC® C011748

ISBN 978-1-4462-7292-3
ISBN 978-1-4462-7293-0 (pbk)

At SAGE we take sustainability seriously. Most of our products are printed in the UK using FSC papers and boards. When we print overseas we ensure sustainable papers are used as measured by the Egmont grading system. We undertake an annual audit to monitor our sustainability.

CONTENTS

About the Authors   vii

Preface   xiii

Introduction   **xv**
Chris Brunsdon and Alex Singleton

Further Resources   xxi

PART I   DESCRIBING HOW THE WORLD LOOKS   **1**

1 **Spatial Data Visualisation with R**   3
   James Cheshire and Robin Lovelace

2 **Geographical Agents in Three Dimensions**   21
   Paul M. Torrens

3 **Scale, Power Laws, and Rank Size in Spatial Analysis**   40
   Michael Batty

PART II   EXPLORING MOVEMENTS IN SPACE   **61**

4 **Agent-Based Modeling and Geographical Information Systems**   63
   Andrew Crooks

5 **Microsimulation Modelling for Social Scientists**   78
   Kirk Harland and Mark Birkin

6 **Spatio-Temporal Knowledge Discovery**   97
   Harvey J. Miller

7 **Circular Statistics**   110
   David Rohde and Jonathan Corcoran

PART III   MAKING GEOGRAPHICAL DECISIONS   **135**

8 **Geodemographic Analysis**   137
   Alexandros Alexiou and Alex Singleton

9 Social Area Analysis and Self-Organizing Maps   152
   Seth Spielman and David C. Folch

10 Kernel Density Estimation and Percent Volume Contours   169
    Daniel Lewis

11 Location–Allocation Models   185
    Melanie Tomintz, Graham Clarke and Nawaf Alfadhli

PART IV EXPLAINING HOW THE WORLD WORKS   199

12 Geographically Weighted Generalised Linear Modelling   201
    Tomoki Nakaya

13 Spatial Interaction Models   221
    Karyn Morrissey

14 Python Spatial Analysis Library (PySAL): An Update
    and Illustration   233
    Sergio J. Rey

15 Reproducible Research: Concepts, Techniques and Issues   254
    Chris Brunsdon and Alex Singleton

PART V ENABLING INTERACTIONS   265

16 Using Crowd-Sourced Information to Analyse Changes
    in the Onset of the North American Spring   267
    Chris Brunsdon and Lex Comber

17 Open Source GIS Software   281
    Oliver O'Brien

18 Public Participation in Geocomputation to Support
    Spatial Decision-Making   301
    Richard Kingston

Conclusion: The Future of Applied Geocomputation   320
    Chris Brunsdon and Alex Singleton

References   327
Index   364

ABOUT THE AUTHORS

Alexandros Alexiou holds a degree in Spatial Planning and Survey Engineering from the University of Thessaly, as well as an MSc in Transport Planning from the Aristotle University of Thessaloniki and an MPhil in Planning from the University of Cambridge. Alexandros is currently a PhD candidate at the University of Liverpool, with research interests in the creation of new models of urban socio-spatial structure that better account for both geographic context and the dynamics of population.

Nawaf Alfadhli is a PhD candidate at the School of Geography, University of Leeds. His PhD is on the subject of the deployment of GIS techniques within the crime field in the local authority of Kuwait. By using a mixture of location-allocation and spatial interaction models Nawaf is building a decision support system for use by local planners in Kuwait to help them monitor, analyse and prevent criminal activity through the better deployment of resources over space and time.

Michael Batty is Professor of Urban Planning at University College London where he is Chair of the Centre for Advanced Spatial Analysis (CASA). His research and the work of his centre is focused on computer models of city systems. In recent years he has been awarded the William Alonso Prize of the Regional Science Association, the University Consortium GIS Research Award and the Lauréat Prix International de Géographie Vautrin Lud. His research has focused on the development of analytical methods and computer models for simulating the structure of cities and regions. His work began with aggregate land use transport models, and he then moved into more visual representations of cities and their models, more recently focusing on complexity theory as reflected in his blogs www.complexcity.info and www.spatialcomplexity.info.

Mark Birkin is Professor of Spatial Analysis and Policy in the School of Geography at the University of Leeds. His major interests are in simulating social and demographic change within cities and regions, and in understanding the impact of these changes on the need for services such as housing, roads and hospitals, using techniques of microsimulation, agent-based modelling and GIS. He is currently the project leader for TALISMAN – the spatial data analysis and simulation node of the ESRC's National Centre for Research Methods.

Chris Brunsdon is Professor of Geocomputation and Director of the National Centre for Geocomputation at the National University of Ireland, Maynooth. He obtained his BSc at the University of Durham and his MSc and PhD at the University of Newcastle upon Tyne. He has been involved in a number of spatial analysis and GIS projects in the UK. His research interests include: reproducible research; exploratory spatial data analysis and visualisation; spatial statistics; Bayesian approaches; health data analysis and the analysis of crime patterns. He is on the editorial boards of a number of journals and has sat on a number of assessment panels for research funding institutions.

James Cheshire completed a BSc in Physical Geography at the University of Southampton before undertaking his PhD in Geographical Information Science at the Department of Geography of University College London (UCL). His topic concerned the spatial analysis and visualisation of large surname databases. Since completing his PhD he has been a Lecturer in Advanced Spatial Analysis and Visualisation at CASA and is now a Lecturer in Quantitative Human Geography at UCL. Aside from his academic outputs, a wide range of his maps and visualisations have been featured in the popular print press (such as the *National Geographic* and the *Guardian*) as well as online. He enjoys blogging both for spatial.ly and mappinglondon.co.uk.

Graham Clarke is Professor of Business Geography at the University of Leeds. He has worked extensively in various areas of GIS and applied spatial modelling, focusing on many applications within urban/social geography. A major research interest has been spatial microsimulation, a technique for estimating 'missing data' and producing detailed household datasets for use in a wealth of social science simulations (especially income, wealth, crime and health applications). Graham has also been very active in retail geography and retail store location research. A major contribution to research in retail location planning has been to rethink the research agenda on store location in relation to the multi-channel growth strategies of retail organisations. An important additional research stream has been work on retail provision, saturation and internationalisation.

Jonathan Corcoran is an Associate Professor in Human Geography within the School of Geography, Planning and Environmental Management at the University of Queensland. His research interests focus around the application of quantitative geographical methods for urban modelling allied with the development and application of analytical, visualisation and prediction techniques. He has worked closely with a range of government agencies to both inform and evaluate operational and strategic planning through the development of geographic-based tools.

Andrew Crooks is an Assistant Professor in the Department of Computational Social Science at George Mason University. His research interests relate to exploring, understanding and the communication of the natural and socio-economic environ-ments using GIS, spatial analysis, Web 2.0 technologies, social network analysis and agent-based modelling methodologies. His web page is http://gisagents.org.

David C. Folch is an Assistant Professor in the Department of Geography at Florida State University. Before joining FSU he earned his PhD in Geography at Arizona State University and worked as a research associate at the University of Colorado Boulder. David's research focuses on geocomputation and spatial ana-lytical methods, with contextual interests in data uncertainty and neighborhood structure and change.

Kirk Harland is a Research Fellow in the Centre for Spatial Analysis and Policy at the University of Leeds. His primary research interests are the investiga-tion of changing demographic trends and the associated impacts on the provision of services in the education and health sectors. He is currently working on a framework for microsimulation and a new version of the GAM cluster hunter, originally developed by Professor Stan Openshaw. Following the completion of his first degree, Kirk worked on large commercial spatial modelling projects for GMAP Consulting Ltd before returning to study for his PhD in spatial modelling and education planning. Having completed his PhD, he joined the Health and Social Care Information Centre of the NHS as a Principal Information Analyst before taking on the role of Head of Clinical Indicator Development.

Richard Kingston is a Senior Lecturer in Urban Planning and Smart Cities in the School of Environment, Education and Development and Deputy Director of the Centre for Urban Policy Studies at the University of Manchester. For the past 18 years he has been researching, developing, testing and implementing web-based planning support systems. Further details can be found on Richard's website at www.ppgis.manchester.ac.uk.

Daniel Lewis is a Research Fellow in the Department of Social and Environmental Health Research at London School of Hygiene and Tropical Medicine. He has a BA in Geography from the London School of Economics, and an MSc in Geographic Information Science and PhD in Geography from University College London. Daniel is a health geographer who is interested in applying the kind of geocomputation techniques expounded in this book to real-world data. Currently, Daniel works on the Olympic Regeneration in East London project, which aims to evaluate the impact of the Olympic legacy of urban regeneration on the health and well-being of young people and their parents in East London.

Robin Lovelace is a quantitative geographer and environmental scientist at the University of Leeds. As research fellow on the TALISMAN project for geocomputation, funded by the National Centre for Research Methods, he has developed teaching materials on R, QGIS and OpenStreetMap. He is currently working on a book that builds on a two-day course, 'An Introduction to Spatial Microsimulation with R'. Robin is committed to applying new data sources and methods to modelling the future including transition away from fossil fuels.

Harvey J. Miller is the Bob and Mary Reusche Chair in Geographic Information Science and Professor of Geography at the Ohio State University in Columbus. His research and teaching focus on the intersection between geographic information science and transportation science. Harvey seeks to understand how people use mobility and communications technologies to allocate scarce time among activities in geographic space – a perspective known as *time geography*. He is also interested in the social dimensions of transportation, and the implications of human mobility and accessibility for sustainable transportation, liveable communities and public health. His main approach to these questions is the development and application of GIS and spatial analytical techniques to extract information from fine-grained mobile objects and spatio-temporal data.

Karyn Morrissey is an economist by training, and her expertise lies in the development of geocomputational methodologies and their application for both health and environmental policy, as reflected by her published work using geocomputational quantitative methods to provide empirical research for public policy-makers and practitioners. With an interest in both health and environmental policy, she is currently embarking on an ESRC-funded collaboration with East Kent Hospital Foundation Trust to establish spatial hotspots for comorbidity across England using spatial microsimulation techniques and geovisualisation tools.

Tomoki Nakaya has research interests that include spatial statistics, geographical mathematical modelling, geovisualisation, spatial analysis of crime and analytical health geography. He joined Ritsumeikan University in 1997, and was made a full Professor there in 2012. He has been the co-director of the Institute of Disaster Mitigation for the Urban Cultural Heritage at the University since 2013. He has held visiting fellowships at the Universities of Leeds, Newcastle, Sheffield, University College London, and at the National University of Ireland, Maynooth. In July 2012 he was a joint winner of the Special Achievement in GIS Award at the Esri International User Conference.

Oliver O'Brien is a researcher and software developer at the Centre for Advanced Spatial Analysis (CASA), a research centre at University College London (UCL). He investigates and implements new ways to visualise spatial data. His current research interests include visualising city transport data, focusing particularly on public transport in London and bike-sharing systems across the world, as well as major UK public demographic datasets such as the Census, the Index of Multiple Deprivation and the Output Area Classification. After an initial career developing financial software, he studied for an MSc in GIS at City University, London, before joining UCL's Geography Department in 2008 and then moving to CASA. He is also a contributor to the OpenStreetMap project. In his leisure time, he competes in orienteering races and has drawn a number of orienteering maps for races in London. He blogs at obrien.com and mappinglondon, highlighting research outputs and the best maps and visualisations of London life.

Sergio Rey's research interests focus on the development, implementation and application of advanced methods of spatial and space–time data analysis. His substantive foci include regional inequality, convergence and growth dynamics as well as neighbourhood change, segregation dynamics, spatial criminology and industrial networks. Recent and current research projects include an analysis of the relationships between spatial linkages and urban economic dynamics (Economic Development Administration), flexible geospatial visual analytics and simulation technologies to enhance criminal justice decision support systems (National Institute of Justice), a spatial analytical framework for examining community sex-offender residency issues over space and time (National Science Foundation), and cyberGIS software integration for sustained geospatial innovation (National Science Foundation). Rey is the creator and lead developer of the open source package *STARS: Space–Time Analysis of Regional Systems* as well as *PySAL: A Python Library for Spatial Analysis*. He is a Fellow of the Spatial Econometrics Association, and since 1999 has served as Editor of the *International Regional Science Review*. In 2014 he became Editor of *Geographical Analysis*. For more information, see http://geoplan.asu.edu/rey.

David Rohde was awarded his PhD at the University of Queensland, applying machine-learning methods for combining information from large astronomical databases in 2009. He also worked extensively in GIS combining information from geographical datasets as both a research assistant and a postdoctoral research fellow also at University of Queensland at various times from 2007 to 2011. He worked as an analyst for the energy consulting company McLennon Magasanik Associates, modelling gas and electricity prices. In 2010–2011 he completed postdoctoral research at Telecom ParisTech on the online EM algorithm. He has previously worked as a Lecturer in Statistics at Universidade Federal do Rio de Janeiro. He is currently a postdoctoral researcher in Geography at the University of Queensland.

Alex Singleton holds an undergraduate degree from the University of Manchester in Geography, along with a PhD in Geography from University

College London (UCL). He previously held research positions at UCL, and is now a Reader in Geographic Information Science at the University of Liverpool. His research interests extend a geographic tradition of area classification and has developed a broad critique of the ways in which geodemographic methods can be refined through modern scientific approaches to data mining, geographic information science and quantitative human geography.

Seth Spielman is an Assistant Professor of Geography and an affiliate of the Institute of Behavioral Science at the University of Colorado at Boulder. As an urban geographer and an urban planner he studies how measures of the urban environment can be used to evaluate policy, assess the impact of urban development, and make inferences about the associations between the character of neighbourhoods and people's well-being – with the aim of informing urban policy. He argues that quantitative geographic descriptions of the urban landscape are inherently 'mappable' and research in this area can be strongly informed by developments in geographic information science and spatial statistics.

Melanie Tomintz is Senior Researcher and Lecturer at the Carinthia University of Applied Sciences (CUAS), Department of Geoinformation and Environmental Technologies, in Austria. She did her undergraduate studies at CUAS in the area of geoinformation that included a funded exchange study at Wageningen University and a funded one year internship at National Centre for Geocomputation, National University of Ireland, Maynooth. At the University of Leeds, School of Geography (joined with the University of Sheffield and University of York) she completed her PhD in the area of spatial health modelling and location–allocation models. At present, Melanie is research leader in health-GIS at CUAS and project leader of SALUD, using spatial microsimulation for health decision support in Austria. Her main research areas include geoinformation, spatial modelling and simulations, especially in the area of health and environment, usability evaluation and health information systems.

Paul M. Torrens is an Associate Professor in the Department of Geographical Sciences and the Institute for Advanced Computer Studies at the University of Maryland in College Park, where he directs the Center for Geospatial Information Science. Paul's work earned him a Faculty Early Career Development Award from the US National Science Foundation in 2007 and he was awarded the Presidential Early Career Award for Scientists and Engineers by President George W. Bush in a ceremony at the White House in 2008. The Presidential Early Career Award is the highest honour that the US government bestows upon young scientists; he is the first geographer to receive the award.

PREFACE

This edited book provides a readable introduction to the subject of geocomputation from an applied perspective. The established model for books on geocomputation in the past has been to focus on methodological progress. However, as techniques have matured, we would argue that there is a role for a more accessible and illustrative text, and specifically for those readers who are new to this area of research. The term 'geocomputation' arose in the mid-1990s to reflect an emerging intersection of research between advanced computational methods and geographical analysis and modelling (Openshaw and Abrahart, 1996). It has also been seen by some as a break away from the geographical information systems (GIS) research agenda. The early days of the 'quantitative revolution' in geography (Chorley and Haggett, 1967) saw geographers creating their own code (in languages such as FORTRAN) to address quantitative modelling and data analysis tasks applied to geographical questions. However, the advent of GIS software saw a trade-off between the flexibility of writing one's own code, and the speed and ease with which more commonly used methods could be applied. Early GIS systems came with a number of key data manipulation and visualisation functions 'out of the box', but lacked the flexibility to implement many other techniques, or to operate as a framework for developing and testing new techniques. Thus, for a period in the late 1980s and early 1990s, the focus in geographical data analysis shifted to the application of the toolkit of methods readily available in GIS, and away from the development of new techniques. On the positive side, this led to the introduction of a number of application areas, and to the development of critical GIS theories. However, to those more concerned with the development of quantitative approaches, this perhaps represented a retrograde step.

Arguably the coining of the term 'geocomputation' arose from a recognition that rather than being a natural progression from the quantitative revolution, it in fact represented a bifurcation, and the need to use more fundamental computing approaches (such as writing code) to solve certain kinds of geographical problems still existed. Thus, the tools of geocomputation are typically orientated towards bespoke software applications developed to implement complex methods, and often for addressing particular problems.

However, over time, the boundaries between the application of GIS and geocomputation have blurred. This in part has been a result of a loosening of the definition of what a GIS is, and the increasingly expansive capabilities of traditional GIS software. For example, we now find many traditional analytical capabilities of GIS are embedded within other software classes; for example, as extensions to relational databases (e.g. PostGIS) or programming languages (e.g. R or Python), as stand-alone libraries to perform specific functionality (e.g. Routino) or as software-based services hosted online and available for remote access (e.g. Google Maps API). Traditional GIS tools such as ArcGIS or the open source QGIS have expanded their analytical capabilities relative to early desktop GIS through the provision of application programming interfaces, enabling third-party developers to utilise both core functionality and expand upon these functions through software extensions. As such, we argue in this book that modern geocomputation can be encapsulated within multiple classes of software, yet will typically involve some element of customisation when utilising desktop GIS.

The book, therefore, aims to provide a portfolio of examples of geocomputation, by considering practical applications in those, and issues (such as reproducibility) that may arise through use. The book is structured into five collections of chapters (which will be outlined in detail in the Introduction) intended to focus in turn on real-world aspects of geocomputation.

INTRODUCTION

Chris Brunsdon and Alex Singleton

From the outset, geocomputation has been an applied and often interdisciplinary area of research, with methodological developments typically embedded within applications seeking to address real-world problems with a spatial dimension. In some sense this empirical focus differentiates geocomputation from more general computer science research, which has a greater prevalence of studying problems through synthetic data about abstract or theoretical problems. The embedding of geocomputation into real-world applications is illustrated well by an enduring example of early geocomputation research that developed the geographic analysis machine (GAM) (Openshaw et al., 1987). Openshaw et al. developed a method of detecting clusters within spatially distributed rare events, and tested it with an exploration of childhood leukaemia cases in the North of England. Through the application of the GAM software, five notable anomalies were identified, only one of which had previously been detected in other more traditional studies. Critically in this example, the analysis technique was developed in parallel with a real-world example utilising spatial data. Over time, the tradition of applied geocomputation has continued; for example, in a study of the Tokyo metropolitan area, Nakaya et al. (2005) used geographically weighted Poisson regression to demonstrate that there are significant spatial variations in the relationships between working-age mortality and both occupational segregation and unemployment; consequently, the application of traditional 'global' models would have yielded misleading results.

Geocomputation research is, however, more than software and methods applied to social science problems, and has developed a strong community, built organically by academics working within the field. In the contemporary period we would echo those earlier comments of Longley (1998: 9) that geocomputation is 'what its researchers and practitioners do, nothing more, nothing less'. Since 1996, geocomputation research has been represented by an international conference series of the same name (http://www.geocomputation.org/), but also, more generally, has outputs found within academic literature that span a wide variety of disciplinary backgrounds – for example, criminology (Hirschfield et al., 2013), transport (T. Cheng et al., 2013) and remote sensing (Comber et al., 2012).

The contemporary significance of applied geocomputation research can be observed in the context of wider developments occurring in cyberinfrastructures and complexity science. Cities are increasingly monitored through a variety of

mechanisms that generate vast quantities of spatio-temporal data about populations and the contexts in which they interact. Such 'big data' are generated
through multiple channels of cyberinfrastructure, including but not limited to
transport systems, administrative data collection, environment sensor networks,
commercial transactional data and volunteered geographic information, such as
social media data. The richness of these new data sources is increasingly utilised
in geocomputation research, addressing complex problems and with the aim of
making cities smarter.

The structure of this book

The topics contained in this book aim to be as comprehensive as possible, although
inevitably in such a text not all potential methods are presented. However, those
that are featured, we would argue, reflect the more prevalent areas of activity that
might be of interest to newcomers to geocomputation. Chapters are pitched
towards a readership of final-year undergraduate and postgraduate quantitative
social science students, and draw contributions from a series of leading international authors. The objective of the text is to present how complex real-world
problems can be solved by the integration of technology, data and established
geocomputational methods. The book is organised around five parts:

- Describing how the world looks
- Exploring movements in space
- Making geographical decisions
- Explaining how the world works
- Enabling interactions

The first stage in most empirical geographical analysis is to explore the data for
any interesting patterns, relationships and anomalies that may help guide the
specification of future explanatory models. However, exploring *how the world looks*
is not without challenges. Data are increasingly highly dimensional, and available
at finer spatial resolutions, which has inevitably created visualisation challenges.
Cheshire and Lovelace (Chapter 1) provide an introduction to the statistical programming language R, which is an increasingly common framework for geovisualisation. Comparisons are made with traditional desktop-based Geographical
Information Systems (GIS) software, citing the benefits of this tool in the context
of applied geocomputation and geographic data visualisation. Whereas Cheshire
and Lovelace focus on two-dimensional visualisation, Torrens (Chapter 2) explores
interactive three-dimensional visualisation linked to agent-based modelling
(ABM). This discussion provides both a theoretical and practical justification for
inclusion of a third dimension vis-à-vis traditional models. A modelling framework

is presented, alongside two illustrative case study applications in the context of dense three-dimensional crowds and the dynamic collapse of buildings. This part of the book concludes with a contribution from Batty (Chapter 3) exploring a 'rank clock' method of visualising change in more aggregate spatio-temporal data about cities.

Time represents a key explanatory component when investigating why certain geographical patterns emerge. As cities have become increasingly connected through enhanced communication and monitoring technologies, so have the opportunities to measure the space–time dynamics of their citizens. Such *movements in space* define cities, be it over short temporal scales such as the shifting population structures attributable to the commute to work, or over longer temporal scales such as the effects of migration trends on future patterns of health. Exploring movements in space is analytically challenging, constrained by issues of individual disclosure, large data volumes and appropriate representation of time in explanatory and exploratory models. Crooks (Chapter 4) introduces the ABM paradigm as a frame-work for investigating the effects of interactions of individuals and any emergent structures that may develop. By linking ABM to GIS and geocomputation software, geographically explicit processes can be simulated and investigated. A number of applications of ABM are outlined, giving examples of how this approach may be used to gain insight into a number of geographical processes. Harland and Birkin (Chapter 5) then present an overview of microsimulation modelling, differentiating between static, dynamic (transition and event) and spatial modelling approaches. In particular, explicit references are made to the differences between microsimulation and ABM techniques which are commonly conflated. Applications are presented over a variety of temporal scales, and the chapter concludes with an overview of those technical and computational considerations that might be considered when applying these modelling techniques. Miller (Chapter 6) introduces a series of methods by which knowledge can be extracted from spatio-temporal data. The chapter begins with an overview of different types of spatial, temporal and mobile objects, providing a high-level discussion of their conceptualisation and potential characteristics. This is followed by discussion of how such data can be stored and processed within spatial data and trajectory data warehouses, concluding with an overview of those main methods of exploring and mining spatio-temporal data.

Within much applied geocomputation, attention is typically paid to data in the form of points, lines or areas, and the attributes attached to these representative forms. However, another important kind of data is directional or circular data that relates to situations where there is some kind of cyclic nature in those quantities being analysed (e.g. time of day, or season in the year), or some kind of circular measure of direction (such as a bearing or direction of travel). Data of this kind have to be considered differently from more conventional linear measures – for example, if time-of-day measurements straddle midnight it makes no sense to compute conventional mean values. In addition, non-standard statistical distributions must be considered. Rohde and Corcoran (Chapter 7) give an overview of the handling of methods for analysing and visualising circular data, together with examples of their usage in practice.

As discussed earlier, applied geocomputation often requires *making geographical decisions* about how best to tackle a real-world problem, be it the location of a new school or where to site a new store for maximum profit. However, when attempting to make an optimised choice from a given set of options, factors influencing these decisions are often multi-dimensional, dynamic and spatially varying, thus stimulating computational challenges related to efficient processing and complexity reduction. Alexiou and Singleton (Chapter 8) present an introduction to geodemographic classification as a method of data reduction that is useful in the mapping of socio-spatial structure of urban populations. This is illustrated by a case study application for the city of Liverpool in the UK. Geodemographic methods are then extended by Spielman and Folch (Chapter 9) through an application of self-organising maps (SOMs) as an alternative method of understanding the multi-dimensional social topology of places. The chapter extends the discussion around the idea of clustering in the context of geographic and attribute space, making explicit the role of distance as a measure of similarity or dissimilarity. It then provides an introduction to the SOM method conceptually and practically through an illustrative case study. Lewis (Chapter 10) makes a case for an *ad hoc* analysis of geographical patterns in the use of particular services – specifically, doctors' surgeries. A key tool in this approach is the application of kernel density estimates, and those visualisation and analysis techniques making use of them. Their use in conjunction with other geographical information is outlined in this chapter. Tomintz, Clarke and Alfadhli (Chapter 11) explore location-allocation methods. The location-allocation problem is an important tool in location analysis, and perhaps one of the longest-standing tools in geocomputation. This chapter provides an outline of the key problem, and those algorithmic approaches used to tackle it, illustrating these more theoretical ideas though a number of practical examples.

Creating a model of *how the world works* enables predictions to be made. For example, if employment were to fall, what effect might this have on crime rates? However, given that the influencing factors of many real-world problems spatially cluster, the parameterisation of these measures into traditional explanatory models can undermine some of their specification assumptions, thus invalidating the outputs. This raises computational challenges for the specification of models that better control for spatially varying relationships. Nakaya (Chapter 12) introduces spatial non-stationarity as an important issue to resolve when attempting to understand how geographical processes operate through explanatory statistical models. There are a number of tools that may be used to explore spatial non-stationarity, and in this chapter the GWR4 software package is introduced.

Although these models are complex, a step-by-step guide to their statistical implementation is given through the use of the GWR4 software to explore spatial non-stationary models of various forms. Morrissey (Chapter 13) then introduces spatial interaction as an alternative framework for explaining human behaviour, which measures flows between an origin and destination location, calibrated by costs attributed to distance travelled, the size (e.g. population) or origins and the attractiveness of destinations. The mathematical basis for these models is explained, together with practical examples using Excel spreadsheet software.

A great deal of geocomputational progress over the last few years may be at least in part attributable to the expansion of the programming languages of Python and R. R was introduced in Chapter 1 by Cheshire and Lovelace, while in Chapter 14 Rey focuses attention on Python, and in particular on the use of the PySAL library as a tool for spatial analysis and modelling. The PySAL library offers a number of modules, covering but not limited to computational geometry, clustering, exploratory spatial data analysis and spatial econometrics. Rey provides an overview of some typical use cases. The availability of R and Python, with the expansion of their spatial analysis capability, alongside increased prevalence of open data repositories, is enabling new methods of conducting applied geocomputation research. Brunsdon and Singleton (Chapter 15) describe some new ways in which applied geocomputation can be presented, linking with data, analytical processes and the presentation of research within transparent and open workflows.

Technical developments in web mapping technology and cyberinfrastructures have significantly changed how can users collect share and interact with geographic information online. Such developments raise challenges related to representation and effective use of non-traditional data sources, but also provide opportunities to disseminate applied geocomputation and engage with end-users. In the final part of the book we deal with three facets of enabling interactions with data, tools for representation and analysis, and public engagement. Brunsdon and Comber (Chapter 16) introduce volunteered geographical information through a case study of expert researchers monitoring the first bloom dates of lilacs from 1956 to 2003 in North America, and investigate changes in the onset of the North American spring. It is argued that care must be taken when analysing data of this kind, with particular focus on the issues of lack of experimental design and Simpson's paradox. Approaches used to overcome this issue make use of a statistical approach termed 'random coefficient modelling'. Using the suggested method, a gradual advance in the onset of spring is suggested by the results of the analysis. Many of the analyses presented in this book have involved the use of open source software, which, as suggested by Brunsdon and Singleton in Chapter 15, are critical to new forms of open and transparent science. Open source software is typically developed by communities of developers, reflecting interactions in applied geocomputation. O'Brien (Chapter 17) reviews some of the main categories of open GIS software and those bodies that have established roles for promotion and standards approval. A case study of using open GIS is given, illustrating the analytical functionality of the desktop GIS software QGIS. Kingston (Chapter 18) discusses how geocomputation can play an important role in democratising spatial decision-making – largely through the provision of websites allowing members of the public to visualise and evaluate the consequences of various planning decisions. Two examples are given – one relating to local climate change and flood planning, and another on the siting and connection of electrical power. The chapter illustrates how new web-based software can be created, but also how existing cyberinfrastructure resources, such as the Google Maps API, may be utilised.

Enjoying the book

We have aimed to present a book that is as accessible as possible, illustrating those complex methods presented with practical examples of implementation or guided analysis using a variety of software. Where possible, technical language and advanced statistical or mathematical presentations are kept to a minimum, although inevitably this is variable over the chapters given the brevity of topics presented. Where appropriate, chapter authors have provided online materials useful for teaching or research; these can be found on the companion website for the book, https://study.sagepub.com/geocomputation.

FURTHER RESOURCES

For examples of the code and data referenced throughout the book, visit
https://study.sagepub.com/geocomputation

PART I

DESCRIBING HOW THE WORLD LOOKS

1

SPATIAL DATA VISUALISATION WITH R

James Cheshire and Robin Lovelace

Introduction

What is R?

R is a free and open source computer program for processing data. It runs on all major operating systems and relies primarily on the *command line* for data input (www.r.project.org). This means that instead of interacting with the program by clicking on different parts of the screen via a *graphical user interface* (GUI), users type commands for the operations they wish to complete. For new users this might seem a little daunting at first, but the approach has a number of benefits, as highlighted by Gary Sherman (2008: 283), developer of the popular geographical information system (GIS) QGIS:

> With the advent of 'modern' GIS software, most people want to point and click their way through life. That's good, but there is a tremendous amount of flexibility and power waiting for you with the command line. Many times you can do something on the command line in a fraction of the time you can do it with a GUI.

A key benefit is that commands sent to R can be stored and repeated from scripts. This facilitates transparent and reproducible research by removing the need for software licences and encouraging documentation of code. Furthermore, access to R's source code and the provision of a framework for extensions has enabled many programmers to improve on the basic, or 'base', R functionality. As a result, there are now more than 5000 official add-on *packages*, allowing R to tackle almost any numerical problem. If there is a useful function that R cannot currently perform, it is likely that someone is working on a solution. One area where extension of R's basic capabilities have been particularly successful in recent years is the addition of a wide variety of spatial analysis and visualisation tools (Bivand et al., 2013). The latter will be the focus of this chapter.

Why R for spatial data visualisation?

R was conceived – and is still primarily known – for its capabilities as a 'statistical programming language' (Bivand and Gebhardt, 2000). Statistical analysis functions remain core to the package, but there is broadening functionality to reflect a growing user base across disciplines. It has become 'an integrated suite of software facilities for data manipulation, calculation and graphical display' (Venables et al., 2013). Spatial data analysis and visualisation is an important growth area within this increased functionality. The map of Facebook friendships produced by Paul Butler, for example, is iconic in this regard and has reached a global audience (Butler, 2010). It shows linkages between friends as lines passing across the curved surface of the Earth (using the geosphere package). The secret to the success of this map was the time taken to select the appropriate colour palette, line widths and transparency for the plot. As we discuss later in this chapter, the importance of such details cannot be overstated. They can be the difference between a stunning graphic and an impenetrable chart.

Arguably Butler's map helped inspire the R community to produce more ambitious graphics, a process fuelled by an increased demand for data visualisation and the development of packages that augment R's preinstalled 'base graphics'. Thus R has become a key tool for analysis and visualisation used by the likes of Twitter, the *New York Times* and Google. Thousands of consultants, design houses and journalists also rely on R – it is not the preserve of academic research, and many graduate jobs now list R as a desirable skill.

It is worth noting that there are a few key differences between R and traditional desktop GIS software. While dedicated GIS programs handle spatial data by default and display the results in a single way, there are various options in R that must be decided by the user.

One example of this is the choice between R's base graphics and a dedicated graphics package such as ggplot2. The former option requires no additional packages and can provide very quick feedback about the nature of the dataset in question with the generic plot() function. The ggplot2 option, by contrast, requires a new package to be loaded but opens up a very wide range of functions for visualising data, beyond the base graphics. ggplot2 also has sensible defaults for grid axes, legends and other features, allowing the user to create complex and beautiful graphics with minimal effort. We encourage users to try both but, following the focus on *visualisation*, have used ggplot2 for all but the first two plots presented in this chapter.

An innovative feature of this chapter is that *all* of the graphics presented in it are reproducible (see the next section for how). We encourage users not only to reproduce the graphics presented here but also to play around with the code, taking advantage of the wide range of visual analysis options opened up by R. Indeed, it is this flexibility, illustrated by the custom map of shipping routes presented later in this chapter, that makes R an attractive visualisation solution.

All of the results presented in this chapter can be reproduced (and modified) by typing the short code snippets that are presented into R. Elsewhere in this book, these principles are extended in the context of reproducible geographic information science.

A practical primer on spatial data in R

This section introduces those steps required to get started with processing spatial data in R. The chapter focuses on the visualisation of so-called vector data (common in socio-economic examples), but R also provides functionality for the analysis and visualisation of raster data (see supporting materials). For users completely new to R, we would recommend beginning with an introductory tutorial, such as Torfs and Brauer (2014) or Lovelace and Cheshire (2014). Both are available free online.

The first stage is to obtain and load the data used for the examples into R. These data have been uploaded into an online repository that also provides a detailed tutorial to accompany this chapter: http://github.com/geocomPP/sdvwR.[1]

In any data analysis project, spatial or otherwise, it is important to have a strong understanding of the dataset before progressing. R is able to import a very wide range of spatial data formats by linking with the Geospatial Data Abstraction Library (GDAL). An interface to this library is contained in the rgdal package: install and load it by entering `install.packages("rgdal")` followed by `library(rgdal)` on separate lines. The former only needs to be typed once to install the package; however, the latter must be run for each new R session that requires use of the functions contained within the package.

The world map that we use is available from the *Natural Earth* website and a slightly modified version of it (entitled 'world') is loaded using the following code (see Figure 1.1).[2]

---

[1]To download the data that will allow the examples to be reproduced, click on the 'Download ZIP' button on the right-hand side of the page, and unzip this to a convenient place on your computer (e.g. the Desktop). This should result in a folder called 'sdvwR-master' being created.
[2]A common problem preventing the data being loaded correctly is that R is not set with the correct working directory. For more information, refer to the online tutorial hosted at http://github.com/geocomPP/sdvwR.

```
library(rgdal) # load the package (this needs to have been installed)
wrld <- readOGR("data/", "world")
plot(wrld)
```

The above block of code loads the rgdal library, creates and then plots a new *object* called `wrld` (Figure 1.1). This operation should be fast on most computers because `wrld` has a small file size. Spatial data can, however, get very large as the number and complexity of zones increases. We recommend keeping track of the size of spatial objects and to simplify them when necessary prior to visualisation.[3]

When spatial data are imported into R, they are saved in a spatial object class using the sp package (Bivand et al., 2013). The spatial data are divided into a series of different *slots*, storing the attribute and geometry data separately.[4] To view the slot names of an object you can use the function `slotNames()`, with the object name written within the brackets.

The contents of 'slots' within spatial data objects can be accessed using the '@' symbol. In the example below, the first two rows of the data slot are displayed, and can be treated as a standard data frame.

```
wrld@data[1:2, 1:5]
```

```
##    scalerank        featurecla labelrank   sovereignt sov_a3
## 0          1 Admin-0  country         3 Afghanistan    AFG
## 1          1 Admin-0  country         3       Angola    AGO
```

Fundamentals of spatial data visualisation

Good maps can have an enormous impact on understanding of spatial patterns, from initial exploratory data analysis through to the communication of results. Graphics do, however, need to be refined and calibrated, and this section describes such considerations. It should be noted that not all good maps and graphics must contain all the features discussed: they should be seen as suggestions rather than firm principles.

Effective map-making is a difficult process. As Krygier and Wood (2011: 6) put it: 'there is a lot to see, think about, and do'. We use ggplot2 as the package of choice to produce most of the maps presented in this chapter because it facilitates good practice in data visualisation. The 'gg' in its name stands for 'Grammar of

---

[3]R makes this easy; see the tutorial that accompanies this chapter (http://github.com/geocomPP/sdvwR).

[4]For more detail on this topic, see 'The structure of spatial data in R' in the online tutorial.

Graphics', a set of rules developed by Wilkinson (2005). Grammar in the context of graphics works in much the same way as it does in language: providing structure to the presented material. The ggplot2 package was developed by Hadley Wickham, and includes a syntax for building graphics in layers using the + symbol (see Wickham, 2010). This layering component is especially useful in the context of spatial data since it is conceptually the same as map layers in a conventional GIS.

In the following analysis, the previously loaded map of the world will be used to demonstrate a series of cartographic principles. This spatial object contains 35 columns of data; however, for our purposes, we are only really interested in population ("pop_est"). Typing summary(wrld$pop_est) provides basic descriptive statistics on population.

Before progressing, we will reproject the data.[5] The coordinate reference system of the world shapefile (named wrld) is WGS84, which is a very common latitude and longitude format. Without projecting the data when plotted, this format distorts the size of countries close to the North and South poles (at the top and bottom of the Figure 1.1 (left)). Instead, the Robinson projection (see Figure 1.1 (right)) can be used to provide a better compromise between areal distortion and shape preservation. Changes of projection can be accomplished using spTransform() with the projection required set with the CRS (coordinate reference system) parameter.

```
library(ggplot2)
wrld.rob <- spTransform(wrld, CRS("+proj=robin"))
  # '+proj=robin' refers to the Robinson projection
plot(wrld.rob)
```

Plotting the reprojected spatial object results in a world map that is better proportioned. As such, when adding detail to the representation, salient patterns will

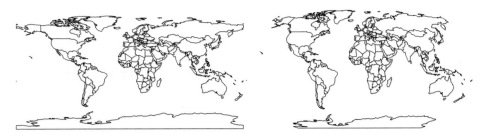

FIGURE 1.1   A basic map of the world in geographic (cartesian) coordinates (left) and the Robinson projection (right)

---

[5]For more information on referencing systems, see the links in the supporting materials.

be presented more clearly to end users. Figure 1.1 was created with R's base graphics. However, the remaining examples will use the ggplot2 package introduced above. This requires the data to be in a slightly different format than base R and should be converted using the `fortify` function. This step discards the attribute data associated with the spatial object and so this needs to be reattached using the `merge` function.

```
wrld.rob.f <- fortify(wrld.rob, region = "sov_a3")
# Use by.x and by.y arguments to specify joining variables
# dataframes together:
wrld.pop.f <- merge(wrld.rob.f, wrld.rob@data, by.x = "id",
  by.y = "sov_a3")
```

Now that the R object is in the correct format to be plotted with ggplot2, the code that follows produces a choropleth map coloured by the population variable. This demonstrates the syntax of ggplot2 by first linking together a series of plot commands, and assigning them to a single R object called map. If you type map into the command line, R will then execute the code and generate the plot shown in Figure 1.2. By specifying the `fill` variable within the aes() (short for 'aesthetics') argument, ggplot2 colours the countries using a default colour palette and automatically generates a legend. geom_polygon() tells ggplot2 to plot polygons. As will be shown later, these defaults can be altered to change a map's appearance.

```
map <- ggplot(wrld.pop.f,
  aes(long, lat, group = group, fill = pop_est/1e+06)) +
  geom_polygon() + coord_equal() +
  labs(x = "Longitude", y = "Latitude",
  fill = "World Population") +
  ggtitle("World Population") +
  scale_fill_continuous(name = "Population\n(millions)")
map
```

Colour has a large impact on how people perceive a graphic. Adjusting a colour palette from yellow to red or from green to blue, for example, can alter the readers' response. In addition, the use of colour to highlight particular regions or de-emphasise others is an important trick in cartography that should not be overlooked. For more information about the importance of different features of a map for its interpretation, see Monmonier (1996).

ggplot2 recognises the difference between continuous and categorical (nominal) variables and will automatically assign an appropriate colour palette accordingly (see Figure 1.3). The default colour palettes are a good place to start, but users can specify them, for example, to print a map in black and white.

The `scale_fill()` (for areas) and `scale_colour()` (for lines and points) family of commands enable such customisation. For categorical data, for example, `scale_fill_discrete()` can be used. The full range of options can be seen within RStudio by typing `scale_fill` followed by the tab key.

```
# Produce a map of continents
map.cont <- ggplot(wrld.pop.f,
  aes(long, lat, group = group, fill = continent)) +
  geom_polygon() + coord_equal() +
  labs(x = "Longitude", y  = "Latitude",
  fill = "World Continents") +
  ggtitle("World Continents")
  # To see the default colours
map.cont
```

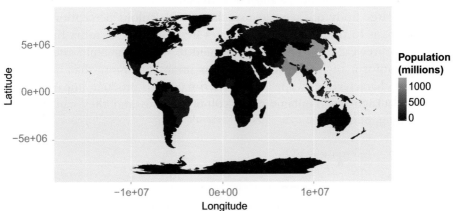

FIGURE 1.2   World population map

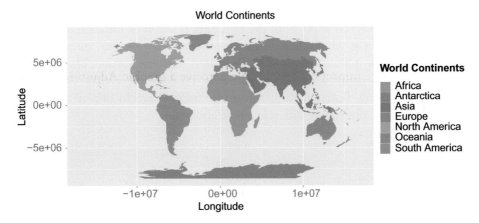

FIGURE 1.3   A map of the continents using default colours

Colours can be specified manually, either using words, as illustrated in the example, or more flexibly, such as through the use of hexadecimal colour codes:

```
map.cont + scale_fill_manual(values = c("yellow", "red", "purple",
  "white", "orange", "blue", "green", "black"))
```

The command `scale_fill_continuous()` is used to set a continuum-based colour scheme:

```
# Note the use of the 'map' object created earler
map + scale_fill_continuous(low = "white", high = "black")
```

Choosing an appropriate colour palette is difficult and there are a variety of considerations, such as the intended destination of a graphic (computer screen, print, etc.), the likely audience and visual impairments such as colour blindness. There is a large body of literature associated with colour perception, and this forms the basis to the *Color Brewer* palettes developed by Cynthia Brewer (see http://colorbrewer2.org). These are designed to be colour-blind safe and perceptually uniform such that no one colour jumps out more than any others. This latter characteristic is important when trying to produce impartial maps. R has a package that contains these colour palettes and they can be easily accessed by ggplot2.

```
library(RColorBrewer)
# Note the use of the scale_fill_gradientn() instead of
# scale_fill_continuous() used above
map + scale_fill_gradientn(colours = brewer.pal(7, "YlGn"))
```

In addition to altering the colour palette used to represent a continuous dataset, it may also be desirable to adjust the breaks at which the colour transitions occur. There are many ways to select both the optimum number of breaks and the locations in the dataset at which they occur. This is important for the comprehension of a graphic since it alters the colours associated with each value. The classINT package contains many ways to automatically create these breaks. We use the `grid.arrange()` function from the gridExtra package to display a series of maps side by side, illustrating different break choices.

```
library(classInt)
```

```
library(gridExtra)
```

```
# Specify the number of breaks - usually less than 7
nbrks <- 6
# Type ?classIntervals for help on this function
brks <- classIntervals(wrld.rob@data$pop_est, n = nbrks,
  style = "quantile")

print(brks)

# Use breaks object (brks) to set breaks in plot
# colour palettes
YlGn <- map + scale_fill_gradientn(colours = brewer.pal(nbrks,
  "YlGn"), breaks = c(brks$brks))

PuBu <- map + scale_fill_gradientn(
  colours = brewer.pal(nbrks, "PuBu"), breaks = c(brks$brks))

grid.arrange(YlGn, PuBu, ncol = 2)

nbrks <- 4
brks <- c(1e+08, 2.5e+08, 5e+07, 1e+09)
map + scale_fill_gradientn(colours = brewer.pal(nbrks,
  "PuBu"), breaks = c(brks))
```

Line colour and width are also important parameters for enhancing the legibility of a graphic (see Figure 1.4). The code below demonstrates it is possible to adjust these using the `colour` and `lwd` arguments. The impact of different line widths will vary depending on screen size and resolution. Also, if you save the plot to pdf (e.g. using the `ggsave()` command), this will also alter the relative line widths. As such, it is often useful to generate and check plots in the desired output format, and then adjust the code until these are appropriate.

```
map3 <- ggplot(wrld.pop.f,
  aes(long, lat, group = group)) + coord_equal() +
  theme(panel.background = element_rect(fill = "light blue"))

yellow <- map3 + geom_polygon(fill = "dark green",
  colour = "yellow")

black <- map3 + geom_polygon(fill = "dark green",
  colour = "black")
```

```
thin <- map3 + geom_polygon(fill = "dark green",
  colour = "black", lwd = 0.1)

thick <- map3 + geom_polygon(fill = "dark green",
  colour = "black", lwd = 1.5)

grid.arrange(yellow, black, thick, thin, ncol = 2)
```

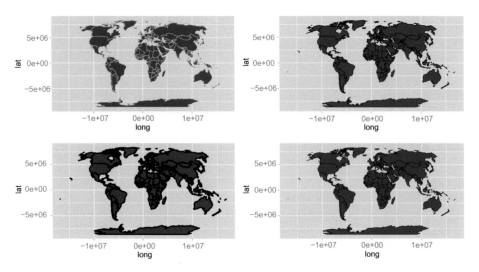

FIGURE 1.4   The impact of line width

There are other parameters, such as layer transparency (use the `alpha` parameter for this), that can be applied to all aspects of the plot – both points, lines and polygons. Space does not permit full exploration here, but more information is available in the ggplot2 package documentation (see http://ggplot2.org).

Map adornments

Map adornments and annotations orientate the viewer and provide context. They include grids (also known as 'graticules'), orientation arrows, scale bars and graphical overlays. None are required on a single map; indeed, it is often best that they are used sparingly to avoid unnecessary clutter (Monkhouse and Wilkinson, 1971). With ggplot2, axes and legends are provided by default, but they can be customised or removed.

The maps created so far have concerned a single dataset. However, it is possible to layer separate datasets together to create a single map. This is something required to create both the north arrow and scale bar since they are, in effect, data that are not stored in the same R object used to produce the plot (in this case only a couple of coordinate pairs). This process can therefore be replicated with additional R objects and not just for map adornments. It first requires an empty plot, meaning that each new layer must be defined with its own dataset, and as such, the syntax is a little different from that presented previously. Although more code is needed, it does enable much greater flexibility with regard to what can be included as new layer content. Another possibility is to use geom_segment() to add a rudimentary arrow (see ?geom_segment for refinements):

```
library(grid)  # needed for arrow
ggplot() + geom_polygon(data = wrld.pop.f,
  aes(long, lat, group = group, fill = pop_est)) +
  geom_line(aes(x = c(-1.3e+07, -1.3e+07),
  y = c(0, 5e+06)), arrow = arrow()) +
coord_fixed()  # correct aspect ratio
```

The scale bar capabilities of ggplot2 are perhaps the least advanced element of the package. To create a scale bar the spatial data will need to be in a projected coordinate system to ensure there are no distortions as a result of the curvature of the earth. In the case of the world map the distances at the equator in terms of degrees east to west are very different from those further north or south. Any line drawn using the the simple approach below would therefore be inaccurate. For maps covering large areas – such as the entire world – leaving the axis labels on will enable them to act as a graticule to indicate distance. The following example uses a shapefile of London's boroughs.

```
load("data/Ind.wgs84.RData")  # load Ind spatial object
load("data/lnd.f.RData")# load Ind fortified object
ggplot() + geom_polygon(data = lnd.f,
  aes(long, lat, hgroup = group)) +
  geom_line(aes(x = c(505000, 515000),
  y = c(158000, 158000)), lwd = 2) +
annotate("text", label = "10km", 510000, 160000) +
coord_fixed()
```

Legends are added automatically, but can be customised in a number of ways. They are an important adornment of any map since they describe what attributes the colours reference. As a general rule, legends with values that go to a large number of significant figures should be avoided. The following code moves the legend from the default position to the top of the map (see Figure 1.5).

```
# Position
map + theme(legend.position = "top")
```

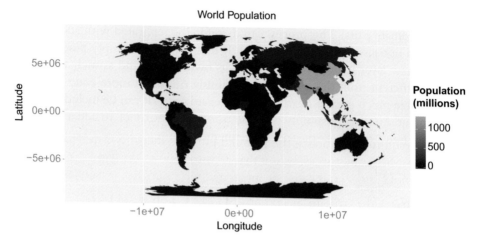

FIGURE 1.5 Formatting the legend

Many more options are available, such as adding a title, adjusting the font size or colour, and controlling other aspects such as the map borders. These are illustrated by the following code.

```
# Title
map + theme(legend.title = element_text(colour = "Red",
    size = 16, face = "bold"))

# Label font size and colour
map + theme(legend.text = element_text(colour = "blue",
    size = 16, face = "italic"))
# Border and background box
map + theme(legend.background = element_rect(fill = "gray90",
    size = 0.5, linetype = "dotted"))
```

Plotting over a base map

The ggmap package extends the ggplot2 package to integrate online mapping services such as Google Maps and OpenStreetMap (OSM) for base cartography. By using image tiles derived from these services, spatial data can be placed in context as users can easily orientate themselves to streets and landmarks. In the following examples, data on London sports participation are used. The data were originally projected in British National Grid, which pertains to a different referencing system than that used in the Google or OSM online map services. This is a common problem and one that is easily overcome using the reprojection function outlined above.

After importing the boundary data and reprojecting, a bounding box of the lnd.wgs84 object was calculated to identify the geographic extent of the map. This information is used to request an appropriate base map from a selected map tile service. The first block of code in the snippet below retrieves the bounding box and then adds 5% so there is a little space around the edges of the data to be plotted. This is then fed into the get_map() function as the location parameter. The code actually utilises two nested functions, ggmap() and get_map,() which are required to produce the plot and provides the base map data. You will notice from the code snippet below that ggmap follows the same syntax structures as ggplot2 and so can easily be integrated with the other examples included here. The example data object contains spatial polygons but spatial points and lines can also be plotted.

```
b <- bbox(lnd.wgs84)
b[1, ] <- (b[1, ] - mean(b[1, ])) * 1.05 + mean(b[1, ])
b[2, ] <- (b[2, ] - mean(b[2, ])) * 1.05 + mean(b[2, ])
# Increase bb by 5% for plot
# Replace 1.05 1.xx for an xx% increase in the plot size

library(ggmap)

lnd.b1 <- ggmap(get_map(location = b))


lnd.wgs84.f <- fortify(lnd.wgs84, region = "ons_label")
lnd.wgs84.f <- merge(lnd.wgs84.f, lnd.wgs84@data, by.x =
 "id", by.y = "ons_label")

# Overlay on base map using the geom_polygon()
lnd.b1 + geom_polygon(data = lnd.wgs84.f,
 aes(x = long, y = lat, group = group,
 fill = Partic_Per), alpha = 0.5)
```

The resulting map looks reasonable, but it would be improved with a simpler base map. A design firm called *stamen* provide the tiles we need and they can be brought into the plot with the get_map() function. This produces a much clearer map and enables readers to focus on the data rather than the cartography of the base map. The integration of data and services from third parties is a growing trend within R and one of its key strengths: users are not constrained to proprietary or paid-for services, they can make full use of the open data sources that are now available.

```
lnd.b2 <- ggmap(get_map(location = b, source = "stamen",
 maptype = "toner", crop = T)) # note addition of maptype

# We can then produce the plot as before
lnd.b2 + geom_polygon(data = lnd.wgs84.f,
aes(long, lat, group = group, fill = Partic_Per), alpha = 0.5)
```

Case Study

As an illustrative example, this final section describes the creation of a map depicting eighteenth-century shipping flows. The data used in this visualisation have been obtained from the Climatological Database for the World's Oceans and represent a sample of digitised ships' logs from the eighteenth century. We are using a very small sample of the full dataset, which is available from http://pendientedemigracion.ucm.es/info/cliwoc/. The example has been chosen to demonstrate a range of plotting capabilities within ggplot2, and illustrate those ways in which they can be applied to produce high-quality maps that are reproducible with only a few lines of code.

The example uses the png package to load in a series of map annotations stored as png graphics files. These have been created in image editing software and will add a historic feel to the map. We are also loading in a world boundary shapefile and the shipping data itself.

```
library(rgdal)
library(ggplot2)
library(png)
wrld <- readOGR("data/", "ne_110m_admin_0_countries")
btitle <- readPNG("figure/brit_titles.png")
compass <- readPNG("figure/windrose.png")
bdata <- read.csv("data/british_shipping_example.csv")
```

The first few lines in the bdata object contain seven columns, with each row reporting a single point on the ship's course. The first step is to specify the format for a number of plot parameters that will remove the axis labels.

```
xquiet <- scale_x_continuous("", breaks = NULL)
yquiet <- scale_y_continuous("", breaks = NULL)
quiet <- list(xquiet, yquiet)
```

The next step is to prepare the world coastlines for input into ggplot2 with the fortify() command, and then combined with background data to create the plot. In the following code, this sets the extents of the plot window and provides a blank canvas on which layers can be built (Figure 1.6). The first layer created is the `wrld` object; the code is wrapped in `c()` to prevent it from executing by simply storing it as the plot's parameters.

```
wrld.f <- fortify(wrld, region = "sov_a3")
base <- ggplot(wrld.f, aes(x = long, y = lat))
wrld <- c(geom_polygon(aes(group = group), size = 0.1,
  colour = "black", fill = "#D6BF86", data = wrld.f, alpha = 1))

base + wrld + coord_fixed()
```

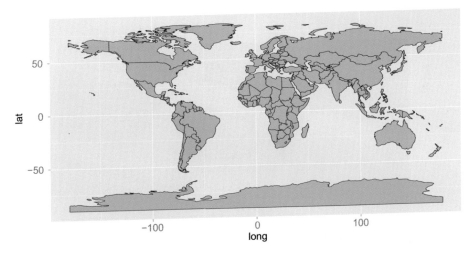

FIGURE 1.6 World map

The code snippet below creates the plot layer containing the shipping routes. The `geom path()` function is used to string together the coordinates to form the routes. You can see within the `aes()` that this specifies the longitude and latitude, plus, pasted together, the `trp` and `group.regroup` variables to identify the unique paths.

```
route <- c(geom_path(aes(long, lat, group = paste(bdata$trp,
  bdata$group.regroup, sep = ".")), colour = "#0F3B5F",
  size = 0.2, data = bdata, alpha = 0.5, lineend = "round"))
```

We now have all we need to generate the final plot, shown in Figure 1.7, by building the layers together with the + sign as shown in the code overleaf. The first three arguments are the plot layers, and the parameters within `theme()` are changing the background colour to sea blue. `annotation_raster()` plots the png map adornments loaded in earlier. This requires the bounding box of each image to be specified. In this case we use latitude and longitude (in WGS84), and we can use these parameters to change the png's position and also its size. The final two arguments fix the aspect ratio of the plot and remove the axis labels.

```
base + route + wrld +
  theme(panel.background = element_rect(fill = "#BAC4B9",
  colour = "black")) +
  annotation_raster(btitle, 30, 140, 51, 87) +
  annotation_raster(compass, 65, 105, 25, 65) +
  coord_equal() + quiet
```

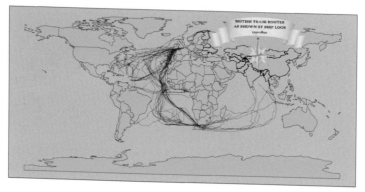

FIGURE 1.7   World shipping

In the plot example we have chosen the colours carefully to give the appearance of a historic map. An alternative approach could be to use a satellite image as a base map. It is possible to use the readPNG() function to import NASA's 'Blue Marble' image for this purpose. Given that the route information is the same projection as the image, it is very straightforward to set the image extent to span −180 to 180 degrees and −90 to 90 degrees and have it align with the shipping data. Producing the plot (Figure 1.8) is accomplished using the code below. This offers a good example of where functionality designed without spatial data in mind can be harnessed for the purposes of producing interesting maps. Once you have produced the plot, alter the code to recolour the shipping routes to make them appear more clearly against the blue marble background.

```
earth <- readPNG("figure/earth_raster.png")
base + annotation_raster(earth, -180, 180, -90, 90) +
  route + theme(panel.background = element_rect(
    fill = "#BAC4B9", colour = "black")) +
```

```
annotation_raster(btitle, 30, 140, 51, 87) +
annotation_raster(compass, 65, 105, 25, 65) +
coord_equal() + quiet
```

FIGURE 1.8   World shipping with raster background

Conclusion

There are infinite combinations of colour, adornments and line widths that could be applied to a map (or any other data visualisation), so do not feel constrained by the examples presented in this chapter. Take inspiration from maps and graphics you have seen and liked, and experiment. The process is iterative, usually taking multiple attempts to arrive at a satisfactory output. To give your maps a final polish you may wish to export them as a pdf using the ggsave() function and then add additional customisations using external graphics programs such as Adobe Illustrator or Inkscape.

The beauty of producing maps in a programming environment as opposed to a GUI, offered by the majority of GIS programs, lies in the fact that each line of code can be easily adapted to a different purpose. Users can create a series of scripts that act as templates and simply reuse them when required. This can save time in the long run and has the added advantage that all outputs can have a consistent style.

This chapter has covered a variety of techniques for the preparation and visualisation of spatial data in R. While this is only the tip of the iceberg in terms of R's spatial capabilities, the simple worked examples lay the foundations for further exploration of spatial data in R, using the multitude of spatial packages. These can be discovered online, through R's internal help (we recommend frequent use of R queries such as ?plot) and other published work on the subject. It is hoped that the techniques and examples covered in this chapter will help communicate the results of spatial data analysis to the target audience in a compelling and effective way.

As the R community grows, so will R's range of spatial applications and functions. The supportive online communities surrounding large open source programs such as R are one of their greatest assets. We recommend you become an active 'open source' citizen rather than merely a passive consumer of new software (Ramsey and Dubovsky, 2013). As R continues its ascent as a spatial analysis and data visualisation platform, the opportunities to create beautiful and useful maps are only set to grow.

FURTHER READING

R is constantly evolving, and as its user community grows the number of online resources is set to increase. An up-to-date resource on R for mapping, with further examples of its spatial functionalities is the online Creating-maps-in-R project: github.com/Robinlovelace/Creating-maps-in-R (Lovelace and Cheshire, 2014). For more on raster datasets, we would recommend the Raster vignette from the raster package. We also recommend "ggmap: Spatial Visualization with ggplot2" (Kahle and Wickham, 2013), available free online, for a more advanced introduction to mapping with ggplot2.

Map projections are a complex and important topic that we only touch on here. The following two web pages offer some further context and technical information: http://spatial.ly/2011/03/flattening-the-earth/ and http://en.wikipedia.org/wiki/Map_projection.

To stay abreast of current developments and tutorials in R we recommend http://www.r-bloggers.com, a feed aggregator about R, and the https://rpubs.com/ code repository.

2

GEOGRAPHICAL AGENTS IN THREE DIMENSIONS

Paul M. Torrens

Introduction

At their foundation, geographical agent-based models (ABMs, see Crooks, Chapter 4) are spatial processors, tasked with collecting and organizing geographic data, relating it to geographical concepts, animating it relative to space–time conditions, and adding contextual value in the meantime. Most geographical ABMs are automata (Turing, 1936; von Neumann, 1951) and follow in the distinguished tradition of finite-state (Minsky, 1967) and intelligent machines (Turing, 1950). As automata, ABMs are highly flexible in their ability to achieve any requirements put to them in computation. One common use of ABMs is in representing geographical things in simulation, and these things could be information, people, cities, biology, cars, snowflakes, and so on. Indeed, much of the appeal of ABMs is a by-product of their flexibility in representing and handling geographical entities, processes, and phenomena, and ABMs have been usefully employed in modeling a range of geographical systems, in public health, urban studies, landscape dynamics, medicine, economics, and other areas.

While they boast considerable inherent and implied extensibility, many geographical ABMs limit their scope to two dimensions of space in practical applications, which often puts them at odds with the realities of systems that they are tasked with modeling. The reasoning for constraining ABMs to two-dimensional spaces is varied, but in short, the addition of a third spatial dimension adds considerable complexity in model-building and simulation, invoking some long-standing challenges in computing three-dimensional spatial processes atop conventional GIS infrastructures. Overcoming the difficulty of adding dimensionality to geographical ABMs could substantially broaden the range of applications to which ABMs could be put, and it could expand their utility in supporting applied geocomputation.

In this chapter, I will review some of the complications that present in building three-dimensional space–time agency and related representations, and I will introduce a novel variation on ABMs – *polyspatial agents* – as a vehicle for addressing those challenges. I will discuss the usefulness of polyspatial agents with reference to simulating human interaction in crowds and earthquake scenarios, where dimensionality is critical to building authentic models, and where the requirements for space–time geocomputation are rather unforgiving for traditional GIS and ABMs.

Why agents need a third dimension of space

When considering a third dimension for geographical ABMs, it is important to assert that this encompasses shifting the theory, representation, applications, processes, data infrastructure, and computing of agents into an immersive *volumetric space*, which is a departure from traditional representation of agents on half-planes or similar two-dimensional manifolds. The potential advantages (and dilemma) of adding a third dimension to geographical ABMs are wide-reaching, but the focus here is on those of particular saliency for geocomputation: *characterization, geographic information, sense and sensing, interaction,* and *meta-simulation.*

Characterization

In many application scenarios, agents are *characters*, and the peculiarities of this characterization are important to the function that they play within the scenario. Characterization affords agents particular attributes, traits, constraints, and tendencies, determining whether (and how) they might be economically motivated (Tesfatsion, 1997), rational (Arthur, 1994), maximizing (Simon, 1956), evolutionary (Holland, 1995), and so on. Agent characteristics also play a (starring, supporting, or bit-part) role in how the dynamics of a system play out, and how agents' individual roles contribute to larger ensembles is critical in assessing the aggregate (i.e. group, population, regional, network, component) characteristics of the system (Janelle et al., 1998; Lee et al., 2007; Parrish and Hamner, 1997; Raafat et al., 2009; Reynolds, 1999; Schmidt and Griffin, 2007; Shaw et al., 1999). This may include system interaction (Doxiadis, 1968; El-Shakhs, 1972), dependencies (Krugman, 2005), complexity (Batty, 2005), allometry (Batty, 2008), and related collective traits for physical, mechanical, biological, chemical, or informational agency. The characterization of the *environments* that house and support agents is also important, as this provides the substrate for much of their agency. For geographical agents, the ambient environment often supplies a significant source of their geographic information (Golledge and Stimson, 1997).

It is therefore important that agents are afforded three-dimensional characteristics when that level of dimensionality is relevant to the role that they are cast in. However, most characteristics of geographical ABMs are represented in one or two dimensions of space, and in many cases this limits their functionality in simulation. Consider some examples. The locomotion of walking agents' bodies is often reduced to vectors cast from artificial centroids of simple point masses (Helbing and Molnár, 1995). In geographical ABMs, urban buildings and streetscapes are usually reduced to two-dimensional polygons (Haklay et al., 2001) or network geometries (Gayle et al., 2009). When terrain is considered in geographical ABMs, it is commonly tessellated and abstracted into an affordance value (Guy et al., 2010). These reduced characterizations are convenient – particularly when ABMs need to be coupled with GIS – but they are also artificially restrictive relative to the real-world systems that modelers *could* represent in simulation.

Geographic information

Reduced dimensionality is quite prevalent in ABMs that rely on geographic information systems (Zlatanova et al., 2002) and spatial data models (Abdul-Rahman and Pilouk, 2008) designed for such systems, in part because GIS has been relatively slow to embrace three dimensions due to lingering technical challenges (Goodchild, 2009).

The usual approach to handling three dimensions of space in a GIS (and in ABMs) is to constrain the system to a 2-manifold and infer a third dimension as *data values* ascribed to objects on the 2-manifold (we can think of this as *2.5-dimensionality*). Yet, *fully* three dimensions of geographic information are critical in understanding and representing a host of systems of interest to geocomputation, and for facilitating geocomputational operators and analyses. By placing agents – and the objects and entities that they might interact with – in a volumetric dimension, one can generate a variety of spatial information that is inaccessible from two-dimensional manifolds (up/down, surmountability, packing and space-filling, shadowing, pitch and roll, morphology of concave (and some convex) polyhedra, 3-manifold structure, etc.). One may also produce more nuanced information about projected two-dimensional conditions by considering a third geometrical dimension (rather than a simple attribute value), and this holds for elevation, relief, collision, visibility, line of sight, occlusion, coverage, parallax, hierarchy, traversability, shape, pose, degrees of freedom, displacement, and so on.

Sense (and sensing) of space and place

Our perception and awareness of environments (as humans or as other objects and entities) is also often informed by three dimensions of space. Many aspects of space and place, for example, simply make little sense on planar representations, such as the shortest path through a multi-story building with stairs (Ferguson and Stentz, 2007) or underwater trajectories (Petres et al., 2007; Ware et al., 2006). Our everyday experiences with sensing ambient geographic information rely on the third dimension, particularly if we must negotiate barriers with ambulatory difficulty (Church and Marston, 2003), for example. The need for a third dimension also translates into the user experience of communicating and working with simulations. In some cases, it may be convenient to reduce the dimensionality of a simulation's output to two dimensions, but in other cases an immersive experience is desirable, and this in part explains the recent popularity of virtual worlds (Bainbridge, 2007; Crooks et al., 2009; Delaney, 2000; *Economist*, 2006) and digital globes (Butler, 2006; Goodchild et al., 2012) in geographic information science. Indeed, the third dimension is likely to become even more important as geographical ABMs move to augmented reality platforms (Feiner et al., 1997; Uricchio, 2011).

Yet, representation of agents' sense of surroundings is often two-dimensional, and this is evident in collision detection routines, synthetic vision, and even economic geography. On the latter point: think about how we usually consider distance decay (Alonso, 1967; Clark and Evans, 1954), for example, and how little sense a two-dimensional (really one-dimensional in the case of bid-rent profiles: Brown and Holmes, 1971) distance decay and friction parameter makes for a real estate market like Hong Kong, which hosts a majority of multi-level structures. To some extent, developments in three-dimensional sensing of geographic information have reinforced the primacy of two-dimensionality, and this is evident for LiDAR (Gamba and Houshmand, 2002), in particular, where height measurements are sampled along a two-dimensional plane, missing overhangs and casting a "melted wax" appearance to many urban environments.

Interaction

It is when we consider interactions of and among things in space and time that denying ourselves a third dimension of space really starts to seem like a shortcut. Touching, brushing, smashing, breaking, separating, covering, fleeing, aligning: these are all interactions that necessitate three-dimensional representation to be treated realistically. This is especially true when we consider *physical processes* among rigid-body objects, particularly on the surface of the Earth, where gravity and friction are crucial in explaining the dynamic of interactions, but other properties between objects also require a third dimension, particularly for soft-body interactions (deformation, elasticity). For human agents, the approximation of movement relative to geographic planes is often problematic when mapped to two dimensions: while walking, for example, people use their feet to contact the ground for acceleration, stopping, rotation, and turning (Winter, 2009), and their center of gravity is a dynamic function of the ambulation of their body and limbs; these details are often reduced to simple point-mass representations (Bouvier et al., 1997; Schweitzer, 1997) or cylindrical extrusions (van den Berg et al., 2008) in geographical ABMs, with the result that well-known artifacts are produced when they interact (Torrens et al., 2012).

Meta-modeling and multi-modeling

Docking – exchange of algorithms and data – between geographical ABMs and other models and simulations is also a challenge in dimensionality. Even though many geographical ABMs are content to sit on two dimensions of space, the models (and model systems) that we often would like them to interact with are treated in three dimensions. For geographical ABMs to participate in meta-model systems (models of models) or multi-models (where the results from one model provide dynamics to another), higher-dimensional data must often be abstracted if they are to have parity. In essence, we must jettison significant detail to allow geographical ABMs to 'talk' to other models that work with higher dimensions, and this is evident in connections between geographical ABMs and voxelized community climate models, many CAD-based building information models, Earth science models that treat sub-surface Earth processes, models of gaseous flow (Epstein et al., 2008), and so on.

Existing approaches

Significant effort is being devoted to changing how the community addresses these challenges, and there *are* many existing implementations of geographical ABMs that adopt three dimensions of space. This is, for example, a prominent topic of research in the *biology* community, where ABMs of cell formations are used to explore dynamic formation of neoplasms (Malleta and De Pillis, 2006), or the dynamics of organ structures such as the epidermis (Chapa et al., 2013). There is also agent-based work on movement of entities within biological systems (Burkitt et al., 2011).

In *physical geography*, automata-based models (of which we can consider agent automata a variant), particularly cellular automata, are popularly used to enhance traditional gridded and lattice models (Malamud and Turcotte, 2000; Rothman and Keller, 1988) or continuum models of particle movement (Avolio et al., 2000; Baxter and Behringer, 1990; Iovine et al., 2003; Kronholm and Birkeland, 2005), and three-dimensional implementations are well represented. In both the biology and physical geography examples, one can rely on the convenience of ascribing well-known laws and rules to the agents (e.g. Navier–Stokes equations or chemotaxis), such that their agency, or behavior, is relatively parsimonious when compared with, say, the human socio-behavioral realm. To some extent, reliance on cellular automata in biology and physical geography examples is illustrative of this point: the automaton cell is used as a container for diffusion of forces and chemical between cellular units and over-arching equations for that geography can be applied across the automata lattice.

Elements of three-dimensional geometry are also used in *human geography* ABMs, particularly when considering landscapes, for example (Dibble and Feldman, 2004). However, these tend to be network or graph geometries with height encoded in vertices or traversal metrics recorded in edges; in other words, the representation of agency in the models does not make full use of the third dimension. Sembolini et al.'s (2004) development of cellular automata models of urban dynamics with extruded building stories is an innovative twist on the idea. Using three-dimensional visualization to *communicate* geographical ABMs is another technique that has been used, particularly in geovisualization of outcomes or parameter spaces from ABMs (Luke et al., 2005). In these instances, the third dimension does not usually feature substantively in the design of agency within the model behavior or algorithms (see Crooks et al., 2009; Dijkstra and Timmermans, 2002).

Some of the most sophisticated work in developing *deeply three-dimensional* geographical ABMs has been in *computer gaming*, where agents present as player or non-player characters (Nareyek, 2002), or massively dynamic elements of the environment in virtual game worlds (Parker and O'Brien, 2009). Several of the toolkits that game developers use have been adopted by the ABM community, for use in construction of serious games (games with a serious educational or research application) (Barnes et al., 2009; Jacobson and Hwang, 2002; Zyda, 2005), or to provide three-dimensional geometry objects and handling for ABMs. As a result, models that were originally developed with the sophistication of their computer graphics as a design goal are now incorporating deliberative agency in their geographical dynamics (Pelechano et al., 2008), in part because of the entertainment value now ascribed to realistic behavior, events, game play, and ambience in gaming (Cass, 2002).

Polyspatial automata as multi-dimensional creatures

In support of this argument for new approaches to how we consider, build, and use geographical ABMs, the concept of *polyspatial automata* will be introduced below. To some extent, this is an extension of earlier work on *geographic automata* (Torrens and Benenson, 2005), but with specific attention to how agents, in particular, *can become many geographical things in many geographical contexts* (Torrens and Nara, 2013). Specifically, with one-agent model structure, the idea is that we could flexibly define as much geography – geographic information, algorithms, data, representation, and perhaps also theory and phenomena – as required of a given scenario, such that the nature of the agents *supports* exploration in simulation, rather than constraining it. In other words, it is perhaps useful to allow the model as much extensibility in supporting the design or experimental goals of the model-builder and user as possible, rather than limiting what the model can be done to a particular strand of modeling, whether that be economic, rational, utility-maximizing, or whatever might be used. The idea of polyspatiality also extends practically, so that polyspatial agents can function nimbly in service of many different models or meta-models (Torrens et al., 2013; Zou et al., 2012). Regarding the central theme of this chapter, polyspatiality should support the design of agents (and of agent behaviors) that can be put to fruitful use in exploring two- *and* three-dimensional spatial scenarios. This is achieved as follows.

First, I provide mechanism for embedding (and therefore also informing) agents in different spaces (Torrens et al., 2012). Often simultaneously, these spaces can take a variety of forms, including mathematical, social, cognitive, physical, urban, architectural, visual, and spaces of the body. These multiplicative representations of space are accommodated by the agents' state variables, which can index conditions relative to a wide variety of spatial references. These references may have semantic meaning relative to one or multiple spaces, thereby providing polyspatial awareness to agents. For example, a geometric point on a GIS representation of a city grid may also have meaning as a goal for a particular agent in its agenda and may have opening hours that refer to a particular time geography, or the goal may refer to the door of a building within an architectural space; the building could be occluded from a particular vantage point in a visual representation; or the door and building might be situated in the corner of a street network in a civic space; and that point on the street network might even serve as the end-effector in a kinematics solver that determines how an agent makes footfalls while walking (Figure 2.1).

FIGURE 2.1 The polyspatial nature of the model can support multiple representations of geography in one simulation. An urban scene can hold representations of (1) social geography within a group of friends; (2) transport geography of road links; (3) vehicular movement on a network space; (4) street infrastructure as way-finding markers for agents' behavioral geography; (5) geometry of buildings for path planning in an informatics space; (6) landmarks and meaning in cultural geography; (7) doors for entry to spaces in a time geography; (8) internal building plans within an architectural space

Second, there must be support for multiple representational geographies for agents in the model, such that their geographic abilities, appearance, and the information that they convey in simulation becomes polyspatial. Of course, this means allowing three-dimensional matrices for agents' geography, replacing one-dimensional representations on two-dimensional planes that are commonly employed. For example, I represent individual agents as fully articulated rigs composed of vertices (joints) and (weighted) edges (bones) that may be dynamically positioned in space and time relative to constraints of realistic human skeletons (Torrens, 2012). This allows agents to articulate their movement through space and time using stride, poise, gait, and body language, rather than simply updating a one-dimensional vector. Agents can also pivot, turn, sidestep, slide, and fall over if necessary. With one rig representation, we can also wrap agents with enveloping meshes to give them corporeal form, and even clothing (Figure 2.2). For collision detection, we may reduce agents to a simple capsule or cylinder to ease calculations of overlap. For animation and communicating results to users, we can texture agents so that they look realistic relative to real-world scenes.

FIGURE 2.2   Multi-representational geographies for agents' bodies. From left to right: a non-uniform rational B-spline envelope for heightened realism relative to real-world scenarios; a block model for collision detection; a three-dimensional texture map; a low-polygon texture-mapped geometrical skin

Third, as a consequence of polyspatiality in spatial reference and representation, agents are afforded a range of interaction schemes. This includes interactions within agents' own cognition, between agents and other agents (Torrens and McDaniel, 2013), and between agents and the environment (in whatever form they consider it) (Torrens, 2014). These interactions are already common for agents as part of multi-agent systems (Ferber, 1999). However, further interactions are required between and within environmental components of the model to factor in these dynamics. The environment, in this context, can take on a variety of forms. And dynamics in one environment can pass via interaction into other model environments during simulation. In this way, interaction is polyspatial in its ability to *translate* interaction signals between agents, systems, and representations. Of most immediate relevance to the agent, agent rigs are ensembles of nodes and links that assume environmental form in considering agent dynamics, when, for example, an agent must determine how a movement in one part of their body should produce an action or reaction in another, or how that system of interactions should be influenced by the streetscape environment while walking, or by other agents' skeletons when avoiding collisions (Torrens, 2012). Similarly, the interactions of building components during a collapse scenario may be handled using a physics model and then translated as dynamic geometries to agents' visual systems and translated into steering interactions via movement (Torrens, 2014).

Fourth, systems (including the objects that populate them) are polyspatial in their processes. Agents – however they are considered or represented in the model – can be driven by (or subject to) a variety of process models, simultaneously in simulation. This allows a model user to better represent the real world in simulation, by building in diverse phenomena for agents to contribute to. It also expands the range of scenarios that can be explored in simulation, as agents can be parameterized and then subjected to different model processes without (necessarily) changing the parameterization (Torrens, 2009). As with interaction, processes can also be translated from one model scheme to another (Torrens et al., 2013; Zou et al., 2012). Indeed, this latter ability is crucial in representing surprising events that agents may encounter, in which agents must try to reason about the dynamics that are unfolding around them, using the model behaviors that they have on hand in their own abilities and cognition (Torrens et al., 2011). In this way agents are able to react to *what* is going on around them, without needing to know *why* or *how* the dynamics are taking place: they can have a *shallow reaction to a deep process* (or vice versa).

Fifth, the systems in which we embed and explore agents are polyspatial in their consideration of scale. We are often interested in the scaling of a system, its form at one phenomenal level relative to another, from one observational range to another, or from dynamics of an individual to that of a population (Batty and Longley, 1994). Similarly, we may be interested in how allometry holds at a given scale or persists across scales as fractal, recursive, or other forms. Indeed, the emergence of phenomena in three dimensions is largely under-explored in many geographical ABMs, although the topic is of critical significance in other geographical domains, including medicine (Coffey, 1998) and biology. Scale and scaling is critical in urban applications of ABMs, and examples are easy to consider (although tough to simulate): how the macro-state of a building relates to its micro-fragments in a collapse scenario; how the macro-kinematics of a body in motion is determined by micro-rotations and translations between bones through a joint (Figure 2.3); how the macro-observation of a crowd collective forms from and informs the microcosm of decisions that crowd members take to step in one place or another, and so on (Batty, 2005). We must also consider temporal scale and scaling in geographical ABMs, even when systems, models, or phenomena operate at varying rates of update and change, from the sub-second awareness and calculations of human movement to many fractions of a second considered in the physics of building interaction during an earthquake.

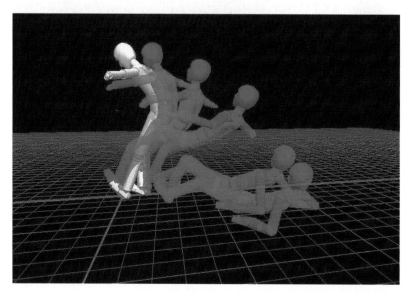

FIGURE 2.3   The model can handle scaling across all entities and objects in the system, as well as considering scaling within entities. The three-dimensional positioning and rotating of an individual joints and main corporeal system (axial skeleton for posture and center of gravity, and the appendicular skeleton for locomotion and ambulation) can be treated in isolation or in groups designed to accomplish a kinematic goal (such as protecting against a fall) relative to space and time. These sorts of dynamics would be incredibly difficult to accomplish in a two-dimensional model

Sixth, there is often some operational need to support polyspatiality in model assumptions for diverse computation, particularly when, say, ABMs need to interface with calculation, estimation, and processing schemes that are conventional to other forms of spatial simulation (and those that are non-spatial). This occurs, in particular, when the results of geographical ABMs need to be translated into numerical simulations (Kevrekidis et al., 2003), which may often be three-dimensional in nature and which handle computation using voxel-based strategies for collecting input and distributing processing. Also, components of

ABM simulation may require diverse computation and calculation to realize different components of simulation, from spatial indexing and geographic information access to rendering and animation (Figure 2.4).

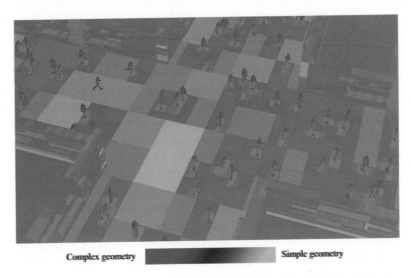

Complex geometry　　　　　　　　　　　Simple geometry

FIGURE 2.4 The three-dimensional (voxelized) view of a frame of animation for an agent-based model of an urban streetscape. The renderer bins the geometry of the scene by triangle count, devoting more in-depth computation (lighting, culling, ray-tracing, reflection, ambient occlusion) to portions with relatively high triangle counts or complex textures. As a result flat and featureless surfaces with no shadows require less computation and assume large voxels, while detailed meshes such as agent envelopes require relatively small voxels and more computing

Some examples

In the examples that follow, I will demonstrate how polyspatial agents can be usefully employed to represent geographical phenomena in three dimensions of space, proceeding beyond the existing capabilities of traditionally favored two-dimensional approaches. The first example tasks agents with handling complicated ensembles of humans in crowded settings; the second example looks at interactions between humans and a highly dynamic built environment.

Three-dimensional crowds of three-dimensional people

Many ABMs have been designed to explore the geography of individuals as mobile entities within crowd aggregates; for an excellent overview of the field, see Pelechano et al. (2008). Indeed, the emergence of crowd-level phenomena from the individual interactions of mobile agent walkers is one of the often-cited exemplars of complex adaptive systems (Ball, 2003; Batty et al., 2003a; Bouvier et al., 1997; Farkas et al., 2002; Helbing et al., 2000, 2005; Helbing and Molnár, 1997; Henein and White, 2007; Hoogendoorn and Bovy, 2000; Moussaïd et al., 2011; Partridge, 1982; Raafat et al., 2009; Stanley, 2000; Strogatz, 2004; Zhang, 2009). Crowd simulations also have significant practical potential, as they may be used to explore dynamics of crowd problems such as congestion, stampedes, and crushes (Harding et al., 2010), in which crowds must assemble and move in limited space. If one considers the experience of being part of a tightly packed crowd, it is perhaps apparent that the complications associated with the phenomenon rely quite significantly on three dimensions of space. Crowd ensembles in these instances are often crowded, that is, people impinge upon each other's personal space (Gérin-Lajoie et al., 2008; Habicht and Braaksma, 1984; Sobel and Lillith, 1975), they interact physically (Ciolek, 1978; Daamen and Hoogendoorn, 2003; Goffmann, 1971), and they must negotiate significant obstacles in their vision and degrees of freedom in movement (Batty, 1997b; Boulic et al., 1994; Foltête and Piombini, 2007). In many cases, the available volume of space around an individual crowd participant becomes so constrained that they must succumb to *reactive* behavior in close quarters – they become locked into small perturbations in the space–time dynamics around them, even when their desires for movement suggest alternative behavior (Figure 2.5).

Generally, many geographical ABMs of crowd congestion are modeled using one-dimensional point masses on two-dimensional planes (Helbing and Molnár, 1995), and they completely miss the third dimension. Variants that use extruded three-dimensional cylinders (van den Berg et al., 2008) or playback animation of agent avatar representations for visualization purposes (Crooks et al., 2009) still often retain this 1D/2D behavioral representation at their cores. Abstracting from the third dimension is useful and convenient, largely because it allows vectors to be mapped to point masses and for physical forces to be calculated between those masses as if they were particles (Henderson, 1971; Treuille et al., 2006). Convenience in this context does not yield realism, and significant details are missed in the abstraction which creates a requirement for results from such models to often be explained in aggregate form (level of service, flow rates, traversal potential, etc.). Furthermore, 1D/2D models often produce problematic artifacts such as spongy compression and expansion in crowd formations (realistically, tightly packed crowds rarely move backwards unless ordered to do so with coordinated help from outside actors), and unrealistic spinning/rotation of point masses (which gives agents unrealistic abilities to see in 360 degrees of vision) (Torrens et al., 2012).

To address these limitations, I extended the representation of synthetic human agents in the model, to give them three-dimensional skeletons (Figure 2.6) and to represent the dimensionality of their skeletons in the simulation. The added dimensionality has consequences for other components of the simulation.

(a)

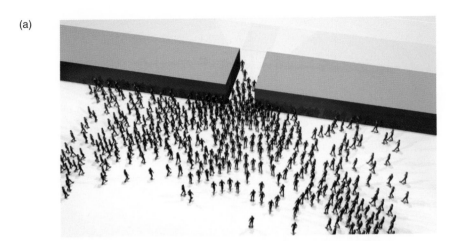

(b)

FIGURE 2.5   A frame from the crowd model from (a) a holistic view and (b) up close within the crowd, showing the significance of three dimensions in constraining the degrees of freedom that walkers have in movement and acquisition of geographic information from their surroundings

Collision detection and resolution must be performed in three dimensions (which is achieved by adding collision spheres to the rig), and locomotion can now consider ambulation of the agents' bodies (which I enabled through inverse and forward kinematics subject to motion-captured constraints). Similarly, I allowed agents to view their surroundings in three dimensions, using frustums in lieu of ray-casting (Figure 2.7). These representations can interact with traditional 1D/2D schemes if needed; for example, vectors for movement can be mapped to agents' center of gravity and projected to the plane.

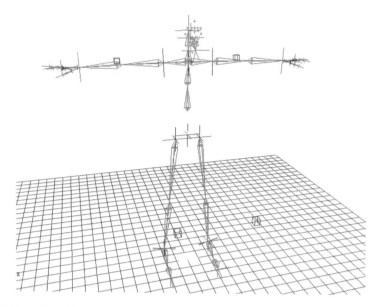

FIGURE 2.6 To allow agent rigs to physically interact in three dimensions during simulation, we added collision spheres (yellow) to key nodes and extremities (b) atop the basic skeletal rig configuration (a)

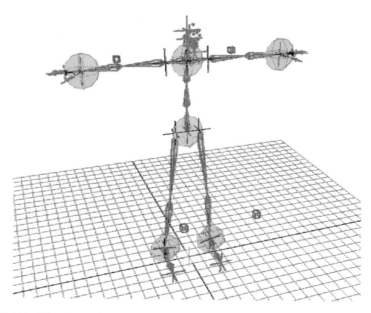

FIGURE 2.7 The three-dimensional characterization of the agents and ambient built environment extends into their vision so that they can collect three dimensions of spatial information as they act, react, and interact in the simulation

The net result is a model with significant added detail, in its dimensionality, and also in its process models. For similar scenarios, such as the well-known challenge of crowd egress through a single-exit room, the three-dimensional model yields substantively different results than standard approaches. Traditional artifacts of simulation disappear and different patterns emerge, such as irregular assembly patterns around congested exits (Figure 2.8). Moreover, we can explore *both* collective

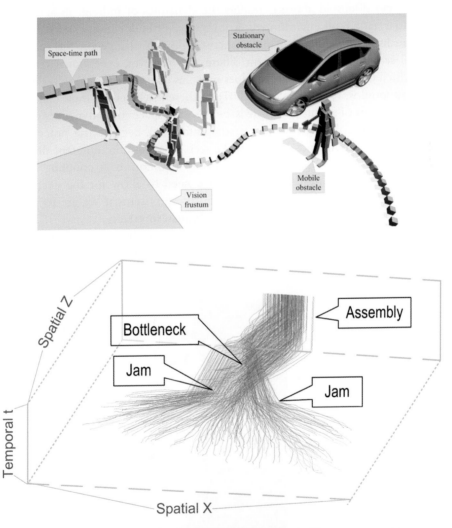

FIGURE 2.8 Space–time paths for individual agents in the room-egress scenario. Collective patterns of jamming, bottleneck congestion, and assembly are evident in the informational signatures of agents' movement, and the contribution of each and every action of the agents to those patterns can be explored in individual or collective form

and individual patterns of movement and interaction in the model as ensemble behaviors and outcomes are appropriately sourced in individual behavior and geographic information (Figure 2.8).

Dynamic three-dimensional built substrates for agent behavior

The scenario discussed above considers the dynamics of individuals and crowds relative to a static built environment. Assumptions of built rigidity and stability are usually realistic, as most urban infrastructures do not change on the spatial and temporal scales of agent movement. There are, however, exceptional conditions under which the built environment does move, as during earthquakes, for example (Anagnostopoulos, 1988; Villaverde, 2007). In these instances, we must consider the three dimensions of agents and crowds, but also the space–time dynamics of building fracture and collapse. This is not straightforward, because the processes that generate building dynamics in earthquake scenarios are quite different than those that drive deliberative human movement and locomotion. A coupling between two model systems is thus required, and we need to allow for the exchange of information, objects, and processes between these systems if we are to generate realistic geographical dynamics at the built–human interface.

To achieve this, I extend the idea of polyspatiality to the built environment representations and processes in the model, by allowing for physical processes of fracture,

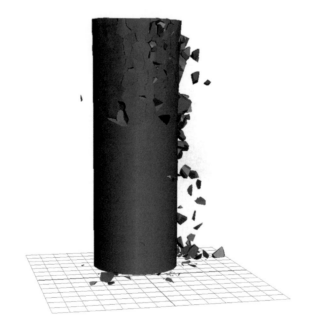

FIGURE 2.9 Dynamic, physical interactions among forming objects as a solid geometry fractures and collapses

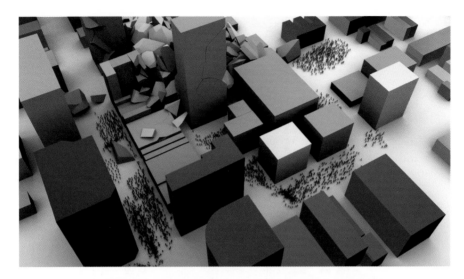

FIGURE 2.10 A frame from the earthquake evacuation model, showing agents reacting both to each other as they form congested crowds, and to the dynamic changes in the built environment following building collapse and shifts in the urban fabric around them

friction, and Newtonian force between the geometry and masses of the built infrastructure. By assuming the passage of energy from seismic activity through modeled objects in simulation, we can determine appropriate fracture patterns and then model the formation of new objects, as well as forces between objects as they collide in space and time. This is shown in Figure 2.9 for a set of simple geometric solids, and for a synthetic representation of an urban area built from LiDAR data in Figure 2.10 (Torrens, 2014). The dynamics from the physical simulation of building collapse pass directly into the crowd simulation as they should – as geometries for human agents to see, avoid, and plan around. In addition, as the built environment changes, so too do the features of the agents' mental map of the city: existing goals may disappear, known paths may vanish, and lines of sight may become obscured. Because the model is built in three dimensions, agents may walk over and pass under debris as they move through the simulated scenarios.

Conclusions

In this chapter, I have argued in favor of increased dimensionality in considering, representing, running, and examining geographical agent-based models. Motivated primarily by a quest for increased detail and realism in geographical ABMs, I have proposed that more useful models and simulations can be built and applied with added dimensionality, owing in large part to the expanded range of questions that can be explored in three dimensions but are simply not accessible in two-dimensional

ABMs. In support of my point, I introduced two simulation scenarios: one involving tightly packed three-dimensional crowds and another that expands on this scenario by including dynamic collapse scenarios in the built infrastructure of a simulated city. More intricate details of how these models work, and the innovation that underpins them, are provided in Torrens (2012, 2014).

It is worth noting that as geocomputational models using agent-based representations of people and things that we experience and sense in the real world begin to add dimensionality and detail, and as our data resources to ground such models in reality expand and refine, we will also begin to move away from traditional ideas of models themselves. In classical spatial simulation, models are usually considered as scaled versions of complicated realities. They have, in the past, been built as pedagogical tools or as estimators, usually with abstraction or coarse appreciation as an aim. In recent years (and perhaps steadily throughout the evolution of such models), an alternative narrative regarding the use and potential of spatial models has developed, one in which the model presents as a synthetic reality, a computational laboratory for playing with ideas, plans, and fancies. While scaled models may always have significant value in the scientific process, it is likely that geographical ABMs will enjoy increasing use as engines for geography in developing virtual worlds and augmented realities in the future.

FURTHER READING

Stan Openshaw's early papers (e.g. Openshaw et al., 1987) set the stage for most of what we now call geographical ABMs, by introducing advanced notions of computation and computability, and by proving that geography is everywhere in the computational realm. Paul Longley's edited book on geocomputation was pivotal in organizing the field around a common research agenda and in introducing its defining concepts (Longley et al., 1998). Readers of this particular chapter may be interested in exploring geosimulation models in Benenson and Torrens (2004), which presents a very detailed overview of the last 50 years or so of geocomputational simulation methods and applications. Andrew Crooks and colleagues have also published a more recent book on geographical agents with contributions from a range of authors in the field (Heppenstall et al., 2012). Michael Batty's (1997a) paper on the computable city is a must-read, as in many ways it predicted the now rapidly appearing fusion of models and data that is coalescing around the notion of a new science of cities, which is largely mediated by geocomputation. His recent book on that very topic is also a useful resource (Batty, 2013). Jean Baudrillard saw the clash of models and reality coming a long time ago, and this is discussed philosophically in *Simulacra and Simulation* (Baudrillard, 1994). Science is always inspired and contextualized by science fiction, and readers may wish to refer to Neal Stephenson's (1993) book on virtual worlds and avatars, *Snow Crash*, for particularly relevant insight. My opinion is that it is all about geography!

3

SCALE, POWER LAWS, AND RANK SIZE IN SPATIAL ANALYSIS

Michael Batty

The properties of spatial probability distributions

Many variables in spatial analysis are distributed as probabilities that reflect competition between the elements, objects or individuals that compose them. The typical form of such distributions is quite unlike those that evolve with no direct competition between their elements, with these being normally distributed, in that there are quite low numbers of both large and small objects and high numbers of objects that are of moderate size. Such distributions have well-defined means and are usually shaped with respect to their frequency and size as bell-shaped curves where most of their elements – at least 95% – lie no more than two standard deviations on either side of their mean. The best examples in the human sciences relate to attributes of ourselves – our height, our weight, and other physical characteristics which evolve slowly with no significant interpersonal competition but according to random mutations that change these attributes over many generations. In contrast, when we examine socio-economic characteristics of ourselves (e.g. our incomes), we find that this variable is distributed with a very heavy skew to the right, where most of the elements or individuals forming the distribution have small incomes and an increasingly small number have large. If we define the size (say, height) of each individual i as x_i, its frequency $f(x_i)$ is distributed according to $f(x_i) \propto \exp[-\lambda (x_i - \mu)^2]$, where λ is a dispersion parameter and μ is its mean. These normal distributions are much rarer in spatial analysis than those that are highly skewed such as income, which can be approximated using power laws which have frequency $f(x_i) \propto x_i^{-\alpha}$, where α is the relevant parameter.

Although skew distributions are ubiquitous in spatial analysis, our detailed understanding of how competition orders the elements in these distributions is quite rudimentary, based on random mechanisms or at best theoretical models that operate under tight constraints on where objects can locate and grow in space (Clauset et al., 2009). For example, the most accessible point in a circular market area is its centre, and the number of locations with lesser accessibilities increase geometrically as one travels further and further from the centre. Thus if we then rank-order these accessibilities by location, the greatest frequencies are the lowest and as accessibility (which is a proxy for size) increases, the number of locations successively decreases (Batty, 2013). We do not need to demonstrate that this is a power law, just as we do not need to show that human characteristics are always exactly normally distributed. All we seek to do is argue that typical spatial distributions are skewed, usually to the right, although we can order these both left and right dependent upon the representation of space that we adopt. These probabilities can thus be approximated by a wide class of skew distribution functions of which the power law is perhaps the simplest exemplar (Simon, 1955). In fact the power law has particular properties that make it even more attractive in that it tends to be applicable to systems that scale, that manifest self-similarity at different scales, and that can be generated as fractals. It is easy to see what this means with a simple power law of the kind we have already noted. If the size variable x_i is scaled by a factor K, then its probability distribution scales as $f(Kx_i) \propto (Kx_i)^{-\alpha} = K^{-\alpha} x_i^{-\alpha} \propto x_i^{-\alpha} \propto f(x_i)$. These are important properties that relate to how we might represent and simulate spatial systems, but they lie beyond the scope of this chapter (Batty and Longley, 1994).

Although most spatial analysis has focused on transforming and searching for variables that approximate normal probability distributions, there is an even more important problem when we examine these distributions over time. Although it would appear that many skew distributions are stable with respect to their skewness when observed at different points in time, even over quite long time intervals, the objects that compose them are seldom fixed. In fact they often change quite radically over quite short intervals of time, but their overall distributions can remain quite stable. We will explore these issues in depth in this chapter, but they pose an enormous conundrum with respect to how we explain the way human and socio-economic phenomena organise and self-organise in space. Spatial distributions that remain comparatively stable with respect to how cities are organised appear to achieve a macro-regularity in their form from one time period to the next, but at the same time they admit rather basic volatility between the elements that make up such patterns. We need not only represent how spatial probability distributions are skewed in a stable and regular fashion, but we will also explore how such distributions continually change in their individual elements while at the same time preserving this macro-regularity.

Let us state this paradox in starker terms: if we rank a set of cities by their population sizes that describe some sort of integrated regional or national system, the distribution tends to follow a power law that has strong regularity across many time periods. In fact this regularity is so unerringly strong that Krugman (1996a) was prompted to say:

> The size distribution of cities in the United States is startlingly well described by a simple power law: the number of cities whose population exceeds P is proportional to $1/P$. This simple regularity is puzzling; even more puzzling is the fact that it has apparently remained true for at least the past century.

However, if we were to examine the size distributions and the cities that compose them at any two times, we would find that cities move quite quickly in terms of their ranks (and of course their sizes). Taking the two distributions of cities in the USA in 1890 and 1990, in 1890 New York City was ranked 1, as it was in 1990. But Houston was not in the list of the top 100 cities in 1890, yet it had reached rank 4 by 1990. In terms of the top 50 cities in the world at the time of the fall of Constantinople in 1453, only six remain today. This micro-volatility in the face of macro-stability implicit in the power law is puzzling, to say the least, in that we do not have good theories of why city systems can maintain their aggregate stability while at the same time shuffling the objects that make up this stability in such a way that the overall scaling appears almost static. It clearly relates to competition between the objects in some way that suggests that the relative size of any object is always constrained by some upper resource limit that remains largely undefined.

In this chapter, we will first state the nature of the power law, examine its properties, and then explore the archetypical example due to Zipf (1965) who in 1949 was the first to draw popular attention to the distribution of US city sizes. We will introduce various visual mnemonics, in particular the rank clock, which will enable us to explore micro-changes in the city ranks that nest within the wider regularities associated with these size distributions. We will illustrate different trajectories and morphologies that compose these visualisations, and this will provide us with the background to attempt a rudimentary classification of rank clocks that details their particular dynamics. We will then extend our analysis of the US urban system using data for the metropolitan statistical areas (MSAs) from 1969 to 2008, and follow this with an examination of scaling in high buildings (skyscrapers) in New York City. High buildings are distinguished by the fact that newer buildings tend to be higher than old, while a building rarely declines in height due to the fact that high buildings tend to be demolished if they are changed at all. This poses a rather different dynamics that produces somewhat different patterns through time. We then examine the change in scaling associated with hubs in a network whose sizes are based on the number of travellers moving through these locations at different times of the working day. Our illustrations of dynamics will all be related to changes in ranks, exploiting the idea of how ranks change over time, where time is displayed as a clock not organised on the 12-hour cycle *per se* but calibrated to the time periods over which the dynamics is considered (Batty, 2006). In conclusion, we will argue that the real puzzle is to unpack the way spatial competition, which organises these patterns into strongly regular size distributions, gives rise to a continual shuffling and mix of cities as they move up and down the size distribution.

Power laws explained and rank clocks defined

As Krugman (1996a) noted, the size distribution of US cities follows the simplest power law where the size of a city P varies in inverse proportion to its rank r as $P \propto r^{-1}$. Expressed in terms of frequencies as a probability distribution, the frequency of the occurrence of a city of size P is $f(P) \propto P^{-2}$, which is the derivative of the previous rank-size expression. This kind of manipulation is quite simple, but it is worth noting that much confusion arises with power laws and rank size because different authors use one or other of these equations and the discussion as to the actual value of the power can become obtuse. In fact the idea that the power of the rank exactly equals 1 or the power of the frequency 2 is an ideal type, although it does appear to be the consequence of a system developing competitively but randomly to a steady state (Gabaix, 1999). A more generic form, however, is to assume that the probability that we introduced above is $f(P) \propto P^{-\alpha}$, with its rank-size form as $P \propto r^{-1/(\alpha-1)}$.

This power law is often contrasted with the Gaussian (or normal) distribution, which is symmetric about its mean and bell-shaped with two very thin tails covering the smallest and largest objects in the size distribution. As we noted above, power laws essentially have long or fat tails that are skewed to the right or left or both (but in this context are usually skewed to the right), where the long tail contains the largest objects for which there is no upper bound. Again there is confusion over fat, thin, long and heavy tails in the literature, but here we will cut through all of this and begin with various city size distributions in the USA for the year 2010. Before we focus on dynamics, we will examine two datasets: the one equivalent to that used by Krugman (1996a) and Zipf (1965) based on cities defined by the US Bureau of the Census; and one based on MSAs defined by the US Office of Management and Budget. We show these distributions as rank (counter-cumulative frequency) size in their untransformed and transformed form (as $\log P(r_i) = \log K - \beta \log r_i$, where $\beta = 1/(\alpha-1)$) in Figure 3.1. The MSA data cover metro areas that are more than four times as large as the US Census cities data, which are based on counties. The plots shown in Figure 3.1 and their estimates have slopes somewhat lower than the pure Zipf parameter of unity, being 0.726 for the cities and 0.862 for the MSAs. Note that Krugman

(1996a) estimates this slope for the top 40 US cities he selected from 1990 data as 1.004, which simply shows that the set of cities chosen and their areal extent can make a substantial difference to these estimates. This suggests that there is no best theoretical value for this parameter with respect to real distributions, and all of them are intrinsically affected by the noise associated with empirical definition (Cristelli et al., 2012).

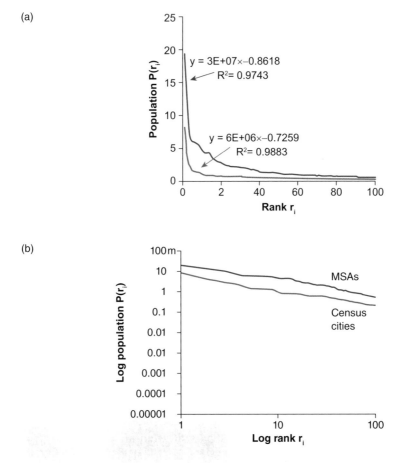

FIGURE 3.1 Rank–size distributions for MSAs and US Census cities 2010 in (a) power law form and (b) logarithmic linear form

If we examine how these rank–size relationships change through time, there is quite remarkable regularity in that the powers vary very little, notwithstanding that there might appear some drift in their values. The best dataset we can use to show this is that on cities from the US Census which we have for the top 100 cities from 1790 to 2000. The rank–size distribution appears extremely stable, as we show in

Figure 3.2(a) and as Krugman (1996a) so clearly remarked in our quote above. But these regularities, remarkable enough in themselves, begin to unravel when we examine the individual cities that make up these ranks. If we plot the shift in ranks, there is considerable movement of cities in terms of their size and rank up and down the hierarchy. In Figure 3.2(a) we also display one measure of this shift by plotting the year 2000 city sizes according to the 1940 ranks; one can see that the smaller sizes tend to shift more than the larger. In fact these shifts are not complete because some of the cities at 1940 are no longer in the top 100 ranked cities in 2000; in fact by then the number of cities that are common to both dates has reduced to 60.

We can enhance this by noting that the cities that are in the top 100 ranks over the 210-year period from 1790 can be displayed individually by plotting their ranks and colouring them according to a spectrum that begins with red and transitions through to yellow, then from green to blue as cities appear in the ranking (using the typical heat map convention). So the first city at the top rank in 1790 is coloured red and the last city to appear in the ranking over the next 210 years is coloured blue, with the transition evenly spaced according to the heat map colour spectrum. We show this for what we call the 'rank space', which is the size versus rank graph (the so-called 'Zipf plot') in Figure 3.2(b), but this is a particularly messy form in which to visualise more than a few objects that comprise the distribution. What this plot does show, however, is that there are several very distinct trajectories defining the space: cities that shoot into the space from outside the top 100 towards the top and vice versa, cities that remain at the same rank defined by vertical lines on the plot, cities that oscillate up and down in terms of rank, and so on.

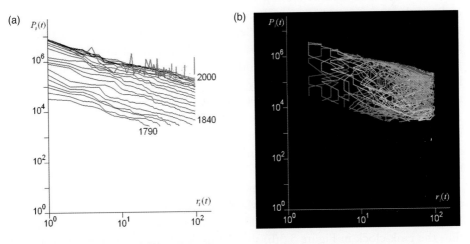

FIGURE 3.2 Rank-size distributions: (a) rank shift, and (b) changes to individual cities in the rank space from 1790 to 2000

A much better mnemonic is to make time explicit and to suppress size; after all, rank is a synonym for size and if the focus is simply on relative position, then rank and time are somewhat more illustrative of volatility in the Zipf plot than rank and size (Batty, 2006, 2010). What we do is to plot time as a regular clock around its circumference, defining the beginning of the time in question (in this case 1790) at the noon–midnight position with the years running in the clockwise direction around the clock until the hand reaches back to noon–midnight at the end of the time period in question (in this case at the year 2000). We can then plot the rank of the city as a radial from the centre of the clock at the appropriate time where we can organise the radial from rank 1 at the centre to rank 100 (or whatever the upper limits of rank) on the circumference, or the other way around, using a linear or logarithmic scale. Here we will use the simplest linear scale with the highest rank at the centre of the clock and the lowest on the circumference. For our 210-year city size distribution taken for the cities defined in the US Census, we show some typical trajectories which compose part of the rank clock in Figure 3.3(a), where it is clear that different cities are associated with quite different trajectories. We will return to this in the next section, where we argue that the clock and its derivatives can be used to think visually about the nature of dynamics in systems that scale.

What is fascinating about this particular clock is that this defines the temporal signature of the development of the US urban system. New York City is the anchor of the clock, being number 1 in rank ever since the Census began in 1790. In some respects, the city is the fulcrum of the entire US system. The opening up of the Midwest, California and the South is also marked out with first Chicago (around 1840), Los Angeles (1890), Houston (1900), then Phoenix (1950) flying into the clock from outside the top 100. Several colonial towns established in the eighteenth century and before, such as Charleston (SC), lose rank and fall out of the top 100, while some colonial settlements in the vicinity of Washington, DC and the northern part of the South lose rank and then begin to stabilise as sprawl makes an impact after the Second World War. Rustbelt cities such as Buffalo (NY) lose rank systematically from the early twentieth century onwards. There are few cities that enter and leave the top 100 in any significant way, but some, such as Atlanta, zoom in only to lose rank as they stabilise, although to some extent boundary changes and suburban sprawl complicate the picture. We have not attempted any classification of different trajectories so far, but the prospect exists for such analysis in future work.

We also illustrate the complete rank clock for all cities that are in the top 100 from 1790 to 2000 in Figure 3.3(b). Many cities enter and leave this exclusive set. Before 1840, there were less than 100 cities catalogued in the US Census, and thus the rank clock in Figure 3.3(b) shows this build-up. We have not normalised any of these cities for boundary changes, so our analysis is inevitably crude.

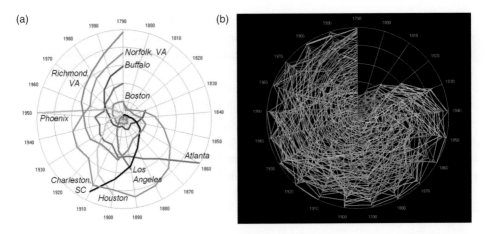

FIGURE 3.3 The US rank clock defining (a) key city trajectories, and (b) cities ranked in the top 100 from 1790 to 2000

Moreover, although after 1840 there are only 100 cities ranked at each time period, in fact over the 210-year period there are 266 cities that are part of the top 100, and this is itself a measure of the volatility of the set, with there being 2.7 times the number of cities appearing in the top 100 over this period. If these cities entered and left the top 100 uniformly, this would mean that on average, about eight cities would enter and leave the top 100 each time period, about 8% per decade. What Figure 3.3(b) clearly illustrates impressionistically through the collage of various trajectories (by their colour) is the substantial volatility of the US system over two centuries. Of the cities in the top 100 in 1840, only 20 cities remained in the top 100 by 2000, consistent with a rate of change of cities being in the top 100 of about 8% per decade.

Trajectories and morphologies of the rank clock

The morphology of the clock is composed of a collage of trajectories of each object, in this case a city, with the trajectories themselves being of different types which we illustrated briefly in Figure 3.3(a), and their intersection forming the overall form of the clock which gives it a distinct morphology as in Figure 3.3(b). Clock-like trajectories can thus be classified into types, although we will only concentrate on the simplest here. First, we show different trajectories. If an object always remains at the same rank, then it will trace out a circular trajectory on the clock, with objects at higher and higher ranks producing smaller and tighter circles to the point where the object is always at rank 1, which is a point at the centre of the clock. Objects that slowly enter the clock and move up towards its centre form inward spirals whose curvature relates to the inverse of the speed at which the objects move up rank. Objects that spiral out of the clock and lose rank perform in the opposite way. The most difficult objects to classify are those that move up and then move down, perhaps even moving in and out of the clock (which always has an upper bound on the number of ranks considered). Objects that oscillate around the clock and stay within it in a regular pattern are more unusual, but, as we will see, when we look at the clocks of transport movements, such oscillations can be seen relating activities to 12-hour, diurnal, weekly and related temporal patterns. In Figure 3.4, we show typical examples of these trajectories.

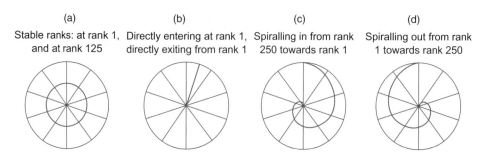

| (a) | (b) | (c) | (d) |

Stable ranks: at rank 1, and at rank 125   Directly entering at rank 1, directly exiting from rank 1   Spiralling in from rank 250 towards rank 1   Spiralling out from rank 1 towards rank 250

FIGURE 3.4   Classic changes in rank defining idealised trajectories

When we generate a collage of these idealised trajectories, we define different morphologies that we can use as baselines to which we can compare real data. First, for objects that remain at the same rank throughout the temporal period, we define a series of concentric circles starting at the pole of the clock and then splaying outwards. To demonstrate this morphology, we have generated an idealised distribution of objects from a power law where $P_t(r) \propto 1 / r$ for all t. This produces a distribution that declines with rank but is identical for all time periods; that is, the object that is ranked r at time t, has the same rank $t + \tau$, for all τ. We show the rank–size distribution in Figure 3.5(a) and its logarithmic form where the slope is equal to 1 in Figure 3.5(b). The rank clock is shown in Figure 3.5(c) where the colouring of these trajectories is based on the rule defined earlier, which reflects the red–yellow–green–blue spectrum ordered according to the time and the rank when the object first appears in the series.

Above we examined the shift in ranks between two points in time by showing the distribution at the first point in time using ranks from another. There are many different statistics that relate to these dynamics, but here we will look at only one, the half-life, so-called because it gives us the number of objects at a given time $t - \tau$ and $t + \tau$ that still exist in the rank-size distribution for any time t. The half-life for the downswing $t + \tau$ is the number of years τ_t^+ when the number of objects is $n / 2$, while for the upswing $t - \tau$ this is τ_t^-, but of course these can be different. The overall half-life of the entire system for the upswings

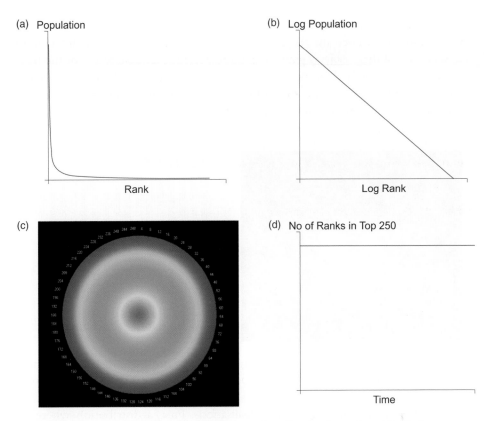

FIGURE 3.5 (a, b) An idealised rank-size distribution based on Zipf's law, where (b) is a log–log transformation of (a); (c) its rank clock; and (d) a plot of its half-life

and the downswings can be defined as the average of the sum of these values over all years T, which are $\tau^+ = \sum_{t=1}^{T} \tau_t^+ / T$ and $\tau^- = \sum_{t=1}^{T} \tau_t^- / T$, while the average for the total system based on upswings and downswings can be formally defined as $\tau = (\tau^+ + \tau^-)/2$. Now in our clock with stable ranks where nothing changes over the entire time period T, the number of objects in the distribution at any time t is the same as the number of objects at $t-\tau$ and $t+\tau$, and thus there is no point where the number of objects drops to one–half of those at a given time. This is illustrated in Figure 3.5(d) where the number of objects in the top ranks is plotted on the vertical axis against time on the horizontal axis. Note that for all these hypothetical explanations of morphologies and trajectories, $n = 250$ and $T = 250$, and the number of objects n is the same at each time t.

Now if we assume that each object in the distribution enters the distribution at top rank 1 at time t and moves to the bottom rank 2 at time $t + 1$, implying that there are only ever two objects in the distribution, then the clock is displayed in

Figure 3.6(a) where it is clear that it consists of a series of spikes. The half-life is quite straightforward because the number of objects at any time in the distribution is 2, with one of these objects generated at the previous time period, thus the half-life (which is only defined for the downswing) is 1 year. We show this in Figure 3.6(b) where the graph shows the number of objects which pertain to any time t, which is 2, and only one of these remains at time $t + 1$.

(a)   (b)

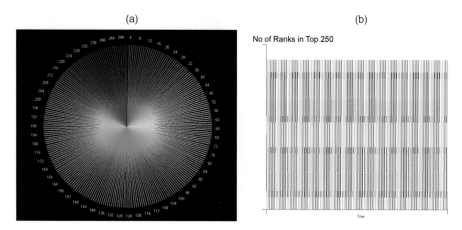

FIGURE 3.6   The rank clock based on (a) an extreme rank-size distribution and (b) a plot of its half-life

We will now examine two other idealisations. First, a distribution where an object enters the rankings at number 1 and then systematically declines until it reaches the lowest rank, one position at a time at the end of its life. The number of objects builds from $n = 1$ to 250 and thus the last object, $n = 250$, which enters at position $r = 1$ at $t = 250$, starts and finishes its life at this point. The rank clock of this structure is shown in Figure 3.7(a) and its half-life plot is a simple set of lines from the time when each object enters. In fact, the number of objects builds up linearly during the 250 time units, and although the first ranked object declines to rank 250 by the end of the time period, all objects remain in the distribution across all times and thus no half-life can be defined as such. The second distribution is one where the object enters at rank 1 and stays there, but new objects then enter, one at each successive time period. The half-lives are the same as all objects remain in the top ranks and the relevant rank clock is shown in Figure 3.7(b).

Our last ideal type involves each object rising from the lowest rank to the highest then declining again to the lowest, and of course dropping out of the mix if the temporal period is beyond the cycle associated with the object. This in fact is a mix of the previous two types of morphology. We begin with the top ranked object at rank 1, and then this loses rank until it disappears at the end of the temporal period. The object with the lowest rank at time 1 gradually rises in rank to

(a)   (b)

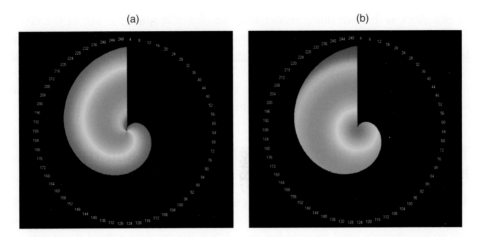

FIGURE 3.7   Rank clocks based on successive objects entering at rank 1: (a) declining according to Zipf's law; (b) remaining in stable orbit

its maximum and then begins to decline. In fact this order from lowest to highest to lowest is associated with each object in sequence, and the way we must illustrate this is according to the clock in Figure 3.8(a). This is complicated because of the way the colours mask the true process, but if we examine Figure 3.8(b), we see that at the beginning of the period the highest rank marked in red declines as an outward spiral to the end of the period. The lowest ranked object at time 1 increases from rank 250 to finish at rank 1 at the end of the time period. This defines the bounds of the trajectories defining the clock. If we take the object which exists at rank 125 (coloured green) at the beginning, this rises to rank 250 in an inward spiral, reaching this at time 125, and then it spirals out to rank 125 again at time 250. We see these trajectories in Figure 3.8(b). One really nice feature

(a)   (b)

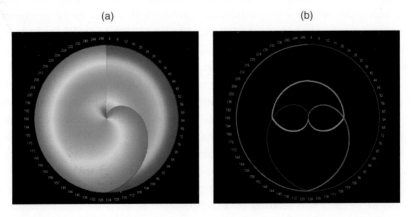

FIGURE 3.8   Rank clocks based on: (a) successive objects entering at all ranks; (b) plots of cities entering at ranks 1 (red), 250 (blue) and 125 (green)

of this clock is that we can consider the first object as spiralling out of the clock from rank 1 to 250 (in red) and then spiralling back into the clock, connecting up to the initial object at rank 1 at the next time 250 (or time 1). This is only clear from the three trajectories in Figure 3.8(b). The half lives of all these objects are not defined as all objects remain in the distribution over all time periods. If we were to plot these, then this would be the same plot as in Figure 3.5(d).

The dynamics of cities, high buildings and transport hubs

To complete this analysis of city systems, we will return to the first dataset that we examined earlier for population associated with the 366 MSAs, and we will extend this to also include total income for each MSA. We have assembled these data in the time series from 1969 to 2008. Its great advantage is that the MSA boundaries are stable and the set of cities is complete, thus implying that the growth dynamics associated with changes in size and rank is clearer than that of the 100 top ranked cities from the US Population Census data. In Figure 3.9 we show the rank-size distributions associated with the population and income measures for the 366 MSAs. These are at a much finer temporal interval than those we examined from the Population Census and thus any measures of shift need to be normalised if comparisons are to be made. In Figures 3.9(a) and (c) we show the rank-size distribution for population and income. These are extremely close to one another, but when we examine their individual clocks in Figures 3.9(b) and (d), these show more volatility. Nothing equivalent to the volatility for the city size distributions from the Census is seen in these distributions, but the time period is much shorter and the urban system is much more mature. In fact the population clock implies that the core cities remain as core, but that there are several smaller cities that rise up the hierarchy, while a few established cities drift down more gradually. Of course this clock implies changes in rank, not size, so although Phoenix, which is ranked 35 in 1969, increases its rank to 12 in 2008, it more than doubles in size. In fact Houston increases its population almost three times while its rank goes from 14 to 6. In the mid-range of the hierarchy, Las Vegas increases its population from about 267,000 to 1.87 million, some 7 times, and its rank from 114 to 30, some 3.8 times. These are substantial shifts given the fact that the system is mature, with a growth rate in metropolitan populations of only around 0.5% per annum.

When we examine the income distribution, there is more volatility, with one very obvious shift in income due to oil being discovered in Fairbanks, Alaska, in the late 1960s. There was a boom in pipeline and related infrastructure construction in 1975–1977, leading to a big increase in income, and then the local economy collapsed back to its former trajectory. This is clearly seen in the income rank clock in Figure 3.9(d). The last distribution relates income to

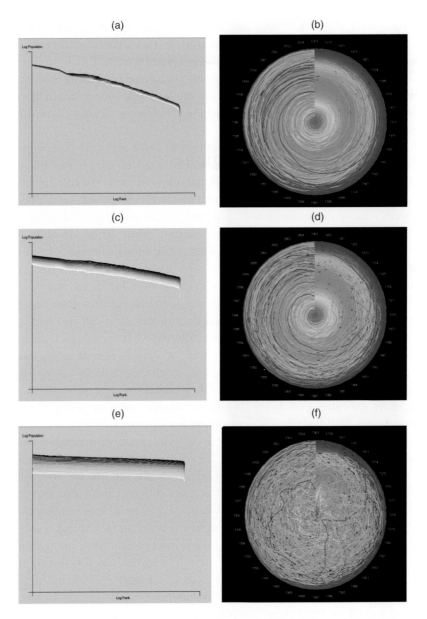

FIGURE 3.9 Population, income and income per capita: (a, c, e) respective rank size and (b, d, f) rank clocks for 366 MSAs from 1969 to 2008

population as income per capita, and the rank size and its clock are shown in Figures 3.9(e) and (f). There is considerable mixing implied by this clock, despite the impressively smooth macro-distributions with respect to their form over time. Quite clearly, as we form composite distributions by taking ratios of more basic data, the noise from one combines with the other and differences (variances) can become magnified. There is in fact little experience of such mixing, and we lack real intuition as to its consequences. But it is enough to cast doubt on many of the more stable and regular relationships that we often begin with, such as power laws, and this suggests that our knowledge of this whole area is primitive and subject to more profound scrutiny than anything that we have attempted so far.

There are many systems where competition in space and time determines how their constituent elements grow or are manufactured in size. If we disaggregate populations and examine the internal distribution of such clusters in cities, we have already seen that these intra-urban elements follow power laws, albeit with considerably more noise associated with their spatial and temporal arrangements than entire cities. If we now divert our attention from actual amounts of such activities to the physical environment which accommodates them – from people to the buildings in which they reside or work – we also find that the sizes of these buildings follow power laws. In fact if we examine high buildings, which are usually defined as being greater than six storeys, certainly greater than ten (which require elevators for their operation), then their distribution can also be shown to follow the rank-size rule. There is a major difference between high buildings and population sizes, in that buildings do not grow or decline, at least in the same way as populations; they are manufactured, and rarely are storeys taken off them through partial demolition and rarely are they added to.

There are exceptions of course, but in our analysis here we will exclude these occasional cases. Buildings do get demolished, but in our analysis we have excluded these too for we will illustrate these ideas only on extant skyscrapers – buildings greater than 40 metres – in New York City from 1909 to 2010, also measuring their height in metres. There are 516 buildings in this set, but we will only ever plot the top 100. In fact the rank clock is quite different from that for cities. As the century progresses, a building that is number 1 in rank does not stay there for long. Skyscrapers have got successively higher as building technologies and materials have progressed, and thus the rank clock is marked by a continuing downward spiral of earlier high buildings, many of them leaving the top 100 during the 101 years that the clock portrays. We show the clock in Figure 3.10(a), and the downward spiral provides the dominant morphology of these dynamics. In fact at the start of the clock in 1909 there is rapid growth to 119 high buildings 'in the top 100' because there are several ties for height, but the rest of the clock is contained within the envelope of 100.

The other feature of this dynamics is that the clock shows quite distinctly the waves of skyscraper building that have dominated New York. At the start of the period in the early twentieth century before the First World War there was a great wave of such building. Then again after the war in the mid-1920s to early 1930s the boom, which

(a)   (b)

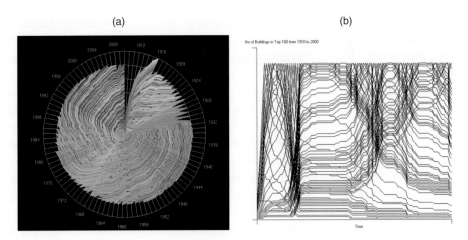

FIGURE 3.10   New York City skyscraper heights, 1910–2000: (a) rank clock;
(b) number of buildings defining the half-lives at each time

preceded the great recession, was a time of massive investment in high buildings. In fact the highest buildings in the city that still dominate the skyline, such as the Empire State and Chrysler buildings, were constructed then and if you look at the core of the clock – the top ten ranks – you will see that these are dominated by buildings (coloured green) that were constructed in the 1930s. In fact earlier buildings are overwritten at the level of resolution used in the clock and some of the early high buildings such as the Woolworth building only reappear, due to the visual limitations of the clock, once the 1930s wave of building subsides. There are waves in the 1960s and 1980s and then more recently in the 2000s, but buildings in general have not been much taller than those built in earlier times. It is elsewhere in the world that the highest buildings have recently been built, in the Middle East and in China.

An even more graphic demonstration of these dynamics is given by the plot of half-lives for the 101 years that comprise these competitive processes, which are firmly linked to boom and bust. In Figure 3.10(b) we show these half-lives, and it is clear that the dynamics produces clusters that relate to specific 'economic events'. Remember that the half-life for the set of 100 buildings that exist at a given point in time is the number of years between the time in question and the time when only half the number of buildings at this time remain in the system. We can of course compute half-lives for buildings that are entering the system, as we indicated earlier, and these ultimately compose the 100 in question. This is often different in time span from those that are leaving the system, being knocked out by higher buildings being constructed where the progression is often slower. The times of rapid building are clearly picked out by the half-life plots in Figure 3.10(b), where the 1910s, 1930s, 1950s, 1960s, 1980s and 2000s are periods of very rapid increase in the size of high buildings whose half-lives on the upswing are much shorter than on the downswing which are more muted. Thus increases in rank during an upswing occur much faster than their relative decrease in rank in the downswing. In fact this figure shows how hard it is to produce an average half-life for the entire series. For periods of rapid growth (boom), the upswing half-life appears to be about four years, whereas the subsequent downswing is about 25 years. Also the downswing half-life seems to be shortening whereas the upswing is less variable. Overall we estimate that the average upswing half-life to be about ten years and the downswing 20 years, but the picture is complicated by the volatility and the dominance of boom and bust.

Our third example refers to the size of hubs in spatial networks. It is very clear that the evolution of networks is governed by competitive forces that enable a limited number of hubs to gain more than proportionate numbers of links. Translated into volumes of activities flowing on such links into hubs, this gives rise to distributions that are similar to power laws. This was first demonstrated by Barabási and Albert (1999) but it pertains to many developments in network science pioneered during the last 20 years. There is some debate as to whether or not these distributions are scaling, for their derivations using laws of proportionate effect tend to generate log-normal distributions, but as most of these distributions are modelled with respect to their heavy tails, power laws can form a good approximation to these. Moreover, there are likely to be more constraints on the form of these distributions due to the fact that spatial networks are constrained in space and cannot generate the numbers of links that are generated in unconstrained network structures. In short, planar graphs which dominate spatial networks do not manifest scaling in their pure form (Barthelemy, 2011).

Our example constitutes the hubs that define the rail stations on the London Underground and Overground (the stations serving heavy, mainly suburban, rail services), where the volumes of travellers entering and exiting these hubs define their size on a typical weekday in November 2010. The dynamics of these hubs relates to the fact that during a typical day, all the hubs only operate for 20 hours, for the system is closed from 1.20 a.m. to 5.20 a.m. each day. The dynamics is also dominated by the morning peak and the evening peak hours, and the volumes reflect this. We have organised the data, available on a second by second basis, into bins of 20 minutes each, of which there are 72 defining the 24-hour day. Twelve of these are empty as there are no trains running. There are a total of 6.2 million entries and 5.4 million exits (the difference is due to open barriers where the RFID (Oyster) card, from which the data are taken, is not used), and we will aggregate these entries and exits to form the volumes for each of the 666 hubs that define the system.

We show the 60 different rank-size distributions in Figure 3.11(a), where it is clear that the differences pertain to different volumes at different times of the day. The dominant cluster of trips is during the two peaks that are clearly evident in Figure 3.11(a), but the shape of these distributions is quite similar at each 20-minute

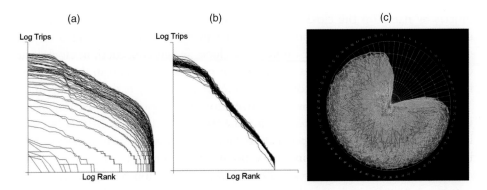

FIGURE 3.11 Sizes of London station trip volumes, November 2010: (a) rank size; (b) collapsed rank size; (c) rank clock

interval, as we show in Figure 3.11(b), where they are collapsed onto one another. We achieve this by taking the distributions from their mean size and normalising by their variance so that they are comparable. It is also clear from Figures 3.11(a) and (b) that the distributions are not scaling but are much closer to log-normal. We show the clock in Figure 3.11(c), from which it is clear there is enormous variability in the way hubs move up and down in the hierarchy of ranks during the day. What we see from this is that some hubs that have low volumes early in the day pick up in the morning peak and then collapse back in the middle of the day to rise again in the evening peak. These are inner suburban hubs, whereas those in the central business and shopping districts tend to remain the biggest during the whole day and close down last. To really explore the meaning of these dynamics, it is necessary to know the actual spatial configuration of rail lines and hubs; to this end, the *Rank Clock Visualiser* that we introduce in the next section indicates how one can make progress in generating much more satisfactory explanations of these spatial network dynamics.

Next steps: the Rank Clock Visualiser

We have developed a visualiser (O'Brien, 2014) for these kinds of scaling distributions that enables the user to link the objects in the clock to their spatial location. The user can generate a rank clock for many different distributions, which so far are confined mainly to city size and building height distributions but also include Fortune 500 data from 1955 to 2010 (with renormalisation in 1994–1995) and the distribution of US baby names. The London Tube hubs are in the set, as are several distributions of UK, US and Japanese populations by small area/cities. Users can plot the rank clock with the highest rank at the centre or the edge, choose any time periods from the maximum available for each dataset and identify specific objects by name on the clock and map. The link to the spatial distributions uses either OpenStreetMap or Google Earth; the user can point to either a city or object on the map or globe or on the rank clock, and its equivalent in clock, map or globe will show up. There is no animation of the clocks so far in this interface, but this will be done in due course for there are many easy extensions like this.

In the dataset, we have world city sizes from the Population Division of the UN Department of Economic and Social Affairs from 1950 to 2010 (some 576 in all, greater than 1 million population each). We show the rank clock of these in Figure 3.12 from the Rank Clock Visualiser, where we have picked out Adelaide in South Australia, a good example of a 1 million population city that is stable in population but declining in rank. The Google Earth display alongside lets the user visualise all these cities and their sizes, and the user can click on a city and see where its trajectory lies on the clock or identify the name of the city from a drop-down list and activate its trace on the clock and on the map or globe. It is not yet possible to zoom into the clock to identify a hot link to a trajectory because the level of resolution is too fine from the static display, but all these extensions are possible and will be explored in future work. The world cities distribution and its rank clock are shown in Figure 3.12.

We have not so far explored the possibility that the morphology of the rank clock itself provides a shorthand for the kinds of dynamics that characterise the system of interest. We noted earlier that the individual trajectories might be classified, but the shape of the clock also varies, and we can see from those illustrated here how different they might be. For example, clocks with no change in range are perfect circular

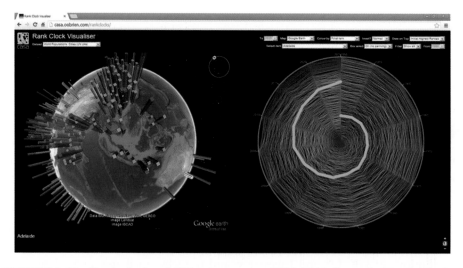

FIGURE 3.12 Rank clocks of UN urban areas, 1950–2010, from the Rank Clock Visualiser, showing the trajectory for Adelaide, Australia

orbits, while those with regular changes in rank up or down or both reflect spirals. We can thus position any clock on this spectrum, getting an immediate picture of its dynamics. Some examples of this kind of classification can be seen in the clocks portrayed for growing cities such as those in Israel, which are dominated by inward spirals (Benguigui et al., 2008). What we need is to tie the dynamics more consistently to this geometry. In fact a purely regular dynamics is not very useful as it is quite unrealistic in city systems whereas the boom–bust structure of economic dynamics is much more likely. Changes in social taste are also likely to be reflected in urban dynamics, and our quest must be to begin to identify different dynamics that are associated with different geometries of clock so that a deeper, more structured picture of the way this world of cities works can be generated.

To conclude, it is worth saying a little more about the notion of scaling in city systems. Our argument began by suggesting that city-size distributions were scaling, following power laws in their heavy tails, although always predicated on the basis that their underlying distribution is more likely to be log-normal. Power laws are thus a good approximation to the heavy tails, but no more than this. And they simply represent our starting point. The rank clock is a good device if we can assume that population is related to rank by a simple logarithmic transformation, as is consistent with a power law, but once we have the idea of the clock, the fact that it relates rank to population can be conveniently forgotten. The clock has its own integrity in that it displays a kind of dynamics that can be explored more generally, and if a rank clock and a size clock are defined, one related to the other, then all kinds of novel animations and explorations suggest themselves. It is in this spirit that the Rank Clock Visualiser has been developed.

We need much better statistics that pertain to the different kinds of dynamics and their variation over time and space. The idea of the half-life needs to be put on a more consistent footing and defined more rigorously. But in the wider perspective, this chapter is as much about space–time dynamics as it is about power laws and scaling. We need to explore the extent to which the kinds of dynamics that define bifurcations, tipping and turning points, and even catastrophes, relate to scaling. Little has been done to date, but there are strong hints in the notion of fractals, self-similarity and hierarchy that need to be exploited in linking this kind of aggregate spatial analysis to the dynamics of spatial modelling. After 25 years or more of consistent but slow development in the field of spatial dynamics, there are now many fertile ideas that will see this field explode intellectually and in terms of applications during the next 25 years.

FURTHER READING

Readers should examine Zipf's (1965) seminal book *Human Behavior and the Principle of Least Effort*, which popularised the rank–size rule, and has led to a renewed interest in distributions that are scale-free. Herbert Simon (1955), winner of the 1978 Nobel Prize in Economics, wrote about power laws focusing on firm size soon after Zipf, and more recently the 2008 winner of the Nobel Prize in Economics, Paul Krugman (1996a), addressed the problem in terms of cities. An excellent summary of the field is in the article by Clauset et al. (2009) in the *SIAM Review*. I have written about rank clocks elsewhere in my 2006 *Nature* paper and in my book *The New Science of Cities* (2013) where I relate these to simple spatial models that generate these distributions. Power laws are ubiquitous in statistical physics and in human systems whose elements are conditioned by competition and evolution, and a good readable summary is by Manfred Schroeder's (1991) book *Fractals, Chaos, Power Laws: Minutes from an Infinite Paradise*.

PART II

EXPLORING MOVEMENTS IN SPACE

4

AGENT-BASED MODELING AND GEOGRAPHICAL INFORMATION SYSTEMS

Andrew Crooks

Introduction

Modeling human behavior is not as simple as it sounds. This is because humans do not just make random decisions, but base actions upon their knowledge and abilities. Moreover, one might think that human behavior is rational, but this is not always the case: decisions can also be based on emotions (e.g. interest, happiness, anger, fear). Emotions can influence decision-making by altering an our perceptions about the environment and future evaluations (Loewenstein and Lerner, 2003).

The question therefore is how we model human behavior. Over the last decade, one of the dominant ways of modeling human behavior in its many shapes and forms has been through agent-based modeling (ABM) (see also Torrens, Chapter 2). The remainder of this chapter provides a general overview of what agents are, why there is a need for agent-based models for studying geographical problems, and how it links to how we believe societies operate through ideas of complexity theory. It sketches out how geographical information can be used to create spatially explicit agent-based models, before reviewing a range of applications where agent-based models have been developed using geographical data. The chapter concludes with an overview of challenges modelers face when using agent-based models to study geographical problems, and identify future avenues of research.

What are agents?

ABM allows us to focus on individuals or groups of individuals and give them diverse knowledge and abilities (Crooks and Heppenstall, 2012). This is possible through the unique properties one can endow upon the agents (people, animals, etc.) within such models (Heppenstall et al., 2012; O'Sullivan and Perry, 2013). The key difference between ABM and other forms of modeling is the ability to focus on the individual and their behaviors. These properties include:

- *Autonomy*: We can model individual autonomous units that are not centrally governed. Through this property agents are able to process and exchange information with other agents in order to make independent decisions
- *Heterogeneity*. Through using autonomous agents the notion of the average individual is redundant. Each agent can have their own properties and it is these unique properties of individuals that cause more aggregate phenomena to develop.
- *Activeness*. As agents are autonomous individuals with a range of heterogeneous properties, they can exert active independent influence within a simulation. There are several ways agents can do this: they can be proactive (goal-directed), as when trying to solve a specific problem; or they can be reactive, in the sense that they can be designed to perceive their surroundings and given prior knowledge based on experiences (e.g. learning) or observation and take actions accordingly.

The primary strength of ABM is as a testing ground for a variety of theoretical assumptions and concepts about human behavior (Stanilov, 2012) within the safe environment of a computer simulation. For example, we know humans process sensory information about the environment, their own current state, and their remembered history to decide what actions to take, all of which can be incorporated with agents (Kennedy, 2012). Through the ability to model heterogeneity within agent-based models we can capture the uniqueness of what makes us human, in the sense that all humans have diverse personality traits (e.g. emotion, risk avoidance) irrational behavior, and complex psychology (Bonabeau, 2002). We also know that human behavior is influenced by others (e.g. Friedkin and Johnsen, 1999), which can introduce positive and negative feedbacks into the system. Through such processes, people form groups, and the results of aggregate behavior can be greater than the sum of the individuals within the group. Such properties again can be captured through the agents heterogeneity and active status.

But what drives humans? What motivates us to take certain actions? By agents being active we can test ideas and theories on what motivates people, and explore why do they do certain things. Perhaps the most cited concept in this regard is Maslow's (1943) 'hierarchy of basic needs'; Maslow discussed how humans order their needs and provided an overview of potentially competing priorities when trying to represent human behavior within an agent-based model. Kennedy (2012) lists three main approaches to capturing such cognitive processes, the first being a *mathematical approach* such as the use of *ad hoc* direct and custom coding of behaviors within the simulation, as when using random number generators to select a predefined possible choice (e.g. to buy or sell; Gode and Sunder, 1993). However, as noted above, people are not always random, which has led researchers to develop other methods such as directly incorporating threshold-based rules. For example, when an environment parameter passes a certain threshold, a specific agent behavior will result (e.g. move to a new location when the neighborhood composition reaches a certain percentage; Crooks, 2010). One could argue that such modeling approaches are appropriate when behavior can be well specified.

The second approach to modeling human behavior is through the use of *conceptual cognitive frameworks*. Within such models, instead of using thresholds, more abstract conceptual frameworks are used, such as beliefs, desires, and intentions (BDI; Rao and Georgeff, 1991), or physical, emotional, cognitive, and social factors (PECS; Schmidt, 2002). Both the BDI and PECS frameworks have been successively applied to modeling crime patterns (see Brantingham et al., 2005; Pint et al., 2010). These conceptual cognitive frameworks are basically decision tree hierarchies with respect to guiding human behavior within agent-based models, and offer a bridge between mathematical representations of behavior and those research-quality cognitive architectures tools that we will turn to next. While the approaches above have been developed with respect to ABM, research-quality *cognitive architectures* focus on abstract or theoretical cognition (Kennedy, 2012), with two of the most widely used architectures being Soar (Laird, 2012) and ACT-R (Anderson and Lebiere, 1998); however, such architectures tend to focus more with artificial intelligence and matching human decision-making over very short periods of time (seconds to minutes) on a small number of agents.

Why agent-based modeling?

The growth of ABM coincides with how our views and thinking about how social systems such as cities are being changed by utilizing ideas from complexity science (see Manson et al., 2012). Rather than adopting a reductionist view of systems, whereby the modeler makes the assumption that cities operate from the top down and results are filtered to the individual components of the system (see Torrens, 2004), people are now adopting a reassembly approach to the system, in the sense of building the system from the bottom up (O'Sullivan, 2004). This change follows the realization that planning and public policy do not always work in a top-down manner; aggregate conditions develop from the bottom up, and from the interaction of a large number of elements at a local scale. Thus there is a move toward individualistic, bottom-up explanations of geographical form and behavior which link to what we know about complex systems (Batty, 2005).

The key characteristics of complexity – such as self-organization, emergence, non-linearity, feedback and path dependence – can all be captured within agent-based models. By their very nature, agent-based models can capture emergent phenomena which are characterized by stable macroscopic patterns arising from local interaction of individual entities (Epstein and Axtell, 1996). A small number of rules or laws, applied at a local level and among many entities, are capable of generating complex global phenomena: collective behaviors, extensive spatial patterns, hierarchies, and so on, which are manifested in such a way that the actions of the parts do not simply sum to the activity of the whole. Thus, emergent phenomena can exhibit properties that are decoupled from (i.e. logically independent of) the properties of the system's parts. For example, a traffic jam often forms in the opposing lane to a traffic accident, a consequence of 'rubber-necking' (Masinick and Teng, 2004).

To give a simple example of how ABM can be used to study such an issue, take Figure 4.1, where each car is an agent and is given two very simple rules of behavior: first, if there is a car in front of it, the car slows down; and second, if there is no car in front, the car speeds up. These two simple rules applied to many agents can demonstrate how traffic jams can form without any serious incident. Moreover, through the modeling of individuals, we can see that there is great variation between individual car speeds that would be lost if we only looked at the aggregate (average) speed of traffic (such behaviors are explored in more detail later). However, if you are wondering if such simple rules can explain the emergence of traffic jams, see Sugiyama et al. (2008).

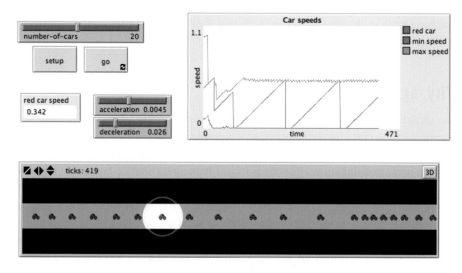

FIGURE 4.1 Simple traffic model where each car is an agent. From top left clockwise, model parameters, a chart of car speeds and the spatial agent environment (source: Wilensky, 1997, 1999)

Studying the behavior of collections of entities focuses attention on relationships between entities (O'Sullivan, 2004) because before change is noticed at the aggregate level, it has already taken place at the micro level. Complexity theory has brought awareness of the subtle, diverse, and interconnected facets common to many phenomena, and continues to contribute many powerful concepts, modeling approaches and techniques especially in relation to ABM. Geographical systems provide many examples of self-organization and emergence (Heppenstall et al., 2012; Batty, 2013); for example, it is the local-scale interactive behavior (commuting, moving) of many individual objects (vehicles, people) from which structured and ordered patterns emerge in the aggregate, such as peak-hour traffic congestion (Nagel et al., 1997) and the agglomeration of firms (Krugman, 1996b).

Why link GIS and agent-based models?

As noted above ABM has relevance to many geographical problems and there is growing interest in the integration of GIS and ABM through coupling and embedding (see Crooks and Castle, 2012, for a review). For agent-based modelers, this integration provides the ability to have agents that are related to actual geographic locations. This is of crucial importance with regard to modeling geographical systems, as everything within a city, region or country is connected to a place. Furthermore, it allows modelers to think about how objects or agents and their aggregations interact and change in space and time (Batty, 2005). For GIS users, it provides the ability to model the emergence of phenomena through individual interactions of features within a GIS over time and space. Moreover, some would consider this linkage highly appealing in the sense that while GIS provides us with the ability to monitor the world, it provides no mechanism to discover new decision-making frameworks such as why people have moved to a new area (Robinson et al., 2007).

The simplest way to visualize the integration of GIS and ABM is by taking the view most GISs do of the world, as shown in Figure 4.2. The world is represented as a series of layers (such as the physical environment, the urban environment) and objects of different types (such as cars or people, which can then be used as the basis for our agents). These layers form the artificial world for our agents to inhabit; they can act as boundaries for our simulations, or fixed layers such as roads provide a means for agents to move from A to B or houses provide them with a place to live. To further illustrate this notion, in Figure 4.3 we show a modified SLEUTH[1] model (Clarke et al., 1997). The model predicts

---

[1]The acronym stands for the raster input datasets needed to run the model: Slope, Land-use (e.g. urban and non-urban), Exclusion (where one cannot build), Urban extent, Transportation (road network) and Hillside.

the extent of urban growth, for example, from blue for the current urban extent, to red for the possible future urban extent after the model has been initialized with the input data and applying specific growth rules controlled by specific growth coefficients (see Clarke et al., 1997, for more details). Aggregate spatial data also allow for model validation: for example, are the land-use patterns we see emerging from a model of urban growth matching that of reality? If they do, it provides us with an independent test of the micro-level processes

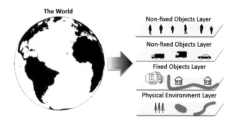

FIGURE 4.2 Representing the world as a series of layers of fixed and non-fixed objects (adapted from Benenson and Torrens, 2004)

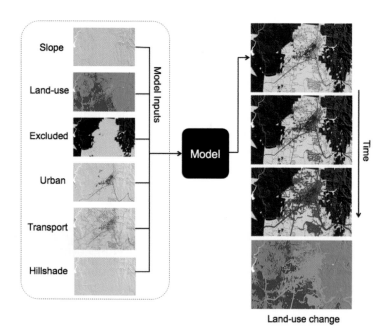

FIGURE 4.3 A basic SLEUTH-type model based on growth rules and coefficients as outlined in Clarke et al. (1997) driving the rates of land-use change in Santa Fe, New Mexico at 30 m^2 resolution

encoded within the model. Taking the example above, the growth coefficients can be calibrated and validated by comparing simulated land-use change with historical geographical information on the area, and if the model matches reality we can be more confident with future growth strategies.

However, it is not just raster data that can be used as the basis for the artificial world, but also vector data. For example, translating vector data into agents and their environment is shown in Figure 4.4. Here the model comprises two vector layers: the urban environment is represented as a series of polygons created directly from the shapefile; and agents are represented as points. It is the information held within fields of the environment layer (Figure 4.4(a)) that is used to create the point agents (Figure 4.4(b)), but also the layer extent defines the boundary of the world (see Crooks, 2010, for further details). The agent-based model uses the data to initialize the model. Note that the underlying color of the polygon can be used to represent the predominant social group in the area (accomplished by counting the number of agents (points) of different types within each polygon). Once the agents have been created they can be given simple rules; for example, in Figure 4.5 we show a model inspired by Schelling's (1971) model of segregation, where each agent has a preference of living in a neighborhoods where 50% of their neighbors are like themselves. Agents then calculate through a buffer operation their neighborhood makeup, and if their preferences are not met (i.e. 50% of their neighbors are not like themselves) they move to a new location where their preferences are met. Through such rules we can explore how the actions of many individual agents result in, say, the emergence of segregation at different levels of aggregation, as shown in Figure 4.6.

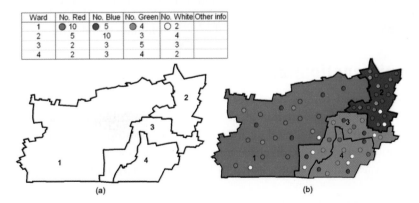

FIGURE 4.4 Populating a model with agents: (a) the environment layer and corresponding attribute fields; (b) reading in the data and creating the environment and the agents based on the attributes of the fields

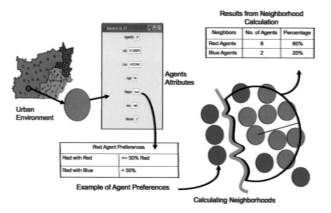

FIGURE 4.5 Basic segregation model structure

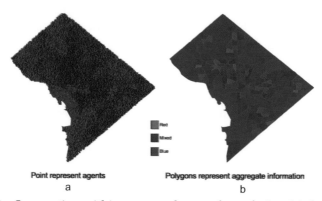

FIGURE 4.6 Segregation within areas and across boundaries: (a) the entire area with individual agents represented; (b) the aggregate information, where each Census block is shaded depending on the most dominate social group

Example applications

Agent-based models have been created to study a diverse range of geographical phenomena (see Heppenstall et al., 2012),[2] and this section provides only a brief review of such applications. The intention is to show how one can use agent-based models to study a diverse range of phenomena at different spatial and temporal scales such as those shown in Figure 4.7. These range in temporal scale

---

[2]A number of ABM toolkits have been developed to allow scientists to focus more on modeling rather than on things such as visualization of model outcomes and charting model progress. Interested readers are advised to see Crooks and Castle (2012) for a discussion on the benefits of ABM platforms which allow for GIS integration.

from the split-second decision-making involving local movements such as people walking, traffic modeling over minutes and hours, residential movement (e.g. Crooks, 2007; Malik et al., 2013) and urban growth over months and years, through to the migration of people over decades (e.g. Gulden et al., 2011; Pumain, 2012). However, it needs to be noted that agent-based models can be used to study a wide variety of fields, such as ecology (e.g. Grimm and Railsback, 2005), biology (e.g. Kreft et al., 1998), and geomorphology (e.g. Reaney, 2008), to name but a few. That being said, we now turn our attention to models that involve humans and space.

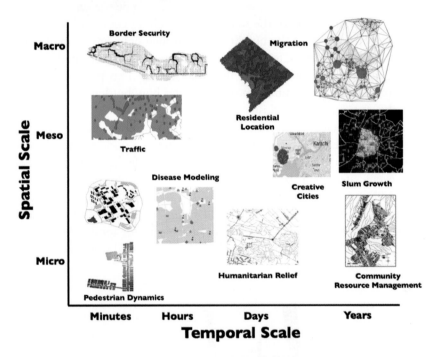

FIGURE 4.7   A sample of application domains for spatial agent-based models

As noted above, agent-based simulations serve as artificial laboratories where we can test ideas and hypotheses about phenomena that are not easy to explore in the 'real world'. One example of this is pedestrian modeling of evacuation or movement in general. For example, without actually setting a building on fire we cannot easily identify people's reactions to such an event. ABM, as with simulations in general, can allow for such experiments. Rather than setting a building on fire, we can re-create the building within an artificial world, populate it with artificial people, start a fire and watch what happens. Such simulations allow the modeler to identify potential problems such as bottlenecks and allow for the testing of numerous scenarios such as the way various room configurations can impact on evacuation time. By building such models we can focus on mitigation and preparedness rather than response and recovery in emergency incident management.

Building the spatial environment of such models is relatively straightforward, as shown in Figure 4.8, for example. From an architect's CAD file of a building we can georeference the building to its actual 'real-world' location. We then take the building and rasterize the space so that each cell represents, say, 50 cm², which approximates the anthropomorphic dimensions of an individual (Pheasant and Haslegrave, 2006). This regular lattice structure is used for our artificial world in a similar way to that of other pedestrian models (e.g. Batty et al., 2003b); however, a continuous (i.e. vector) space representation could also be used if so desired (see Castle, 2007, for a discussion of the role of space within pedestrian models). Once we have the spatial layout of the building, we can then populate the building with agents that have simple rules; for example, once the alarm is activated, agents follow emergency signage (or move to the nearest exit). This process is shown in Figure 4.8(c) and (d). The simulation demonstrates how bottlenecks form at exits and how this causes agents to cluster around such obstacles.[3]

It is not just evacuation models that can benefit from utilizing agent-based models. Such models have been developed to explore a wide range of phenomena where the movement of people plays a critical role (see Torrens, 2012), for example, event safety (e.g. Batty et al., 2003b), and understanding how people move around town centers (e.g. Haklay et al., 2001) or theme parks (e.g. S.-F. Cheng et al., 2013). Geographically explicit pedestrian models have also been developed to explore how people find out and search for food in times of crisis (e.g. Crooks and Wise, 2013). and to explore the spread of diseases within refugee camps (e.g. Crooks and Hailegiorgis, 2013).

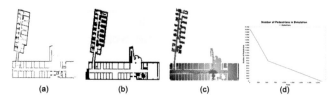

FIGURE 4.8 Simple pedestrian model: (a) CAD floor plans of a building are converted into (b) a raster layer and are used as the environment for the agent-based model. (c) shows the simulation running with agents (red) who are exiting the building and leaving behind walking traces (yellow). (d) shows a time series plot of agents evacuating from the building

---

[3]Interested readers can see an animation of this model and download the source code and data along with other spatially explicit models from http://www.cs.gmu.edu/~eclab/projects/mason/extensions/geomason/.

Moving from the small-scale interactions of pedestrians, agent-based models have also been developed to explore how people navigate through space, and the results of these behaviors in aggregate. Such models have been developed to understand and potentially reduce traffic congestion or pollution, for example (Helbing and Balietti, 2011). ABM is ideal for studying the effects of traffic, as each vehicle and potentially each entity of the simulation (e.g. traffic lights) can be represented as individual objects or 'agents' which make independent decisions about their actions. Behavior can be incorporated into such models; for example, how many individuals can cause traffic jams (Nagel and Schreckenberg, 1992), how different road-pricing scenarios influence usage (e.g. Takama and Preston, 2008), how traffic conditions can be simulated over entire countries (e.g. Raney et al., 2003), and how people find a place to park (e.g. Benenson et al., 2008).

Geographical information plays a critical rule in such models. For example, in the model presented in Figure 4.9, we explore commuters who are working within the Tyson's Corner area of Virginia on the border of Washington, DC. We take road and travel-to-work data from the US Census as the basis for our spatial agent-based model. The road data act as a basis for our agents to move from their homes to Tyson's Corner, and the Census data provides us with the number of agents who travel to the area on a daily basis. The agents attempt to find the shortest path from their home to their destination, with preferential attachment to highways and freeways over smaller country roads. By running the model, cars start at homes and travel towards Tyson's Corner, and as more cars join certain sections of roads, traffic jams start to form (as speed is a function of the number of cars on a specific section of road). For example, in Figure 4.9(b), individual cars can be distinguished when they are not clustered, but when traffic density increases, larger clusters develop. Such a model could easily be extended to incorporating a range of route choice behavior (e.g. Manley et al., 2014).

Researchers have also combined traffic models of cars with those of pedestrians. For example, Banos et al. (2005) explored the interaction of pedestrians and vehicles within the urban street network, with the aim of understanding how one can reduce the number of casualties from vehicles. In another model, Łatek et al. (2012) explored how an agent-based model could be used to understand dynamics along the US–Mexico border – specifically, how border security infrastructure (e.g. integrated fixed towers) can impact on the prevalence of smuggling (e.g. of drugs or people), by modeling both pedestrians and vehicles. Figure 4.10(a) outlines the data layers needed to explore this question.

Within the model, land cover and terrain relief are used to calculate viewsheds for the placement of security infrastructure via line-of-sight analysis. As the model is also interested in routing across the border and for border patrol agents, roads and terrain roughness are also needed, as this enhances or impedes movement. However, as the model incorporates the movement of smugglers,

FIGURE 4.9   Using agent-based models to explore rush-hour congestion: (a) road and Census data used for model inputs; (b) zoomed-in section of (a) with agents (red circles) moving towards Tyson's Corner and causing traffic jams

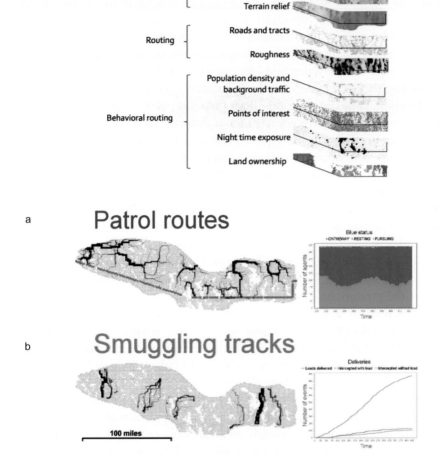

FIGURE 4.10 Modeling border security along the Arizona–Sonora borderland:
(a) data layers and analysis type for the simulation; (b) visualization of the
model, showing both patrol routes and smuggling tracks

one also needs to incorporate geographical information that impacts on behavioral routing decisions. This includes population density, in the sense that smugglers initially do not want to be seen and therefore avoid highly populated areas, along with land ownership that elicits different access constraints. Once such information is obtained, we can then simulate how smugglers and border protection agents might interact on this terrain, as is shown in Figure 4.10(b). We can also validate such movement patterns based on real-life drug seizures whose locations are made publically available.

Cities have continued to grow over the last 200 years and there is little sign of such growth slowing down. Physically, they can grow in two different ways: expansion and compaction. The process of expansion leads to more space being occupied (e.g. sprawl); compaction leads to the same amount of space being occupied by more people, resulting in increased population density. Not only do cities grow physically, but they also change their resourcing demands; for example, they may require more water, building materials, food, goods, and services from the surrounding region, which results in population, economic activities and technology diffusion. The motivations for modeling urban growth processes should be obvious. If we can understand and estimate such growth, we can not only raise awareness of the consequences of urban growth, but also test out policy ideas (e.g. restricting growth on specific land-use types or promoting higher-density development) to see how these policies might impact future land-use patterns.

While urban modeling has a long tradition (e.g. Batty, 1976), recently researchers have started to explore such issues through agent-based models due to specific limitations of past modeling endeavors (see Benenson and Torrens, 2004). For example, Torrens (2006) combined GIS data and ABM to explore urban sprawl in Michigan, while Xie et al. (2007) explored urban growth in China, specifically focusing on the pressure on land from developers. Others have explored the evolution of land markets at the urban–rural fringe. For example, Magliocca (2012) and Wise and Crooks (2012) have explored how heterogeneous agents representing farmers, developers, and buyers could influence the spatial pattern of residential development through interactions in the land market. The results from such models are in accord with the classical urban theory (e.g. Alonso, 1964), in the sense that as the distance from the central business district increases, land prices and housing density decrease, a pattern that is also empirically observed in many US cities. By building on such models, one can explore how various planning scenarios such as lot-size zoning and municipal land acquisition strategies could reduce the impact of urban sprawl (e.g. Robinson and Brown, 2009).

Coinciding with urban growth is the rise of slums, especially in developing countries (Patel et al., 2014). Over 900 million people live in either slums or squatter settlements, a number that is projected to increase to approximately 2 billion by 2030 (United Nations Human Settlements Programme, 2003). Just as agent-based models have proved useful in exploring urban growth, they can also be a useful tool to study questions about how slums come into existence, how they expand, and which processes may make some slums disappear. For this reason several researchers have started to explore slum formation from the bottom up through ABM (Vincent, 2009; Augustijn-Beckers et al., 2011; Barros, 2012). Patel et al. (2012), however, notes that there is still much to be done with respect to modeling and improving our understanding of slums.

Conclusions

This chapter has presented a brief introduction to agent-based modeling and how such models can be linked to GIS (both in terms of raster and vector data). The applications have demonstrated how through such a linkage a wide variety of problems can be analyzed at various spatial and temporal scales. This generative (or bottom-up) approach allows us to explore how a small number of rules or laws, applied at a local level and among many entities, are capable of generating complex global phenomena at different spatial resolutions. These collective behaviors and extensive spatial patterns are manifested in such a way that the actions of the parts do not simply sum to the activity of the whole. They provide us with a new way of thinking about problems, taking ideas and insights from complexity science.

However, in conclusion, while there is great potential for ABM, there are several additional challenges that ABM faces, ranging across the spectrum of theory to practice, hypothesis to application (see Crooks et al., 2008). Validation schemes are a classic example of this. One reason why validation poses such a challenge is the degree to which the true micro-geography of any geographical system (e.g. a city) is still largely unknown in many situations. Nevertheless, this style of modeling provides a tool for testing the impact of changes in, say, room configurations, land-use type or the transportation networks in dense metropolitan areas via a simulation approach. This approach is less focused on predicting the future accurately, than on understanding and exploring the system from the bottom up.

FURTHER READING

Readers wishing to know more about ABM and GIS might be interested in the following works. For an introduction to ABM and complexity more generally, Miller's and Page's (2007) book *Complex Adaptive Systems* provides a good overview of why one might to use ABM along with considerations about how to model more generally. Gimblett's (2002) edited book provides one of the first introductions to ABM and GIS, especially how the integration of the two can be used to study social and ecological processes, and to this day provides the foundations to many agent-based models. A more recent comprehensive edited volume of ABM applications applied at various spatial and temporal scales can be found in Heppenstall's *et al.* (2012) edited volume *Agent-Based Models of Geographical Systems*. A detailed review of agent-based models, how they have evolved with respect to GIS and their relationship to other modeling approaches can be found in Benenson and Torrens (2004), while those interested in cities might like to read Batty's (2005) *Cities and Complexity*.

5

MICROSIMULATION MODELLING FOR SOCIAL SCIENTISTS

Kirk Harland and Mark Birkin

Introduction

Microsimulation modelling has increased in popularity across a variety of disciplines over the last 60 years. The approach has become particularly prevalent in policy analysis with an emphasis on demographic and geographical implications of policy decisions. The work presented here will introduce the different 'flavours' of microsimulation models, including static, dynamic and spatial. Furthermore, the interface with another emergent modelling technique, agent-based modelling (ABM) (see Crooks, Chapter 4), will be discussed, followed by some examples of recent microsimulation applications. A more technical discussion will be the focus of the final section, which will suggest an architecture for a microsimulation model framework, with the trade-off between structure and efficiency considered.

What is microsimulation?

Ballas et al. (2006: 66) define microsimulation as

> a methodology that is concerned with the creation of large-scale simulated population microdata sets for the analysis of policy impacts at the micro level. In particular, microsimulation methods aim to examine changes in the life of individuals within households and to analyse the impact of government policy changes for each simulated individual and each household.

Microsimulation models can be categorised into static or dynamic and whether they are spatially explicit. A static microsimulation model generates a population at a snapshot in time. A dynamic microsimulation model produces a population that is subsequently moved through time and optionally space. A microsimulation model becomes spatially explicit when population characteristics are simulated to represent realistic social heterogeneity across geographical areas.

Static microsimulation

Static microsimulation models generate a synthetic population from aggregate data, with the resulting population being immobile in both space and time. Population generation is normally a process of sampling or 'cloning' candidate individuals from a sample survey to create a population consistent with a series of aggregate constraint values. Where a survey population is not available for sampling, a population can be generated using the conditional probability distributions derived from the aggregate data, as demonstrated by Birkin and Clarke (1988, 1989). There are a variety of approaches for producing synthetic populations from aggregate data sources that fall into three broad categories: deterministic, statistical estimation and combinatorial optimisation. A detailed discussion of the different population generation approaches is beyond the scope of this chapter; further information on a selection of techniques can be found in Harland et al. (2012).

Dynamic microsimulation

Dynamic microsimulation models forecast past trends into the future to estimate changes in a micro–unit population (Ballas et al., 2006). As the population is moved forward through time, the likelihood of events occurring based on the characteristics of the micro–unit is assessed against probabilities derived from available data. As identified by Orcutt (1957), micro–units of the same 'type' have the same characteristics, but outputs for identical units need not be the same. For example, two synthetic individuals in a population aged 65, married and retired, may both have identical characteristics and the same probability (say, 0.4) of a death event occurring. However, if individual 1 has a random draw (random number generated) of 0.7 the mortality event does not occur, whereas individual 2 may have a random draw of 0.3 that results in a mortality event occurring. This demonstrates that even though many synthetic individuals can share identical input characteristics the output is dependent on the random draw and probability distribution of an event occurring, reflecting the uncertainties and decision-making processes observed in real life. Orcutt (1957) reflected that despite the variation observed, in reality a significant proportion has a relatively regular set of outputs dependent on the input characteristics of the micro–unit. This is one of the central considerations used to derive statistical risk models for managing risk exposure during the issue of personal loan and mortgages by large financial institutions and it is also, as Orcutt (1957: 118) notes, 'because of this that insurance companies do so well'.

Probabilistic approaches may be suitable in situations where a known probability distribution for an event occurring can be derived, such as the use of life tables for mortality events. However, in some circumstances an output can be considered as a direct result of an input. Ageing may be considered this type of characteristic: an individual gets one year older, not two or three, on a particular date and only on that date. There may also be second-order implications of an event occurring. For example, if individual 1 is married to individual 2 and individual 1 has a mortality event, then a direct impact of this is that the marriage status of individual 2 will change from married to widowed.

Transitions and events

Dynamic microsimulation models can generally be categorised into either transition-based or event-based models. The difference between the two approaches centres around the way time is handled. Each approach is described briefly below.

Time-steps are the temporal currency of transition-based dynamic microsimulation models. At each time-step, events are evaluated based on the input characteristics of each micro-unit to ascertain which events take place and what the outcomes of those events are. A time-step is a discrete unit of time, normally a week, a month or a year, but any discrete measure of time can theoretically be used. Orcutt (1957) recommends relatively short periods such as a week or a month; however, the time-step of the model should be considered in the context of the intended use and the data available to derive the required event probability distributions. A model run ends when the last specified time-step is completed.

Time is treated as continuous in an event-based model. From an initial starting point, all events considered by the model have a lapse time randomly generated. The lapse time is the time to that event occurring for a particular synthetic individual. The event with the shortest lapse time is executed first. This point then becomes the starting point, and all events that are influenced by the outcome of first event have new lapse times calculated.

When events can be considered in isolation the event-based approach does have the advantage that causality can be observed (Spielauer, 2009b). The execution of an event may adjust the probabilities of another event happening, bringing the events execution forward (or perhaps pushing it back) in time. In a transition-based model, events happen between two fixed points in time and therefore causality between two events happening in the same time-step cannot be linked. It is only when the time-steps within a transition-based model become small enough for multiple events not to happen within one time-step that causality between events becomes apparent (Galler, 1997). When this is the case the transition-based model is effectively treating time as continuous. Orcutt (1957) emphasised the fact that time-steps within the transition-based architecture he described had to be of a fine enough resolution to enable the outputs of events to be considered only within the next iteration of the model and not by events within the same iteration.

Events in reality rarely happen in isolation, and therefore the relative simplicity and elegance of the event-based model is quickly lost in complex equations to account for the interdependence of events (Galler, 1997). Galler concludes that transition-based models with small enough time-steps to preclude multiple event occurrences provide the simplest and most promising architecture for dynamic microsimulation models, echoing the recommendation of Orcutt in his original paper some 40 years earlier. However, this presents another problem: the issue of model scale against processing efficiency. Despite the increases in processing power of computers and the significant fall in the cost of digital storage, synthesising and moving a large population, such as one of a whole country, is still a significant task. As noted by Spielauer (2009b), when individual micro-units can be simulated throughout their life course in isolation, an event-based approach offers the ability to process each micro-unit's life in parallel on distributed computing networks. This reduces processing time and also offers the efficiency that only event points in the life of the micro-unit are considered, not every time-step where on many occasions nothing may happen. Again there is a conceptual trade-off with this approach. In the social sciences microsimulation models are generally used to replicate the interrelationships between events and individuals that are either not captured or ineffectively represented by aggregate approaches. To consider the life course of each individual micro-unit in isolation undermines the interrelated nature of events and individuals. Event-based models can process individual life courses simultaneously across the population to preserve relationships between individuals; however, executing the model over a distributed computing networks becomes a very complicated task.

Spatial microsimulation

Most microsimulation models have a spatial element. However, models that are spatially explicit, examining policy or demographic change at a small geographical scale (typically consisting of several hundred households), are described as spatial microsimulation models (Ballas et al., 2006). The distinction between spatial and non-spatial models is not as well defined as the difference between static and dynamic models. As noted above, most models have a spatial element, even if this is implied. The extent to which small-area demographic detail is captured within the underlying population of a microsimulation model defines the spatial resolution of the model and whether that model is described as explicitly spatial.

The explicit preservation of relationships between survey attributes at the expense of variation in small-area characteristics, such as the approach of Kao et al. (2012), assumes spatial homogeneity over small geographical areas. This approach generates a population that is attribute-rich, reflecting the underlying survey sample well, but lacks the detailed underlying spatial variation present in reality. Spatial models preserve the geographical detail by constraining the model to multiple known aggregate relationships at a small-area level, usually extracted from population census data. The spatial detail and small-area heterogeneity are included at the expense of explicitly preserving the relationships between the attributes within the sample population. Work by Smith et al. (2009) and Birkin and Clarke (2012) suggests that tailoring a microsimulation model structure, and incorporating additional information such as a geodemographic classification between survey sample and aggregate constraints, can alleviate some of these issues, thus improving population reconstruction.

Microsimulation modelling is a tool to construct theoretical arguments, aid understanding and ultimately solve complex problems. To achieve these aims, it is crucial that the researcher select the most appropriate approach for the research question under study and those data available which can feed into models. The goal of the research should inform the spatial resolution of the model and the acceptable trade-off between spatial granularity and preservation of survey detail. It should also influence whether a dynamic microsimulation model is event-based or transition-based, or indeed whether a dynamic model is required at all. It is possible that a series of static models could provide the required insights.

Agent-based modelling and microsimulation

Agent-based modelling approaches the modelling of complex systems from the activity of constituent micro-units, agents, to explore emergent macro-patterns from the micro level. Agent-based models move individual agents, expressed as discrete objects in a digital environment, through time and optionally space. The movement or behaviour of each agent is dependent on its internal state and its interactions with other agents and the surrounding environment. These inputs are processed through a set of rules or a behavioural framework contained within the agent to provide a reaction to the input stimuli (see Crooks and Heppenstall, 2012, for a good introduction to ABM concepts).

There are some obvious similarities between dynamic microsimulation and ABM approaches within the social sciences. Both approaches move a population through time and optionally through space, and both additionally require a base population to begin the simulation. This presents obvious synergies between static microsimulation's population generation capabilities and both agent-based models and dynamic microsimulation models. However, a novel approach to integrating static spatial microsimulation with an agent-based model was undertaken by Malleson and Birkin (2012). Here the generated agents were used to enhance the environment within which burglar agents operated rather than the agents themselves, producing more realistic opportunities for crimes to be committed.

Another similarity between microsimulation and ABM approaches is the degree to which they lend themselves to being programmed using object-oriented computer programming languages such as C++, Java or Visual Basic .Net. As noted by Ballas et al. (2006), individuals, households or firms can easily be represented and stored as objects within a computer program which provides a useful and intuitive abstraction of reality. Although Crooks and Heppenstall (2012: 89) note the distinction between the object-oriented programming paradigm and ABM structure, they do note that 'the object-oriented paradigm provides a suitable medium for the development of agent-based models. For this reason, ABM systems are invariably object-oriented.'

Despite the similarities between dynamic microsimulation and ABM, there are differences. The time-steps taken in each model and the total simulation period are normally much shorter in ABM, simulations being representations of days, weeks or months, whereas dynamic microsimulation models tend to consider longer time periods spanning years and even decades. Within both modelling approaches, events occur that are influenced by both the internal state of the micro-unit (age, etc.), inputs from other micro-units (e.g. marriage proposal) and inputs from the digital environment (e.g. amount of rainfall) (Orcutt, 1957; Crooks and Heppenstall, 2012). However, in microsimulation models, these events tend to be unidirectional, policy changes being modelled influencing the micro-units, whereas in agent-based models the interaction is bidirectional, with policy influencing the behaviours of micro-units but also the resulting behaviour of micro-units having the potential of influencing policy within the model (Crooks and Heppenstall, 2012). Furthermore, agents in an agent-based model interact with the digital environment around them; the environment can be regarded as an input to the decisions each agent makes, but also the agents can influence the environment. Consider the crime simulation of Malleson et al. (2010); when a burglary occurs, the house burgled in the digital environment may have its security upgraded; the environment has changed, which will reduce the likelihood of another criminal activity taking place in the same location. In microsimulation models the interaction between the environment and the micro-unit is unidirectional, with the environment influencing the micro-unit but no reciprocal impacts taking place.

Considering the way that events occur in both modelling approaches, differences are also apparent. Outcomes of probabilistic events in microsimulation models will likely be designed to conform to a derived probability distribution usually informed by real-world observations. One of the attractive features of ABM is the ability to include rich behavioural models such as the Beliefs, Desires and Intentions (Bratman et al., 1988), Behaviour Based Artificial Intelligence (Brooks, 1986) or Physical Conditions, Emotional State, Cognitive Capabilities and Social

Status (Schmidt, 2000; Urban, 2000) frameworks. Behavioural frameworks such as these enable each agent to deliberate over courses of action and formulate action plans to satisfy their goals, providing an intricate level of detail absent from microsimulation models.

It is clear that both dynamic microsimulation and ABM approaches have much in common conceptually (both are time-based with micro-unit actors), and architecturally (both share a natural synergy with the object-oriented programming paradigm). However, there are also significant conceptual and functional differences between the approaches, possibly the most important of which being the importance of detailed behavioural abilities in ABM and the bidirectional nature of interactions. However, the high level of behavioural detail and complexity of interaction are computationally time-consuming, and, despite significant forward progress in computing power, still limit the size of the models that can be constructed.

Examples of dynamic microsimulation applications

In this section we present three examples of dynamic spatial microsimulation which are applied to different problem domains, but more importantly, indicate the relevance of the technique over three different timescales – the short, medium and long term. First, we discuss a case study that reflects decision support for medium-term planning. This type of study has been most prominent in the literature for studies involving infrastructure investment or service location planning, often of the 'what if?' variety. The second application extends this to a much longer time-horizon, in which case the planning process is likely to be oriented much more towards the strategic evaluation of options rather than specific projects and proposals. In the last of the examples, some recent work looking at the fast dynamics of urban mobility is reviewed.

Medium-term decision support and impact analysis

The work of Jordan (2012) presents an interesting application of microsimulation principles to the problem of house-building, home ownership and tenure management. Similar examples can be found in the literature relating, for example, to labour markets (Ballas and Clarke, 2001), education (Kavroudakis et al., 2013), retailing (Nakaya et al., 2007) and health care (Smith et al., 2009).

The construction of the model combines reconstruction of the population of small geographical areas with a two-stage process in which the desire to move house is evaluated. The 'mover model' (first stage) is regulated by a decision tree, which includes key household attributes such as age, family composition, social grade and ethnicity. Figure 5.1 shows the structure of the decision tree, while Table 5.1 identifies the attributes that contribute to branching at each of the levels shown in Figure 5.1. When the final choice in the decision tree has been reached, a movement probability is identified. The 'choice model' (second stage) mimics an evaluation process in which the influences of seven destination properties are combined. These elements are drawn using a combination of empirical research and evidence from the literature, comprising access to workplaces and schools, neighbourhood preferences for both ethnicity and social status, property size, tenure and distance (for more details, see Jordan, 2012; Jordan et al., 2012).

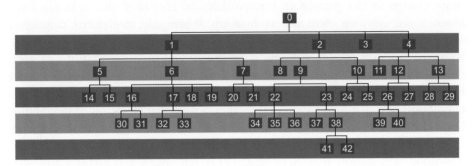

FIGURE 5.1   Decision tree structure

Source: Adapted from Jordan et al. (2012)

TABLE 5.1   Attributes used to differentiate the mover probability

Level	Attributes used in level
1	Age
2	Tenure Rooms required Accommodation type
3	Social class Family type Tenure Rooms required/rooms occupied in house
4	Number of residents in house Ethnicity Number of rooms in house Rooms required Accommodation type
5	Qualifications Accommodation type/family type

The simulation is applied to a substantial part of Leeds with more than 100,000 individuals in which a social housing (publically owned and rented) programme has sought to create improved housing conditions for those on low income through a combination of new investment and changing tenure (e.g. stimulating private ownership of public housing stock). A number of scenarios were investigated, including the provision of new schools (found to provide a positive boost to the objective of increased mixing in the community) and the creation of a new transport corridor (found to have a negligible impact on social diversity).

Many of the extant applications of microsimulation to 'what if?' policy scenarios have a comparative static flavour – for example, they might compare equity of service provision to a population before and after the introduction of some change in the pattern of service delivery. However, this example has a much more genuine dynamic as housing moves are moderated through a 'vacancy chain' in which the decision to move simultaneously reduces the availability of property (as a new home is occupied) and increases it (through vacation of the current home). The decision rules introduce an interdependence between the decision-making units (e.g. if an affluent household moves into a neighbourhood of a lower social grade, then a marginal elevation of neighbourhood status takes place, and vice versa). This also starts to suggest an ABM flavour to the simulation, which has some similarities to well-known theories of neighbourhood segregation (e.g. Schelling, 1971) but much less stylised and with greater realism and practical potential.

Long-term strategic planning of infrastructure provision

A recent example of dynamic microsimulation applied to a strategic planning problem looking ahead more than 80 years has been provided in the work of Hall et al. (2014). This work has been undertaken as part of a bigger programme to explore alternative approaches to long-term transitions of infrastructure systems comprising transport, energy, water, waste and information technology. The long cycles of investment required for these crucial networks require that robust frameworks for future changes can be provided in both demographics and infrastructure demand, notwithstanding the obvious uncertainties and difficulties associated with this process (Beaven et al., 2014).

In practice, the only sensible way to approach this problem is to adopt a scenario planning format. The work of Zuo et al. (2013) attempts systematic construction of a portfolio of scenarios across a full range of plausible future projections in the key demographic components of fertility, mortality and migration. The presentation of the model projections as a suite of microsimulated individuals and households provides a very flexible foundation for further modelling of the demand for infrastructure and the associated commodities – in fact, these authors implement a decision tree for attributes driving consumption which bears more than a superficial relation to the migrant generation process in Jordan's housing model. As an example, Figure 5.2 shows how a long-term demographic projection, combined with baseline assumptions about energy consumption, can be used as a benchmark for changing patterns in future energy demand. The demographic

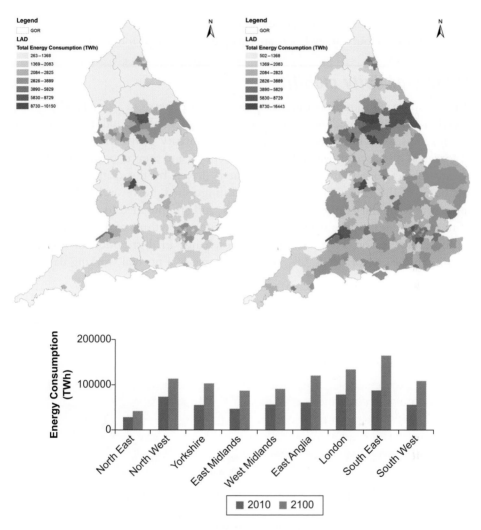

FIGURE 5.2   Long-term energy demand projection

projections can then be used as a basis for evaluation of alternative options and strategies for long-term infrastructure enhancement against a backcloth of both spatial and temporal disaggregation (Tran et al., 2014).

The process mechanisms for long-term demographic microsimulation of this type have been laid down in the work of Birkin et al. (2009). The flexibility of representing households, their constituent individuals and associated attributes is exploited as a means for driving transitions from one time period to another, through key sub-models which advance members of the population through cycles of migration, fertility, ageing, survivorship, household formation and fragmentation. In terms of the earlier discussion, the conceptual clarity and ease of a transition-based approach to the dynamic modelling task are preferred to the greater efficiency but somewhat more opaque event-based models.

Models of daily mobility

A novel approach to representing daily movement patterns in a simulated population has been proposed recently in the work of Harland and Birkin (2013a, 2013b). This work has some representational similarity to the movement models of Jordan as discussed above. Individual households and their constituent individuals are subject to a transition process that is driven by the spatial location and attributes of the entity. This example is distinguished, however, by the fact that moves are temporary rather than permanent, and are intended to represent cycles of movement against the daily, weekly and seasonal rhythms of demographic flux within a city region. Figure 5.3 shows the change in distribution of disposable income in the city of Leeds, UK, over the course of a normal working day.

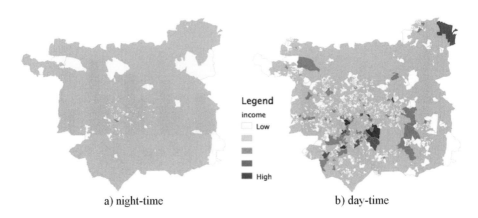

a) night-time   b) day-time

FIGURE 5.3 Simulated distribution of income

The potential value in model applications of this type is both evident and substantial (Birkin et al., 2013), for example in relation to problems of emergency planning, disease transmission, crime prevention or retail service provision. The procedures for capturing short-term movements in the population are not inherently difficult, although they do introduce some computational challenges in the execution of large numbers of decisions and also in monitoring and storing frequent shifts in the spatial position of each actor. Naturally simulating behaviour in a way that absorbs enough of the richness and complexity of movement and interactions in the 'real world' is a challenging programme. To date, the possibilities that have been adopted within the models combine elements of the mapping of different agents onto activity containers such as schools, hospitals and workplaces (in turn building on foundations laid in work such as that of Cockings et al., 2010); or consider the potential of social media for tracing movements and behaviours (Birkin et al., 2013). The greater potential of more wide-ranging datasets such as those maintained by mobile-phone-service providers such as Telefónica or Vodafone remain a tantalising but exciting proposition at the time of writing.

Technical architecture considerations

Abstraction

Ballas et al. (2006) point out synergies between the object-oriented programming paradigm and the conceptualisation of a dynamic microsimulation model into individual objects representing each individual micro-unit. However, representing individual micro-units as individual encapsulated programming objects has a trade-off against the memory and processing power required to run a model of considerable scale, such as for the whole of the UK. The Office for National Statistics (ONS) in the UK estimated the 2012 mid-year population to be approximately 63.7 million (ONS, 2013). Representing each UK individual with 63.7 million objects within a computer program would be a challenge, even allowing for the power of current technologies for computation.

Designing agent-based models, with their complex behavioural models, so that each individual is represented by an encapsulated object is logical. Each agent object can interact with other agent objects or the environment and react to input stimuli from either other agent objects or the environment independently. However, dynamic microsimulation models seldom require such detailed behavioural frameworks, with the outputs of events normally being defined by probabilistic or deterministic reactions derived from real-world observation. Taking the conceptual differences between ABM and dynamic microsimulation into consideration, it is potentially beneficial to consider the object currency (or level of abstraction) for a dynamic microsimulation model to be the processing framework, with the micro-units simply being data items passing through the framework. Figure 5.4 shows the difference between the two conceptual designs in diagrammatic form. Each object represents an individual micro-unit in Figure 5.4(a), with the behavioural framework and data reflecting the state of the micro-unit contained (or encapsulated) within each object. In contrast, Figure 5.4(b) shows a conceptual design where the micro-unit data are stored separately from the event-processing objects and are cycled through for events to take place.

To exemplify the difference in processing efficiency, a simple test has been constructed. For benchmarking purposes the specification of the test computer is a MacBook Pro with 16 GB of random access memory and 500 GB solid state

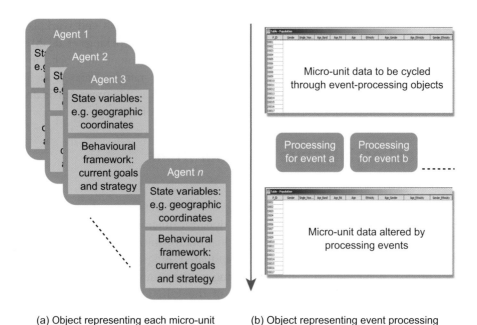

(a) Object representing each micro-unit   (b) Object representing event processing

FIGURE 5.4   Alternative model architectures

storage space running OS X Lion Version 10.7.5 on a 2.6 GHz Intel Core i7 processor. The test was run using the Java programming language through Netbeans version 7.3.1 and simply consists of storing 64 million randomly generated double-precision values in a single array. In the first test the numbers are stored as a 'Double' core language object in the Java programming language. In the second test the numbers are generated and stored as primitive values in the Java programming language (for a more detailed discussion of primitive values and objects in the Java programming language, see Schildt 2002). The first two rows in Table 5.2 show the amount of memory used and time taken to create and store the 64 million values for both tests. The bottom two rows show the extra memory used and time taken to cycle through the 64 million values, multiply the stored value by another randomly generated number and adjust the original value to be the result of the calculation. It is clear from the results in Table 5.2 that although considering the modelling approach from an event-processing perspective sacrifices the inherent logic of representing individuals as encapsulated objects, it has considerable advantages in storage capacity and processing time.

The tests executed here are very simple representations of the two architectures considered but serve to highlight the significant differences between the two approaches. Adjusting the level of abstraction from the micro-unit to the event processing level provides significant scalability benefits.

TABLE 5.2   Test results for storage and processing efficiency

	Test 1: objects	Test 2: primitives
Memory use	1.79 GB	0.52 GB
Time to create	31 seconds	1 second
Memory used to cycle and adjust	100 MB	0 MB
Time to cycle and adjust	66 seconds	2 seconds

Events and actions

Orcutt (1957) outlined the need to represent interactions between demographic attributes from the micro level to effectively estimate and simulate how relationships between attributes develop and change over time and potentially space. The interrelated nature of events and attributes means that the dynamic microsimulation approach needs to have the flexibility to represent new events or interactions between events as they become apparent or important for the needs of the researcher. It is also important that the inclusion of new or modified events and relationships result in as little change to existing model architecture as possible, while considering that two seemingly separate events may result in very similar results. For example, consider a child leaving home for the first time and the breakdown of a marriage. These are two very different events potentially happening at different life stages of the individual. However, the action resulting from these two events is very similar, as one household unit effectively splits into two.

For these reasons, the programming architecture recommended here to represent events and actions is one similar to the architecture used for controlling user interaction with a graphical user interface. The graphical user interface typically consists of two parts: the first is something that the user interacts with, such as a digital button that generates events; the second is something that performs an action when notified to, an event handler. A third object is also created to hold information called the event. When the button is created, code is written to create an event object. The event handler is written to only handle the event type that it is concerned with, and it is attached to the button. When the button is pressed it broadcasts to anything listening that it has been pressed, creates the event object containing all the information important for the event and passes that to any objects listening. Any event handlers listening can then process the event. Figure 5.5 demonstrates this process diagrammatically.

This architecture promotes loose coupling between event generation and actions taken through event handling. As demonstrated in Figure 5.5, multiple event handlers can listen and action events created by a single generator, but also a single event handler can listen and action events created by multiple event generators. Additionally, the actions taken by the event handler may be dependent on

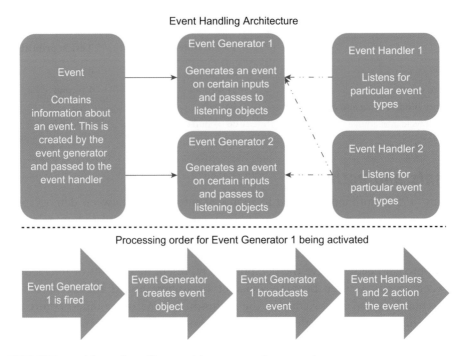

Event Handling Architecture

FIGURE 5.5 Event handling architecture and processing

the information contained in the event object which will be generated with spe-cific information about the event, and likely information relevant to the micro-unit from which this event has been generated. Event handler 2 in Figure 5.5 could represent the example provided above where a new household is formed from both a marriage breakdown and a child leaving home. Event handler 2 actions the for-mation of a new household based on the information contained within the event object. Event generator 1 may be fired by a marriage breakdown, whereas a child leaving home may fire event generator 2. However, event handler 2 is 'listening' to both of these event generators and will action an event when either is fired.

Handling time

The advantages and disadvantages of transition-based and event-based models have been discussed above. It is proposed here that both of these approaches are accom-modated. As Spielauer (2009a, 2009b) notes, dynamic microsimulation models are likely to require some elements of both approaches and that they are not mutually exclusive, citing the Australian DYNAMOD model as an example of a hybrid approach. It is proposed here that time is handled as a schedule broken down into time-steps. This is a similar approach to that adopted at George Mason University in the development of the MASON ABM toolkit:[1] time-steps are the heart of the model, but there is also a scheduler that can be used to schedule events or actions. Here the model time-frame (time from start point to the end of the modelling time period) is specified alongside the time-step; for example, the time-frame is maybe 30 years and the time-step one month, resulting in a schedule of $30 \times 12 = 360$ time-steps. Events are then scheduled by the length of time to the event and placed onto the scheduler. However, the time-step mechanism also implements the event handling architecture, and it is therefore a straightforward matter to attach event listeners to event generators called at each time-step or at particular time-steps. In this way the model can work as a purely event-based model, purely as a transition-based model or as a hybrid, with some actions working from a schedule and some at each and every time-step.

Simple example

A simple example has been developed to test the architectural principles of event and time handling. The first stage of the process constructs a random population of 63.7 million individuals with a random age between 0 and 100 and a gender of either male or female, again randomly assigned. A schedule is created consisting of a time-frame of 30 years with a time-step of one week, which therefore consists of a schedule with 1560 time-steps. Births and deaths are randomly assigned to individuals and added to the schedule at a point in the future consistent with the numbers reported by ONS (2013) for 2012, 812,970 births and 569,024 deaths. Births are restricted to only occur to females aged between 15 and 45, while deaths randomly occur throughout the population.

Figure 5.6 shows a diagrammatic representation of events scheduled for each micro-unit. Once the schedule is formulated for each micro-unit, events are only processed at the scheduled points in time. However, if the interrelationship between events demands adjustment to future events, this can also be undertaken. Likewise, if an event occurring to one individual impacts another, the schedule for the secondary individual can be adjusted.

In our simple example, and Figure 5.6, the whole population is aged by one year after 52 time-steps, representing the 52 weeks in a year. Running this simulation uses 1 GB of memory and takes nine seconds to complete. The estimated population of the UK by 2042 using this very simple model is 73,114,216. The main aim of this example is to demonstrate the proposed model architecture functioning. It does not include critical features such as age-related mortality probabilities, fertility rates by age or any real representation of interactions between

---

[1]http://cs.gmu.edu/~eclab/projects/mason/.

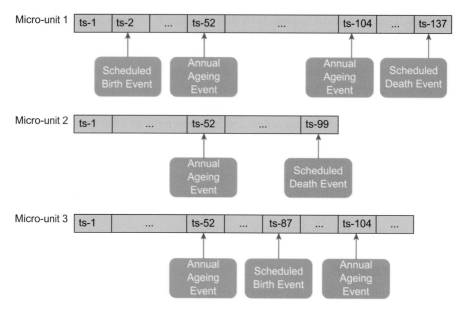

FIGURE 5.6 Diagrammatic representation of time handling

individuals or space. Parameterisation of the model is minimal, and it is accepted that as additional parameters and interactions are added, processing time and memory use will increase substantially. Despite these serious limitations, comparing the simple model output with a linear extrapolation of the ONS figures for births and deaths highlights an interesting point. Projecting the births and deaths 30 years into the future results in 569,024 × 30 = 17,070,720 deaths and 812,970 × 30 = 24,389,100 births. Combining these figures with the initial start population 63,700,000 + 24,389,100 − 17,070,720 results in a population estimate of 71,018,380. The simple dynamic microsimulation has captured a potential increase in the population greater than that resulting from a linear extrapolation. This is simply the result of births outstripping deaths and an expanding number of females eligible to give birth, producing a non-linear expansion of the population. This supports Orcutt's (1957) suggestion that modelling at the micro-unit level naturally captures relationships between attributes that are difficult or even impossible to represent using aggregate data alone.

Discussion

Advances in both theory and computation have allowed significant extension of the ideas originally presented by Orcutt (1957). One powerful innovation that has been facilitated is the representation of individual micro-unit actors as software objects within social science models (Ballas et al., 2006). Nevertheless, alternative modelling architectures are open to researchers, as Crooks and Heppenstall (2012) note. An exemplification of this point was demonstrated above. Representing individuals as computer objects proved to be computationally much more onerous than holding and processing the same information as a series of indexed individual data items. Evidently, a trade-off exists between model efficiency and the logical representation of the individual micro-unit that needs to be considered when a model is in the design phase. Design questions should be asked including:

1. What are the scalability requirements for this simulation exercise?

   a. the number of actors in the model;
   b. the time-horizon for the simulation.

2. Is complex behaviour an integral part of the research?
3. What level of interaction is required?

   a. actor to actor;
   b. actor to environment;
   c. environment to actor.

These types of questions help ascertain the modelling approach most suitable for the research exercise. Dynamic microsimulation and ABM have many similarities, and it is becoming increasingly difficult to definitively select one approach to use in a research project. Some suggestions have been made above relating to where the interface between dynamic microsimulation and ABM lies. One suggestion is that agent-based models have bidirectional interaction between the digital environment and the micro-unit actors, whereas dynamic microsimulation has unidirectional interaction with actors unable to influence the environment. However, this distinction may not always be clear-cut. The work of Jordan (2012) represented migrations in a dynamic microsimulation model through interactions with housing stock in the digital environment. The migration of one micro-unit is seen to impact on the decisions of other micro-unit actors on where to locate as the available housing stock in the environment has changed. The interaction here was clearly bidirectional, so elements of this model might equally well be described as agent-based. Since the model also clearly has many of the traits of a transition-based dynamic microsimulation model, perhaps an overall characterisation as an hybrid model would be most appropriate. Increasingly it may be less critical to think in terms of model definitions as ABM and dynamic microsimulation, but rather as a continuum between approaches (e.g. Wu et al., 2008).

Recently microsimulation, ABM and other modelling approaches not discussed here such as cellular automata have been described under the umbrella term 'individual-level' models. As research progresses and computational power increases, the distinctions between different modelling approaches will become ever more difficult to discern and arguably less important. In 50 years, perhaps less, computational power may well be at a level to enable models with millions of actors each with complex behavioural frameworks and multi-directional interactions. Such models could be relatively easily achievable for researchers to access as modelling toolkits become more advanced.

A future scenario of this kind is not inconceivable, considering that 30 years ago the desktop computer and the internet were still embryonic research projects. The pace of progress is increasing, not slowing; therefore it is entirely likely that such complex models of human society may well be achievable within the working lifetimes of the current generation of researchers. Examining the differences between modelling approaches to enable closer integration is perhaps one of the more exciting prospects for researchers in the social science modelling arena. There is every likelihood that increasingly massive portfolios of behavioural data at a variety of spatial and temporal scales ('Big Data': Birkin, 2013) will provide a further impetus to the pace of change. However, in a research world where 'big data' and individual level models may well proliferate, considerations about model evaluation and validation will become increasingly important and significantly more challenging than they already are. Where and how these important requirements for social science modelling will fit in with individual-level models is a question yet to be resolved. And how sense can be made of the huge volumes of data that will be produced by such models is an equally significant challenge for future research. These questions may be the real future challenge as the integration of modelling approaches appears to be naturally occurring as computational power and model accessibility progress.

FURTHER READING

A recent overview of microsimulation approaches is provided by Tanton and Edwards (2013). For a a real and detailed example of the use of agents within a dynamic microsimulation, see Wu et al. (2011). Harland (2013) is a practical manual which presents software codes, sample data sets and easy-to-follow instructions for readers wishing to develop their own microsimulation models. Finally, for an authoritative introduction to the application of microsimulation as a technique for demographic modelling, see Van Imhoff and Post (1998).

6

SPATIO-TEMPORAL KNOWLEDGE DISCOVERY

Harvey J. Miller

Introduction

The transformation from a data-poor to a data-rich environment is the most profound event in the history of geography and earth science. For most of history, we invested a tremendous amount of resources and time, and quite often human lives, to tease small trickles of geographic data from the environment. Those trickles have become floods due to the rise of sensing systems and location-aware technologies such as satellite and airborne remote sensing, the global positioning system, mobile phones, sensor networks and georeferenced social media.

Much of the data torrent is spatially referenced and time-stamped (Rohde and Corcoran, Chapter 7). These ubiquitous, ongoing spatially and temporally referenced data flows are potentially revolutionary since they allow us to capture spatio-temporal dynamics more directly (rather than inferring them from periodic snapshots) and at multiple spatial and temporal scales. Also, since the data are collected on an ongoing basis, they can capture both mundane and unplanned events, allowing data-rich, naturalistic analysis that can lead to new discoveries about the dynamics of people, cities, societies and the environment (Miller and Goodchild, 2014).

Our statistical and spatial analysis techniques are to a great extent not designed to find patterns and information that may be hidden in spatio-temporal data streams and mobile objects data. These techniques are *confirmatory*: they require the researchers to hypothesize variables tied together in a highly specific relationship. Statistical testing can only tell if the postulated model does not fit the data; it cannot suggest alternatives (Guo and Mennis, 2009). While confirmatory methods have their role, they are not well suited to exploring massive spatio-temporal data: the number of possibilities is too large to explore in this tedious manner.

Massive spatio-temporal datasets require new *exploratory* methods: techniques that can efficiently discover patterns hidden deep within their stuctures. While this information may be tentative, it can enhance the initial stages of the scientific process by generating new insights and novel hypotheses to be tested using confirmatory techniques (Gahegan, 2009; Gahegan et al., 2001; Miller, 2010)

It is common to use the term 'data mining' to describe the process of exploring databases for novel, unexpected patterns. Data mining is the application of low-level algorithms and techniques to search for hidden patterns in data (Alexiou and Singleton, Chapter 8). However, this is only one part of a broader, higher-level *knowledge discovery* process that requires interlinked decisions about how to prepare and manage data, the identification of those properties and features to be explored, the selection of data-mining techniques and interpretation of the results in light of the analyst's background knowledge about the real-world domain. Knowledge discovery is a complex, intricate process that requires human-level intelligence for guidance.

This chapter discusses processes and techniques involved in discovering new spatio-temporal knowledge from spatially referenced and time-stamped data, including data on moving objects. It presents background material, and includes discussion of those issues surrounding the processing of spatial, temporal and moving objects data to extract properties of interest. It also includes strategies for building and populating a data warehouse or the database designed to support such analysis and exploration. It discusses major methods for exploring and mining spatio-temporal data, including data cubes, clustering, association rules, sequence mining, mining collective patterns and visual analytics. The chapter concludes with some summary comments.

Processing spatial and temporal data

Spatial, temporal, and mobile objects data are potentially interesting since proximity in space and coincidence in time affect many human and physical phenomena. *Spatial relations* are measures of proximity in space. Major classes of spatial relations are set-oriented, topological, directional and metric relations (Worboys and Duckham, 2004).

Set-oriented relations conceptualize spatial relations between objects using set theory concepts such as intersection, union, difference and complement. *Topological* relations focus on connectivity, interior, exterior and boundary relationships between spatial objects (Egenhofer and Franzosa, 1991). *Directional* relations between spatial objects include cardinal (e.g. 'the factory is west of the neighborhood'), object-centered ('the neighborhood is behind the factory') and ego-centric directions ('the factory is on your left'). The major type of *metric* spatial relation is distance, such as Euclidean or network-based distance.

Temporal relations are measures of coincidence. The number of possible temporal relations is also surprisingly large: there are 13 possible temporal relations between two time intervals, including *meets, equals, starts* and *finishes* (Allen, 1984).

Time also does not need to be linear: cyclical and branching real-world timelines are also possible. Natural cycles such as seasons are cyclical, and branching time can represent different versions of the past or possible futures.

With respect to a moving object, we can distinguish among movement properties based on temporal scale. *Moment properties* include its time, location, direction, speed, acceleration/deceleration and accumulated travel time and distance at a given instant in time. *Interval-based properties* include its geometric share, traveled distance, duration, the dynamics and distribution of speed and direction, and the ordering of these properties over a sub-segment of a moving object's trajectory in space with respect to time. *Episodic* properties are movement properties associated with an external event in space or time; an example is people within a park during lunch. *Global* properties are those defined for the entire space–time trajectory traced by an object (Andrienko et al., 2008b; Laube et al., 2007).

A challenge in exploring spatio-temporal data is that spatial and temporal relations and properties are usually not explicitly stored in the database. A database will only store the locations and footprints of the objects in a spatial reference system, and time-stamps corresponding to sample times, events in the real world and database transactions (e.g. when a change was recorded in the source database). Spatial and temporal properties must be computed from the data based on the location and geometry of the spatial objects as well as time-stamps.

Processing spatial data is computationally demanding because it is often voluminous due to its geometry. Furthermore, many spatial algorithms require computational geometry procedures that are time- and storage-consuming. A common database and tool design question is whether to conduct operations involving spatial data before the data-mining process (as a so-called *precomputation stage*) or 'on the fly' during the process (Andrienko et al., 2006). The former case requires a great deal of overhead and requires these calculations to be updated if the database is updated. Some of these extracted properties and relations may not be needed. The latter strategy can, however, slow the data mining, causing great harm to the exploratory process which needs to be fast and nimble to maximize the human–computer interaction experience.

Another issue in mining spatio-temporal and moving objects data is the spatial resolution or temporal granularity of the data. There is a trade-off between the interest level and the support strength for patterns at different levels of resolution and granularity. More interesting patterns are likely to be discovered at the highest resolution and granularity levels, but the smaller number of database cases at these levels means that data support may be low, meaning that confidence in these patterns will also be low. Conversely, there is likely to be large data support at coarser resolutions and granularity, but these patterns are not likely to be interesting (Andrienko et al., 2006). It is also critical to keep in mind that the problem of artificial or 'modifiable' units of analysis in both space and time can also greatly affect the results of exploratory analysis just as it does in confirmatory analysis (Long and Nelson, 2013).

A challenge associated with processing moving objects data is that the object's movement pattern is not directly available: location-aware technologies only capture a sequence of locations at discrete moments in time. There are several ways to capture this location sequence (Andrienko et al., 2008b; Ratti et al., 2006). The simplest and most common manner to reconstruct an object's movement from a temporal sequence of recorded locations is through linear interpolation. This assumes that the object followed a straight line between recorded locations. We can deal with the location uncertainty that results from undersampling by constructing uncertainty regions that delimit the locations where the object could be, based on how fast the object could move and the length of the time interval (Kuijpers and Othman, 2009; Pfoser and Jensen, 1999).

Data warehousing

Data explored and analyzed during the knowledge discovery process are often integrated from disparate sources and stored in a specialized database known as a *data warehouse* (DW). This is usually an integrated archive or repository of other databases. It may be drawn from a project or organization's operational database that supports day-to-day activities. However, it also need not be internal to a project or organization: it may consist of views into external databases existing on remote servers and accessed using the internet. Data accessed remotely may be published versions of internal databases, such as data selectively released on the web for public consumption by governments, business and organizations. It also may be data 'scraped' from the web and social media through application programming interfaces, including volunteered and shared data from citizens. Therefore, a DW may be a heterogeneous collection of authoritative, clean data and informal, messy data.

Extraction, transformation, and load (ETL) functions perform the tasks required to copy data from source databases, making them more usable, and store them in a DW. ETL functions include decoding and encoding data, merging/splitting of attributes, data aggregation and summarization, cleansing through removal of errors and outliers, and data smoothing. ETL for spatial and temporal data requires wide-ranging expertise. For example, the system must ensure that source data are topologically correct and respect spatial and temporal integrity constraints, and that the overlay of these data with other spatial data is also topological correct and coherent with respect to updates. There are also problems of georeferenced data with different map scales and referencing systems, objects and timelines represented with different spatial and temporal granularities and semantic heterogeneity (e.g. the definition of 'city' or 'mountain' may vary with location). Mobility data may be collected using different methods and at different levels of granularity; these data may need to be resampled prior to integration. Given this range of expertise, it is unclear that spatial ETL functions could be executed automatically without human intervention; for example, software 'wizards' that ask users questions at crucial decision points in the ETL process (Bédard and Han, 2009).

A DW has different design objectives than standard databases. Operational databases are designed to support *transactional processing*: editing, updating and other database activities by multiple users. This usually involves avoiding repeated data and minimizing the number of logical connections among data items in the database design. Since a DW is an archive we do not need to worry about trans-actions and instead can design it to support *analytical processing*. Therefore a DW will have repeated and highly interconnected data to reduce the effort required to search and link data during exploratory analysis. This means that a DW may be much larger than the combined volume of the individual source databases (Miller and Han, 2009).

Spatial data warehouses (SDW) and *trajectory data warehouses* (TDW) must have support for spatial and temporal data storage and access methods such as spatial, spatio-temporal and mobile objects database indexes (Bédard and Han, 2009; Pelekis et al., 2008). Given the complexity and volume of spatial, temporal and mobility data, a design issue is also which measures to pre-compute and store in the database, as discussed above. There are three major strategies for SDW. First, we could simply store spatial data with no pre-computed spatial measures. A second strategy is to pre-compute and store a rough approximation of the spatial measures, such as those based on minimum bounding rectangles rather than the spatial objects themselves. A third strategy is to selectively pre-compute some spatial measures (Papadias et al., 2002; Shekhar et al., 2001; Stefanovic et al., 2000). With TDW we also have the choice of whether to not pre-compute any trajectories, pre-compute a limited portion of an individual trajectory, an entire trajectory or all trajectories (Pelekis et al., 2008).

Exploring and mining spatio-temporal data

The above discussion concerned how spatio-temporal data can be integrated within SDW and TDW. In this section, this discussion is expanded to consider a series of common methods that are used to both explore and mine spatio-temporal data. The methods discussed have been selected as they are commonly used; however, within this topology, there is diversity both within classes, and at the boundaries between different methods.

Space–time cubes

A *data cube* is method for organizing and managing database summary measures for querying, visualization, or inputs into data mining techniques. A data cube is a multidimensional extension of a spreadsheet: it is a logical arrangement of data measures that allows quick summaries and cross-tabulations at different levels of aggregation. Summary measures can include *distributive functions* (e.g. count, minimum, maximum, sum), *algebraic functions* (e.g. average, standard deviation), or *holistic functions* (e.g. median, mode, rank). These summary measures can be pre-computed and stored to increase query performance (Gray et al., 1997; Han and Kamber, 2012; Rivest et al., 2005).

Figure 6.1 illustrates a data cube. Given a measure (such as 'sales') and dimensions of interest (such as 'item', 'location' and 'time'), the corresponding data cube is the three-dimensional data lattice shown. A data cube will store the measures as well as their summaries at all possible levels of aggregation (e.g. total sales by item and location, total sales by location and total sales). We can also manipulate and query the data cube in several ways. A *slice* of the data cube is a selection based on a single dimension: for example, we could select all sales data for 30 November 2012. *Dicing* creates a sub-cube by selecting data based on conditions for two or more dimensions: for example, we could select all sales for all items in Chicago on 30 November 2012. We can also perform a *roll-up* (aggregation) operation that removes a dimension from the cube. For example, if we rolled up the cube in Figure 6.1 by removing the time dimension, we would have sales for all items by location for all times. We can also disaggregate or *drill down* the data cube by adding an additional dimension, such as 'customer type' (Han and Kamber, 2012).

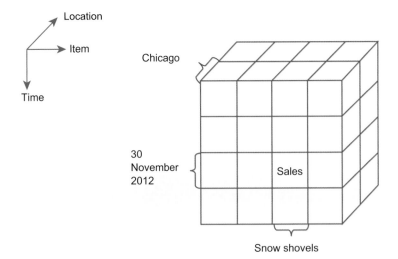

FIGURE 6.1   A three-dimensional data cube

Concept hierarchies can enhance roll-up and drill-down operations on data cubes. For example, we might provide a concept hierarchy for the location dimension with the levels (from lowest to highest) *City < State/Province <*

Country. Similarly, we could have a concept hierarchy for the time dimension with the levels *Date* < *Week* < *Month* < *Season* < *Year*. Concept hierarchies provide a different and often richer way to aggregate and disaggregate data for exploration (Han and Kamber, 2012).

A *spatial data cube* or *map cube* is a data cube for spatially referenced data. A map cube can have three types of spatial dimensions: *non-geometric* (nominal categories, such as 'Chicago' in Figure 6.1), *geometric* (having spatial objects at all levels) and *mixed* (spatial objects associated with some levels of aggregation but nominal spatial references at other levels; an example is spatial objects at the state/province level but nominal spatial references at the national level). A map cube can also contain spatial aggregation operations, including spatial distributive functions (such as a minimum bounding box for an object or object collection, union and intersection among spatial objects), spatial algebraic functions (such as centroid and center of gravity) and spatial holistic functions (such as nearest neighbor) (Bédard and Han, 2009; Rivest et al., 2005).

A *space–time attribute cube* contains spatial, temporal and non-spatial dimensions (Guo et al., 2006). *Space–time cubes* contain only spatial and temporal dimensions with different levels of granularity. An example of a space–time cube is a *traffic cube*: a space–time cube for traffic data (Lu et al., 2009). Figure 6.2 shows a space–time plot for traffic counts based on slicing the space–time cube for day of week by time of day for the week corresponding to the American Thanksgiving holiday in the year 2003. The impact of the holiday on Thursday can clearly be seen in the plot, as well as the transition into the unofficial shopping holiday on Friday (Song and Miller, 2012).

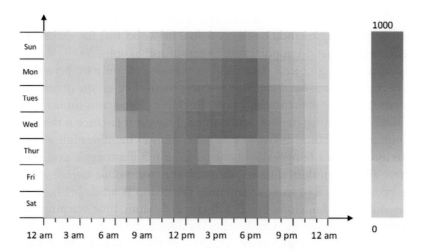

FIGURE 6.2   Visualizing a slice of a space–time cube: time of day versus day of week for traffic count data

A large number of data-mining techniques can be applied to spatio-temporal data after they have been prepared, stored and organized. Major classes of data-mining techniques include clustering, association analysis, classification, generalizations, trend detection and outlier analysis (see Andrienko et al., 2008a; Fayyad et al., 1996; Miller and Han, 2009). Most of these techniques have been extended to spatio-temporal data, typically focusing on algorithmic aspects and fast approximation techniques for calculating spatial and temporal relations and properties for inclusion into data mining techniques (Nanni et al., 2008). In the remainder of this methods section, we will focus on techniques that are particularly well suited for spatio-temporal and mobility data.

Spatio-temporal clustering

Cluster analysis involves sorting a set of objects into a smaller number of groups consisting of members that are similar (see Alexiou and Singleton, Chapter 8). In standard (non-spatial) clustering, similarity is based on attribute values. With spatial and temporal data, similarity may also include spatial proximity and/or temporal coincidence. There are numerous cluster analysis techniques. *Partitioning* methods divide a dataset into a smaller set of groups such as that each group has at least one member and each item belongs to exactly one group. *Hierarchical methods* start from the top down (*divisive*) or bottom up (*agglomerative*) to build a multi-level system of clusters. *Density* methods grow clusters by adding objects as long as the density of objects within a defined neighborhood meets a minimum threshold. *Grid* methods divide the object space into a grid and cluster based on that structure. Density- and grid-based methods can be effective for spatial and temporal data since they can discover arbitrarily shaped clusters as opposed to compact, spherical shaped clusters as in partitioning and hierarchical methods (Han et al., 2009). *Self-organizing maps* are another clustering method based on neural networks that has also been shown to be effective for spatio-temporal data (Agarwal and Skupin, 2008).

Clustering moving objects data requires measuring similarity between space–time trajectories. *Shape-based similarity measures* focus only on the geometry of the trajectories, ignoring sequence and time. Euclidean distance is intuitive and easy to calculate, but sensitive to noise and outliers. Hausdorff distance is the maximum of the minimum distances between two paths but can be misleading since it does not take into account temporal sequence. Figure 6.3 illustrates a case where two paths have a low Hausdorff distance although the two mobile objects are unlikely to have been proximal in space at any time (Yuan and Raubal, 2014).

Time-based methods include Fréchet distance, dynamic time warping and longest common subsequences. Unlike the Hausdorff distance, the *Fréchet distance* takes into account the sequence of locations in the path and is therefore better suited to mobile objects (Maheshwari et al., 2011). *Dynamic time warping* measures the similarity between two sequences or trajectories, based on the effort

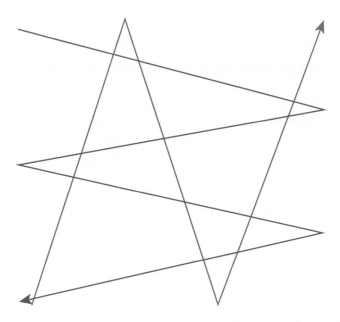

FIGURE 6.3 Two trajectories with a low Hausdorff distance despite the lack of spatio-temporal proximity between the mobile objects

involved to stretch or compress time to get the sequences to match. *Least common subsequence* measures similarity based on the length of least common subsequence between two sequences (Dodge et al., 2012; Nanni et al., 2008; Yuan and Raubal, 2014).

Edit-distance functions measure similarity between sequential patterns based on the cost of the insertion, deletion and substitution operations required to transform one sequence into the other. In traditional sequence pattern similarity measurement the cost of each operation is set constant, but for spatio-temporal sequences we can measure the operations costs based on the spatial and temporal relations between the corresponding trajectory data points. Yuan and Raubal (2014) develop edit-distance similarity measures for comparing mobile-phone trajectories based on spatial, temporal and spatio-temporal distance of each point from the trajectory's centroid. They also develop an edit-distance function constrained by time-windows corresponding to semantic time periods (such as 'morning' and 'evening').

Spatio-temporal associations

Association rule mining involves searching data to discover conditions occurring together frequently. Association rules are of the form $X \Rightarrow Y(s\%, c\%)$ where X and Y are conditions that occur together and $(s\%, c\%)$ are the levels of *support* and *confidence* for the rule; these correspond to the empirical probabilities $P(X \cup Y)$ and $P(Y | X)$, respectively. For example, the spatial association rule

$$\text{Is } (School) \wedge \text{near } (Sports\ Center) \Rightarrow \text{near } (Park)\ (2\%, 55\%)$$

means that 55% of schools that are near a sports center are also near a park, and this occurs in 2% of the items in the database (Han and Kamber, 2012).

Association rule mining requires finding all itemsets in the database that occur together based on a user-defined support threshold and then generating association rules from that subset based on a user-defined confidence level. Mining for spatial, temporal, and spatio-temporal association rules is complex since finding associations requires computing spatial proximity and temporal coincidence relations from the data (Mennis and Liu, 2003). A strategy for managing this complexity is *filter-and-refine*: first mine the full database using approximations of the spatial objects such as minimum bounding rectangles or convex hulls; then refine these results by mining the detailed spatial objects in the smaller candidate dataset, removing the false positives (Han and Kamber, 2012).

Verhein and Chawla (2008) develop *space–time association rules* (STARs) that describe how objects move among a set of regions over time. STARs apply to specific time intervals and describe how objects move among a set of regions during those intervals. Regions can be any shape or size. An example is the rule:

$$\text{Is}(Office\ worker) \wedge \text{outside } (City) \wedge \text{time}(Night) \Rightarrow \text{inside } (City) \wedge \text{time } (Morning)$$

In other words, office workers who are outside of the city at night tend to head into the city in the morning. They also extend the concept of coverage and support to space and time by considering the lengths and sizes of the time intervals, spatial regions and objects described by the rules. They develop rules for empirical concepts such as *stationary regions* and *high traffic regions*, with the latter consisting of *sources*, *sinks* and *thoroughfares*.

Sequence mining

Closely related to association rule mining is *sequence mining*: searching for patterns in time or other sequences. Most sequential mining techniques concentrate on symbolic patterns, searching the database for events (symbols) that occur frequently together. The search is similar to association rule mining, but is based on three search parameters: the *duration* of the time sequence, the *event window* or time-horizon for considering events as temporally coincident, and the time *interval* between events.

Periodic pattern mining is a type of sequence mining that searches for recurrent patterns. These can include: (i) *full periodic patterns* where every point in time contributes either exactly or approximately to the cyclic pattern (e.g. all days of the year approximately contribute to the season pattern); (ii) *partial periodic patterns* that are true for some but not all points in time (e.g. a person may visit a particular coffee house at 7:00–7:30 a.m. every Monday through Friday but display no other regular activity patterns); (iii) *cyclic or periodic association rules* that associate events that occur periodically (e.g. a busy afternoon in a restaurant implies a busy evening rush). Both sequential and periodic mining can be conducted as a database search in a manner similar to association rule mining (Han and Kamber, 2012).

Extending sequential and periodic mining techniques to mobile objects data requires solving two sub-problems: (i) how to detect the periods in complex movement patterns; (ii) how to mine the period behavior. A strategy is to search for *reference spots* or locations that are repeatedly visited by the objects and then mine for recurrent movement patterns between these reference spots (Li et al., 2012). Bleisch et al. (2014) combine these techniques with a conceptual framework regarding the possible causal relations between states and events in the world to infer causal and causal-like relationships between movement patterns of mobile objects with associated measures of support and confidence.

Mining collective patterns

Another important set of patterns associated with moving objects data are *collective motion patterns*. These focus on individual object movement patterns within the context of a larger group of mobile objects (Long and Nelson, 2013). Distance-based measures search for collective patterns such as flocks by searching for moving objects that are densely connected in space given user-defined distance and time thresholds (Gudmundsson and van Kreveld, 2006). Laube et al. (2005) develop a relative motion (REMO) approach that considers the individual directions in a group of moving objects to detect collective patterns such as flocking, leadership, convergence and encounter. Gudmundsson et al. (2007) improve the efficiency of the REMO approach using techniques from computational geometry.

Visualization and visual analytics

Visualization is a powerful strategy for leveraging human visual acuity to help make sense of data and discover patterns and anomalies in spatio-temporal data. Visualization is particularly essential for analyzing processes unfolding in geographical space with respect to time. The heterogeneity of geographic space and the variety of properties and relationships within it cannot be adequately represented in algorithmic processing, exploration, and analysis of spatio-temporal data. Visualization facilitates the derivation of knowledge from spatio-temporal data by leveraging the human analyst's sense of space and place, tacit knowledge of their inherent properties and relationships, and space/place-related experiences (Andrienko et al., 2008a).

More than just another exploratory technique, visualization is also a powerful method for integrating and enhancing all aspects of the spatio-temporal knowledge discovery process from data management through to guiding the exploration process, interpreting the results and constructing new knowledge (Gahegan, 2009). *Visual analytics* is the science of data-driven reasoning facilitated by interactive visual interfaces to data management and analysis techniques (Thomas and Cook, 2004). Visual analytics is more than just information visualization: while visualization provides insights into data, visual analytics provides insights into how we *process* data during exploration and analysis (Keim et al., 2008).

Visual analytics for spatio-temporal data includes techniques such as attribute plots, time plots, space–time cube operations, methods for data selection and data-mining techniques such as clustering; all linked together though interactive visual interfaces with the map as the central metaphor (Guo, 2003, 2009; Guo et al., 2006). Visual analytics for moving objects data includes linked techniques for analyzing and comparing entire trajectories, variations of properties within trajectories, synoptic visualization and analysis of the moving objects as a whole, and investigating movement within the spatio-temporal context in which it occurs (Andrienko et al., 2013).

Conclusion

We have entered an unprecedented era for research in geography and earth sciences. The world is awash with spatially and temporally referenced data flowing from remote sensing systems, location-aware technologies, embedded sensors and social media. Confirmatory techniques are not well equipped to explore these massive spatio-temporal databases and to discover information hidden in them.

Spatio-temporal data-mining techniques applied within the broader knowledge discovery process have the potential to generate new insights into human, physical and linked human–physical systems. Spatio-temporal knowledge discovery is the intricate process of preparing, managing and exploring massive spatio-temporal databases to generate novel geographic knowledge. Spatio-temporal data offer unique challenges to knowledge discovery due to the volume and complexity of these data as well as the implicit nature of spatial and temporal relations and properties within spatially and temporally referenced data. However, recognizing the value of spatio-temporal data, a large and vibrant interdisciplinary research community has emerged over the past two decades to meet these challenges. New methods and techniques for integrating and managing spatio-temporal and moving objects data are being developed, as well as techniques that can exploit the spatial and temporal properties of these data to generate new geographic knowledge.

A new data-driven scientific geography is emerging in response to the wealth of spatially and temporally referenced data, flowing from sensors and people in the environment, as well as capabilities for extracting new knowledge from these data. While this is very promising, perhaps revolutionary, there are some cautions. Data-driven knowledge is idiographic – contingent on particular places and times – while science seeks nomothetic (general, law-like) knowledge. We must be cautious about where this research is occurring – in the open light of scientific peer review and reproducibility, or behind the closed doors of private-sector companies and government agencies. Privacy is a concern not only as a human right, but also as a potential source of political and societal backlash that will curtail data-driven research. Finally, we must remember that data should not make decisions for us. Data should support but not replace decision-making by intelligent and skeptical scientists (Miller and Goodchild, 2014).

FURTHER READING

Hey et al. (2009) is a collection of short, empirically oriented essays about how data-driven science will change fields ranging from earth science to medicine. Miller and Han (2009) is an edited collection of chapters describing spatial and spatio-temporal data-mining methods. Several chapters provide a basic review of methods such as spatial data warehouses, map cubes and spatio-temporal clustering.

7

CIRCULAR STATISTICS

David Rohde and Jonathan Corcoran

Introduction

Geocomputation involves the analysis of data in many forms, including real numbers that might include information on distances or times, counts derived from census tables, or categorical variables drawn from survey data, to name only a select few. The analysis of these datasets requires the use of statistical methods based upon distributions with appropriate support, for example the *normal distribution* which has support over the real numbers (i.e. negative or positive values) and the *Poisson distribution* which has support over the natural numbers (i.e. count data, see Nakaya, Chapter 12). In this chapter we consider another type of data of interest to geographers: namely, *angular* or *directional* data.

Angular quantities are real numbers but with bounded support on the interval $[0,2\pi]$ (radians) or $[0,360]$ (degrees) and the unusual property that quantities near the extremes of 0 and 2π are in fact close. The consequences of these unusual properties are that statistical methods that apply to real data can give very misleading results when applied to angular datasets. It is from these unusual properties that the sub-discipline of circular statistics was born; Fisher (1995: 1) described it as a 'curious byway of statistics', with origins dating back to the mid-eighteenth century (Bernoulli, 1808). Early applications of circular measures were employed to demonstrate that the orbital planes of planets in the solar system could not be aligned by chance (Mardia, 1975), and Florence Nightingale (1858) developed the coxcomb (or rose diagram) to visually depict the efficacy of improved sanitation in hospitals during the Crimean War. A number of books have since been devoted to the topic of circular statistics; of particular note are Fisher (1995) and Jammalamadaka and Sengupta (2001).

The importance of circular statistics to analytical geography in the examination of spatial phenomena is twofold: first, as a technique to analyse direction (e.g. direction of travel); and second, to investigate the temporal dynamics of phenomena (e.g. time of day or day of week, see Figure 6.2 on page 103 for a visual approach).

In both cases circular statistics have a potentially important role to play in uncovering directional and temporal dynamics of spatial phenomena. The focus of this chapter will be on the examination of directional dynamics. In general, the previous research on using directional data has tended to focus on data with no explicit spatial frame in which applications in geology and biology have been particularly prevalent. Despite the general lack of application of circular statistics in geography, the potential that such techniques hold in terms of their capacity to augment the geographer's analytical toolbox is significant in permitting new ways to analyse and visualise spatial data. In bringing this new approach to the study of geographical datasets, new questions can be posed that may include derivatives of the following (taking commuting data derived from the census by way of example):

1. What is the mean direction of commuter travel, and how does this mean direction vary from one region to the next?
2. Does the mean direction of commuter travel of any one travel zone differ significantly by gender and mode of travel?
3. Have there been observable shifts in mean directions of commuter travel between census waves, and, if so, how do these relate to changes in urban form?

With the general thrust of these questions in mind, this chapter aims to encourage readers to think about their own datasets and how these questions may be reformulated with these different contexts. Furthermore, this chapter aims to encourage readers to experiment with the analysis of circular data using pre-existing software libraries such as Berens (2009) for MATLAB or the circular package for R (Agostinelli and Lund, 2011).

The remainder of this chapter is organised as follows. First, we outline how descriptive statistics analogous to the mean and variance can be used to summarise circular or angular data. We then present the von Mises distribution as one of the most useful circular distributions, but also wrapped distributions as another very flexible way to define a probability distribution of a circular quantity. These discussions are then extended to develop semi-parametric and non-parametric statistical models, alongside the statistical testing of circular quantities. The remainder of the chapter is concerned with a case study analysing the direction of travel utilising data derived from a bus electronic ticketing information system in Brisbane, Australia. Concluding remarks are then made alongside further reading.

Descriptive statistics

There are two ways to represent an angle, using either polar coordinates θ or a vector $\mathbf{r}$ constrained such that $|\mathbf{r}| = 1$; the latter method is also useful in considering angles on spheres, a topic that is beyond the scope of this chapter. The two methods are related in the following way:

$$\mathbf{r} = \begin{pmatrix} c \\ s \end{pmatrix} = \begin{pmatrix} \cos\theta \\ \sin\theta \end{pmatrix}$$

The most basic of questions we might consider for a collection of angular data $\theta_1,\ldots,\theta_n$ is to compute the mean. It is rather obvious that the conventional mean $\frac{1}{n}\sum_{i=1}^{n}\theta_n$ will fail to take account of the fact that 0 and 2π are identical and that values around this region are in fact close, the so-called cross-over problem. An alternative approach is to compute the average based upon rectangular coordinates $\mathbf{r}_1,\ldots,\mathbf{r}_n$, i.e. $\bar{\mathbf{R}} = \begin{pmatrix} \bar{S} \\ \bar{C} \end{pmatrix} = \frac{1}{n}\sum_{i=1}^{n}\mathbf{r}_n$, through which the angle can be recovered by converting this quantity back to polar coordinates. This is achieved using

$$\arctan\left(\frac{\bar{S}}{\bar{C}}\right) \text{ if } \bar{C} > 0,\ \bar{S} \geq 0,$$

$$\frac{\pi}{2} \quad \text{ if } \bar{C} = 0,\ \bar{S} > 0,$$

$$\arctan\left(\frac{\bar{S}}{\bar{C}}\right) + \pi \text{ if } \bar{C} < 0,$$

$$\arctan\left(\frac{\bar{S}}{\bar{C}}\right) + 2\pi \text{ if } \bar{C} \geq 0,\ \bar{S} < 0,$$

$$\text{undefined if } \bar{C} = 0,\ \bar{S} = 0.$$

The rather complicated conditions in this equation are due to the fact that the ratio $\bar{S}/\bar{C}$ cannot distinguish between cases such as $\bar{C}$ and $\bar{S}$ both negative and $\bar{C}$ and $\bar{S}$ both positive (or one positive and the other negative). The operation is common enough that the function $\arctan2(\bar{C},\bar{S})$ is often defined as above (or more commonly, as a slight variant with support over $[-\pi,\pi]$) and is available in many programming languages, including MATLAB and R.

It is noteworthy that unlike the individual observations $\mathbf{r}_1,\ldots,\mathbf{r}_n$ which all have magnitude 1, the statistic $\left|\bar{\mathbf{R}}\right|$ usually does not; in fact, a quantity that is useful in measuring the spread is given by $1-\left|\bar{\mathbf{R}}\right|$, and is sometimes referred to as the circular variance. The circular variance ranges from 0 to 1. An intuition for this statistic can be gained by imagining that $\mathbf{r}_1,\ldots,\mathbf{r}_n$ are all identical, in which case $\mathbf{R}$ lies on the unit circle and $\left|\bar{\mathbf{R}}\right| = 1$, and as a consequence the circular variance is 0. Alternatively, if $\mathbf{r}_1,\ldots,\mathbf{r}_n$ are uniformly spread, these cancel each other out exactly, then $\bar{\mathbf{R}} = \begin{pmatrix} 0 \\ 0 \end{pmatrix}$ and $\left|\bar{\mathbf{R}}\right| = 0$ and the circular variance is 1; also note that in polar coordinates the mean angle is in this case undefined. An undefined mean is an interesting special case that occurs with circular statistics, but not linear statistics. While it might seem strange to have the mean undefined, it does makes intuitive sense that a distribution that is uniform over all angles has no mean.

Circular distributions

The normal or Gaussian distribution is fundamental to linear statistics and has support on the real numbers. In order develop statistical methods for circular data it is useful to consider distributions with support on the interval $[0,2\pi]$. One of the important properties of the normal distribution is that it has *sufficient statistics*, i.e. it is possible to summarise all of the information about the data likelihood from a sample from a normal distribution using just the sample mean and sample variance. The distribution for circular quantities that has sufficient statistics is the von Mises distribution, which has the form

$$f_\theta\left(\theta \mid \mu,\kappa\right) = \frac{e^{\kappa \cos(x-\mu)}}{2\pi I_0(\kappa)}$$

for polar coordinates in radians, or

$$f_r\left(r \mid \mu,\kappa\right) = \frac{e^{\kappa \mu^T r}}{2\pi I_0(\kappa)}$$

for rectangular coordinates (often referred to as the Fisher–von Mises distribution). The latter distribution can easily be generalised to spheres and hyperspheres. The normalisation of these probability densities requires the use of a Bessel function which is available in most programming libraries, although direct use of this function is often made unnecessary by the use of circular statistics libraries available for many programming languages. A series of scripts accompany this text in MATLAB, and are provided in order to demonstrate how to reproduce these results.

A histogram of circular data is shown in Figure 7.1. For this dataset, $\bar{\mathbf{R}} = \begin{pmatrix} 0.5211 \\ 0.5270 \end{pmatrix}$ and $\bar{\theta} = 0.2518\pi$ or 45.3 degrees. This estimate of the average angle gives a good representation of the peak in Figure 7.1. The linear mean is 0.6459π or 116.3 degrees. The failure of the linear mean to identify the mode of the distribution gives an illustration of the change-point problem. The change-point problem refers to the fact that 0 and 2π are identical and points on either side of these values are in fact close.

A fit of the von Mises distribution found using numerical methods results in the parameter estimates $\mu = 0.2518\pi$ and $K = 2.2914$.

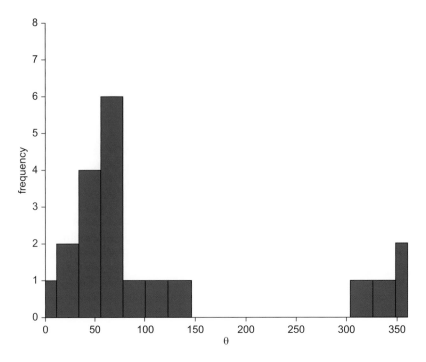

FIGURE 7.1 Histogram of angular data

Note that the estimate of the mean is identical to the sample circular mean, but there is no simple conversion between the circular variance and K. A normal or Gaussian distribution has a parameter representing the mean and the variance of the distribution, and in the case of the von Mises distribution there is a parameter for the mean, but the K parameter does not have a simple relationship with the circular variance. While $1/\kappa$ might reasonably be seen as analogous to the variance, $1/\kappa = 0.4364$ is quite different from the computed circular variance, which is $1 - |\bar{\mathbf{R}}| = 0.2588$.

The probability density function (pdf) of the fitted von Mises distribution is shown in Figure 7.2.

It can be useful to draw a rose diagram, which is the circular analogue of a histogram, or a plot of the pdf on polar coordinates; these are given in Figures 7.3 and 7.4. The main advantage of the circular plot is that there is no arbitrary edge at 0 or 2π radians or at 0 or 360 degrees. The possible disadvantage is that intuition may be worse in a circular plot as they are less familiar, and it may be difficult to read the magnitude on a rotated axis.

The fact that the von Mises distribution has sufficient statistics makes it a very convenient form to analyse circular datasets. However, other distributions are also of interest.

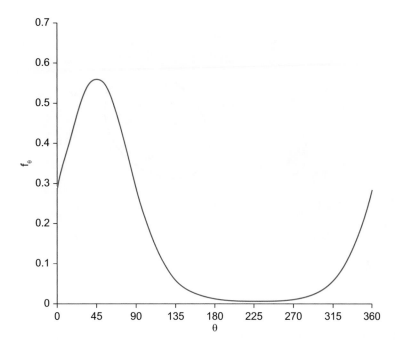

FIGURE 7.2 The pdf of a von Mises distribution fitted to the data shown in Figure 7.1

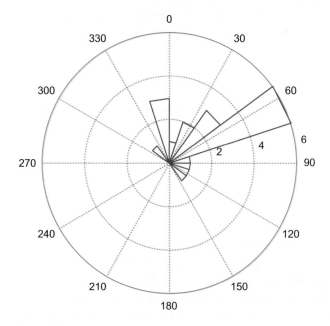

FIGURE 7.3 Rose diagram

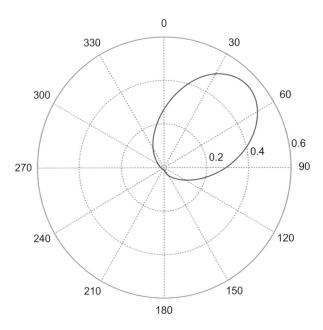

FIGURE 7.4   The pdf of a von Mises diagram fitted to the data shown in Figures 7.1 and 7.3

One way to construct a circular distribution from a distribution with real support $G()$ is to use modular division (Ferrari, 2009); recall that taking modular division with 2π is equivalent to taking the remainder after dividing by 2π and then multiplying this number by 2π. Thus

$$\omega \sim G()$$

$$\theta = \omega \bmod 2\pi$$

The pdf can then be constructed with an infinite sum

$$f_\theta\left(\theta\right) = \sum_{k=-\infty}^{\infty} f_\omega(\theta + 2\pi k)$$

where $f_\omega\left(\omega\right)$ is the density of $G()$.

The infinite sum is inconvenient, although it can be removed in special cases such as the wrapped Cauchy distribution. Unfortunately, it cannot be removed in the more interesting case of the wrapped normal distribution. The wrapped normal is of interest because it arises when applying the central limit theorem on the circle; also, the convenient properties of the multivariate normal are useful in multivariate statistics and the wrapped normal also enjoys other convenient mathematical properties: e.g. the sum of two independent wrapped normal distributions is also wrapped normal. While for linear statistics the normal distribution has a variety of convenient analytical properties, for circular statistics these convenient properties are awkwardly shared between the von Mises and the wrapped normal distributions.

The absence of sufficient statistics requires the use of numerical methods such as the expectation–maximisation (EM) algorithm in order to fit wrapped distributions; the rather technical details are given in Fisher and Lee (1994) with more background given in Jammalamadaka and Sengupta (2001), and a Markov chain Monte Carlo approach is outlined for fully Bayesian inference in Ravindran (2003) and Ferrari (2009). Fortunately, modern software libraries, particularly those provided in the R circular package, mean that implementing these complicated algorithms is usually unnecessary for the practitioner. A fit applied to the same data gives $\mu = 0.2549\pi$ or 45.3 degrees and $\sigma = 0.5162$, and a circular plot of the pdf is shown in Figure 7.5, which shows it to be a very similar shape to the von Mises. For completeness a wrapped Cauchy is shown in Figure 7.6 which also has a very similar shape, but is a little more concentrated and as a result the mean is slightly shifted.

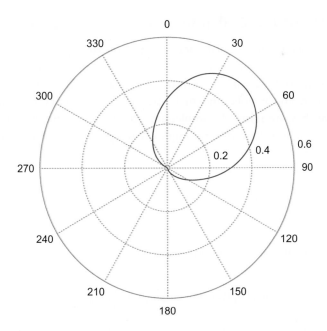

FIGURE 7.5 The pdf of a wrapped normal fitted to the dataset shown in Figure 7.1

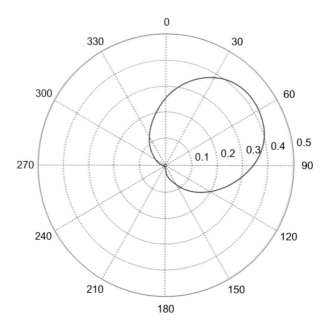

FIGURE 7.6 The pdf of a wrapped Cauchy fitted to the dataset in Figure 7.1

Semi-parametric and non-parametric methods

So far we have seen how circular data can be summarised by descriptive statistics that correspond to the sufficient statistics of the von Mises distribution, and we have also shown how the von Mises distribution can be fitted to data as well as the wrapped Cauchy and wrapped normal distributions. A general limitation of these models is that they are relatively simple; nevertheless these simple distributions are useful in order to produce more complex semi-parametric and non-parametric statistical methods.

Mixture models as semi-parametric models

An effective way to construct a more complex distribution, is by the use of a mixture model with K components which has the following pdf $m_\theta(\theta)$ and is constructed from a number of individual components $f_\theta(\theta)$

$$m_\theta\left(\theta \mid \alpha, \mu, k\right) = \sum_{k=1}^{K} \alpha_k f_\theta(\theta \mid \mu_k, \kappa_k).$$

Efficient algorithms exist for the estimation of such models, including the widely used EM algorithm.

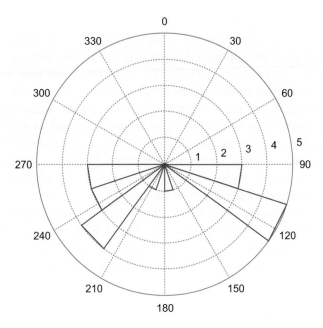

FIGURE 7.7   Rose diagram of a multimodal dataset

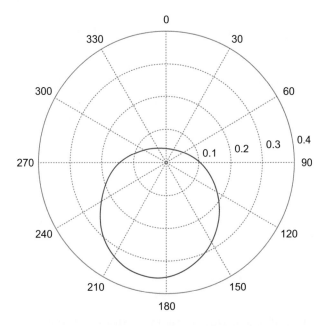

FIGURE 7.8   The pdf of a von Mises distribution fitted to the dataset in Figure 7.7

In order to consider why it might be necessary or desirable to use a mixture model, consider the dataset displayed in Figure 7.7. It is fairly easy to see that all of the methods we have considered so far will fail in one way or another. The average angle is -0.0184π or -3.3187 degrees, which is nowhere near the main modes observed at 50 degrees and 240 degrees. The variance for the data is quite large at 0.8645 (recall that the maximum value is 1), which, assuming a von Mises distribution, suggests that the data is near uniform instead of the two distinct modes observed. A polar plot of the pdf of the von Mises distribution is shown in Figure 7.8. In contrast, when a mixture of two von Mises distributions is fitted to the data, the two modes are located at 51.1 degrees and 246.6 degrees; a plot of the pdf of the mixture model is shown in Figure 7.9.

There are some possible difficulties with this methodology. The first is that the number of mixture components K must be selected. While it seems straightforward in the given case study that $K = 2$ is preferred, this is not always clear for real datasets. Moreover there is no easy statistical solution to this type of problem, and traditionally heuristics are employed such as the use of cross-validation or human intuition to select K. The EM algorithm also occasionally shows numerical problems failing to converge to the optimal solution or, in rare cases, finding a singularity in the likelihood. Failure to converge can usually be mitigated by rerunning the algorithm, while singularities can be avoided by preventing k becoming too large and thereby

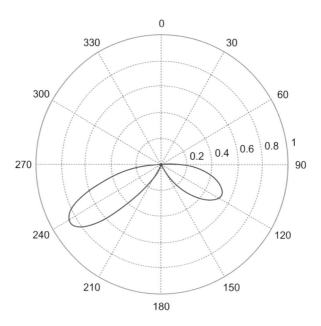

FIGURE 7.9   The pdf of a mixture of two von Mises distributions fitted to the dataset in Figure 7.7

preventing one of the mixture components being a sharp spike of probability mass centred on a single data point.

The technical details of the EM algorithm for the Fisher–von Mises distribution are given in Banerjee et al. (2005); again, fortunately, libraries such as Hornik and Grün (2014) in R or the simple MATLAB implementation included in the code accompanying this chapter reduce the need for practitioners to be overly concerned with implementation details.

Kernel smoothing as a non-parametric method

An alternative and very flexible approach to estimating a pdf is to use kernel density estimation, also known as Parzen's window (Hastie et al., 2005), which amounts to using the following expression as an estimate of the predictive distribution:

$$\bar{f}_\theta(\theta) \approx \frac{1}{n} \sum_{i=1}^{n} f_\theta(\theta_{n+1} \mid \theta_i, \kappa).$$

Here $f_\theta(\cdot \mid \theta, \kappa)$ is the kernel being used, and for circular problems, this is often set to be the von Mises distribution, which makes sure the estimated distribution has correct support and correctly handles the change-point problem at 0 or 2π. The bandwidth, given by $1/\sqrt{\kappa}$, is an important free parameter that must be set carefully, although no generic solution is known, and again, heuristics such as cross-validation or human judgement must be employed.

A demonstration of the technique using the dataset from Figure 7.7 and a von Mises kernel with $k = 20$ and $k = 100$ is presented in Figures 7.10 and 7.11, respectively. When k is smaller and consequently the bandwidth is larger, a much smoother estimate results, as is shown in Figure 7.10; and when k is larger and the bandwidth is smaller, a more delicate estimate results, more vulnerable to noise, as in Figure 7.11.

Testing

Testing is a vexed subject in statistics, and perhaps no topic more so than the use of p-values. In a relatively positive paper on the subject, Senn (2001) recommends not relying on p-values alone, using likelihood as well as well as reporting point estimates and standard errors, and considering the use of Bayesian methods (amongst other things!).

With these qualifications noted, here we introduce a circular analogue for testing whether two quantities are independent. A test of this form involves the computation of a statistic $T(x)$, a function of the two datasets that returns a real number; furthermore, we need to know what the distribution of T under the null

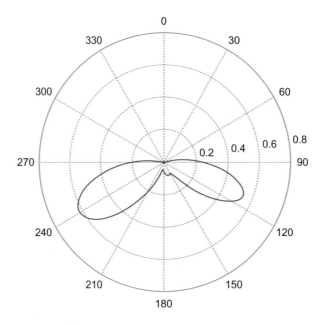

FIGURE 7.10   Kernel density estimate of pdf using the data shown in Figure 7.7 with a von Mises kernel with $\kappa = 20$

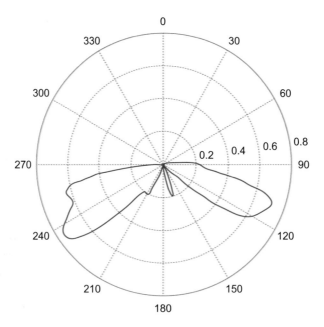

FIGURE 7.11   Kernel density estimate of pdf using the data shown in Figure 7.7 with a von Mises kernel with $\kappa = 100$

hypothesis is. The statistic $T(x)$ is computed on the real data, and then the probability of the statistic, or a more extreme value, is computed as the p-value. If the p-value is suitably low, then it is reasonable to say that either the null hypothesis is wrong or something unusual happened. The peculiarities of circular statistics enter when finding a reasonable definition of the statistic $T(x)$. In cases where the data consist of real numbers a simple statistic such as the difference in means might suffice, but, as noted, computing differences with circular quantities constitute a problem.

In order to give an example, a bivariate problem will be considered where the first example dataset θ^1 will be used. As noted earlier, this details the direction of a passenger travelling on the bus network, and the second datset θ^2 will be considered, which relates to the direction of travel for the journey to work for the same passenger. A question of interest might be whether the model should utilise the joint distribution $f_{\theta^1\theta^2}(\theta_i^1, \theta_i^2)$ or whether an independence assumption can reasonably be made by implying $f_{\theta^1\theta^2}(\theta_i^1, \theta_i^2) = f_{\theta^1}(\theta_i^1)f_{\theta^2}(\theta_i^2)$.

In order to make this approach as flexible as possible, Brunsdon and Corcoran (2006) propose as a statistic the mutual information between the two densities, estimated using the above kernel smoothing approach. This results in the statistic

$$
T\left(\theta^1, \theta^2\right) = \frac{1}{n}\sum_{i=1}^{n}\log\frac{\bar{f}_{\theta^1\theta^2}(\theta_i^1, \theta_i^2)}{\bar{f}_{\theta^1}(\theta_i^1)\,\bar{f}_{\theta^2}(\theta_i^2)}
$$

where θ^1 refers to the first sample which has estimated density $\bar{f}_{\theta^1}(\cdot)$, θ^2 refers to the second sample which has estimated density $\bar{f}_{\theta^2}(\cdot)$, the estimated joint density is $\bar{f}_{\theta^1\theta^2}\left(\theta_i^1, \theta_i^2\right)$ and the number of data points is n. If the two distributions are independent then the mutual information is 0.

As this is a complicated statistic, there is no known form for the sampling distribution, and instead computational methods such as drawing random permutations of mismatching θ_i^1, θ_j^2 $(i \neq j)$ can be used in order to sample from the statistic conditional on the null hypothesis, or alternatively the very general bootstrap procedures proposed by Efron and Tibshirani (1993) can also be adopted, as it is here. This methodology, as stated, is quite a generic method for carrying out a test of independence of two distributions; the adaption to circular statistics occurs by using a von Mises kernel in order to carry out the density estimation.

In order to demonstrate this method, the test is applied to the two presented datasets using $k = 20$. A colour map representing the joint pdf of the joint estimate is shown in Figure 7.12(a), and one representing the product of the two densities separately estimated is shown in Figure 7.12(b); the similarity suggests that the two distributions might be close to being independent.

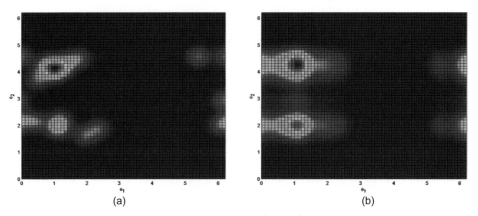

(a)   (b)

FIGURE 7.12   (a) The joint estimate of $\bar{f}_{\theta^1\theta^2}\left(\theta_i^1,\theta_j^2\right)$. (b) The product of the marginal estimates $\bar{f}_{\theta^1}\left(\theta_i^1\right)\bar{f}_{\theta^2}\left(\theta_j^2\right)$

The statistic takes the value $T = 113.8$; when we bootstrap samples from the distribution of the statistic we find that the probability of this value being more extreme (i.e. of greater value) is 0.666, which indicates that the null hypothesis of independence is not rejected. This finding is consistent with the observation that the two estimates in Figure 7.12 are quite similar. Also note that the value 113.8

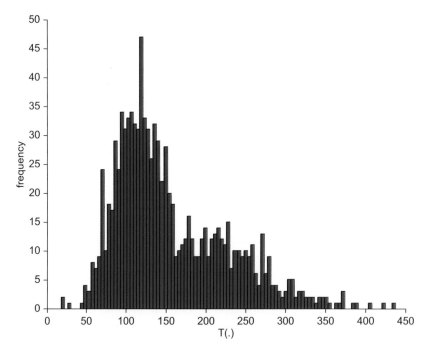

FIGURE 7.13   Bootstrap simulations of the sampling distribution

is a fairly typical mid-range value in the sampling distribution (see Figure 7.13: the higher the value, the less compatible the statistic is with independence).

While the test found that the data were consistent with the hypothesis that the two datasets are independent, in practical situations there are often compelling a priori reasons to believe that there will be some sort of dependence between distributions, and in a situation like these, with fairly small sample sizes, the failure to find a difference between the joint distribution and the product of the marginal may be because the dataset is too small to detect such differences. Users of these tests should bear in mind the impact of effect size and sample size on the outcome of the test. Testing within the context of circular data is a large well-studied topic and much more information about circular statistical testing can be found in Mardia and Jupp (2000).

A case study modelling direction of travel of bus commuters

Brisbane, Australia's third largest city and Queensland's state capital, has a relatively extensive bus network which includes over 400 bus routes and 10,000 bus stops. The origin and destination of hundreds of thousands of trips per day can be tracked by means of electronic ticketing or smart cards. An important problem from a transport planning perspective is to determine the volume of people travelling between given origins and destinations and how this varies from location to location.

A simplified extract from the 150,000 record smart card dataset is shown in Table 7.1, which for the purposes of this study includes information capturing the location (in (x, y) coordinates) of each trip origin and destination, or, in other words, the bus stop at which the passenger began and completed their journey.

In order to conduct the circular analysis of Brisbane, the city is divided into a 14 by 14 grid, and for each cell in the grid, the set of trips that originate from that cell are aggregated for analysis. The amount of data available per cell is variable and in some cases away from the bus network the cells are empty.

TABLE 7.1 Smart card dataset

Record_ID	Origin x	Origin y	Destination x	Destination y
1	502,330	6,961,700	494,400	6,964,500
2	502,390	6,961,400	505,780	6,958,800
3	506,100	6,958,900	503,140	6,962,700
4	502,490	6,961,400	508,520	6,949,000
5	503,870	6,958,500	502,790	6,959,900
6	496,850	6,962,100	502,520	6,961,700
150,000	493,530	6,964,300	501,870	6,961,100

Bus Stop

Statistical Area 2

(1) Queensland University of Technology
(2) CBD
(3) New Farm Park
(4) Indooroophilly Shopping Centre
(5) University of Queensland
(6) Cooparoo Shops
(7) Carindale Shopping Centre
(8) Griffith University
(9) Sunnybank Hills
(10) Eight Mile Plains
(11) The Gap
(12) Buranda
(13) Tingalpa
(14) Inala

N

5

Kilometres

FIGURE 7.14   Case study region

First, a simple method to model the data is to compute the circular mean for each of these locations, the results of which are shown in Figure 7.15. Perhaps unsurprisingly, the main direction of travel from almost the entire area of the city is towards the central business district (CBD, number 2 on the map). While the circular mean is obviously identifying the strongest trend, it is neglecting to detect when passengers move from a non-central location to some other non-central location. It seems reasonable to suppose that the circular distributions obtained at a number of locations around the city will have multiple modes, indicating multiple important lines of direction. This can be confirmed by plotting rose diagrams, or alternatively by using kernel density estimation for some of the locations. Consider the location highlighted in Figure 7.16 which is on the outskirts of the

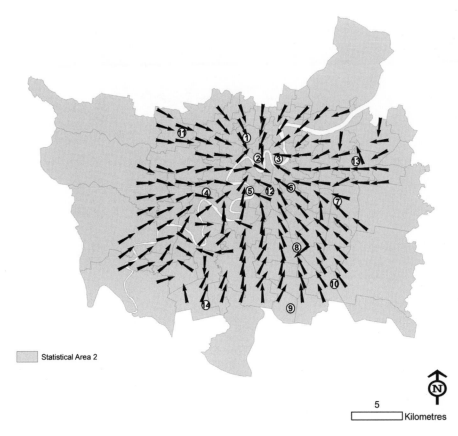

Statistical Area 2

5 Kilometres

N

FIGURE 7.15 Mean directions of travel for bus passengers in Brisbane

city in the South West; a rose diagram for the angular data is shown in Figure 7.17, a graphical representation of the mean angle in Figure 7.18 and a kernel density estimate in Figure 7.19. From these plots it is confirmed that there are several directions of flows of differing importance from passengers in this region, and summarising this with the mean direction of flow shown in Figure 7.18 is unsatisfactory. A much better fit is given by a four-component mixture of von Mises distributions, with the pdf of such a fit shown in Figure 7.20, and a plot of the direction and strength of the four modes shown in Figure 7.21. The significance of Figure 7.21 is that the four preferred directions of flow can be summarised by four circular parameters representing the means of the four components of the mixture model. The choice of four mixture components makes intuitive sense in this case as the rose diagram shows four clear modes, but in other regions it may make sense to have fewer or more modes.

FIGURE 7.16 Selected region (in yellow) for detailed analysis of directions of travel

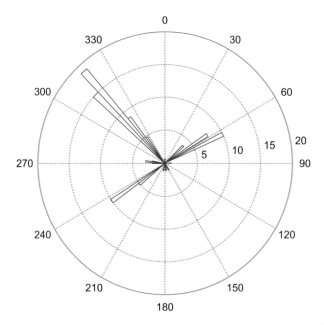

FIGURE 7.17 Rose diagram of the direction of travel on the selected data in Figure 7.16

FIGURE 7.22 The direction and strength of travel on a 14 by 14 lattice estimated using a two-component mixture model

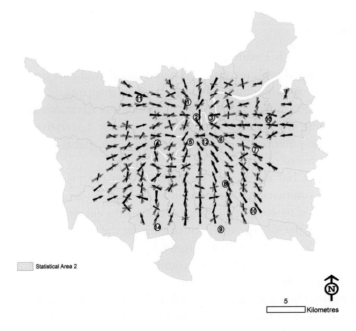

FIGURE 7.23 The direction and strength of travel on a 14 by 14 lattice estimated using a three-component mixture model

In order to build up a picture of secondary flows around a city, a two-component mixture model was fitted to the same dataset, and the two arrows plotted in Figure 7.19. A similar procedure was followed using a three-component mixture model in Figure 7.20 and a four-component one in Figure 7.21. The advantages of using the mixture model approach are immediately apparent as it is now possible to identify corridors of movement that are not moving towards the centre of the city but rather run across it. Without the semi-parametric mixture model approach only the most obvious city-bound trips could be extracted. The question of how many components should be used in the mixture model is perhaps more vexed; interesting details become apparent as the number of components is increased, but also in many places only one or two components seem adequate. Ideally the number of components could be varied depending on the data and the particular location. Another issue of concern is that the EM algorithm is a local optimiser that usually does not find global solutions to the maximum likelihood problem; slightly different results will therefore be obtained due to different initial conditions on subsequent runs of the algorithm. The use of mixture models allows Figure 7.15 to be improved by plotting secondary directions of travel using two mixture model components as in Figure 7.22, or three mixture components as in Figure 7.23. This methodology appears useful in highlighting secondary corridors of travel flows to other significant locations distinct from the CBD.

Conclusions

Geographers deal with a wide range of different data types, and while there are many well-developed tools for real, count and categorical data based upon distributions with appropriate support, angular or circular quantities which are also of great interest to geographers receive considerably less attention. This chapter has addressed how circular statistics can be used for both descriptive and modelling analyses with the potential to offer new insights into spatial datasets and extend the geocomputational toolbox.

The most basic but useful idea presented is that the circular mean is a convenient way of summarising simple unimodal angular or circular datasets. The critical idea here is to convert the angles to a vector quantity on the unit circle, before computing the average and then converting the output back to polar coordinates. It was demonstrated that this simple idea mitigates the change-point problem, i.e. the closeness of points around $[0, 2\pi]$ or $[0, 360]$. Another simple statistic based upon the magnitude of this vector quantity that was shown to be a useful analogue of linear variance was referred to as circular variance.

In order to consider more sophisticated statistical methods it was necessary to introduce some distributions with support for angular or circular quantities. The easiest of these to use in practice is the von Mises distribution which has the mean computed on the rectangular representation of the angle, as used previously for descriptive statistics. The other common and useful way to use distributions with real support (i.e. the distribution applies to regular numbers including fractions and negative numbers) is to convert the model to one with the more limited support used for angles by wrapping the distribution around the interval $[0, 2\pi]$ or $[0, 360]$, thus giving the wrapped normal and wrapped Cauchy distributions amongst others. Both of these distributions can be fitted to datasets and therefore used for relatively simple univariate models. More complex modelling can be achieved using semi-parametric methods such as a mixture of von Mises distributions or non-parametric methods using kernel smoothing methods.

A case study of electronic smart card data was used in order to identify the main routes of travel in Brisbane, Australia. A simple analysis based upon applying circular means to a lattice of data found that the main directions of travel are predominantly towards the CBD. In order to further investigate the data, a particular region in the lattice was shown to contain multimodal structure that required either non-parametric or semi-parametric approaches. The mixture model approach was demonstrated to be particularly useful in identifying corridors of movement as each mixture component could be plotted as an individual arrow showing a general direction of movement. This allowed the identification of more subtle movements within the city.

In summary, the frequencies with which geographers might consider the analysis of direction suggest that there are many opportunities to employ a circular statistical and graphical approach to a wide variety of spatial datasets. It is hoped that this chapter highlights and promotes the ways in which circular statistics might be more routinely embedded within geographical analyses.

FURTHER READING

This chapter outlines the fundamental differences with analysing and modelling circular data, offers some techniques and demonstrates their usefulness with a case study. There has, however, been a great deal more development in this area. In this section we point the reader to some of the literature on more advanced topics of interest for the examination of geographical problems.

ACKNOWLEDGEMENTS

We would like to thank Translink for access to the data on which this chapter is based. However, the interpretations of the analysis are solely those of the authors and do not necessarily reflect the views and opinions of Translink or any of their employees.

Fisher, N.I. (1995) *Statistical Analysis of Circular Data.* Cambridge: Cambridge University Press.

A generic tool that has proven to be of great utility to geographers is the use of correlation or covariance to measure the relationship between two real variables, such as the determination of a relationship between income and level of education across a region. The use of correlation is complicated when one or more of the variables is an angular quantity. In order to deal with these issues useful empirical measures of correlation between two circular quantities or between a circular quantity and a linear quantity have been proposed in this text. Taking our earlier questions based on the commuting example, one could now explore whether the mean direction of commuter travel is correlated with the distance travelled across a region.

Lee, A. (2010) Circular data. Wiley Interdiciplinary Reviews: *Computational Statistics*, 2: 477–486.

A second topic of interest is that of regression modelling where the dependent variable is either a count or a real number. Again modifications are needed if the variable represents a circular or angular quantity. Common to the application of correlation and covariance, regression modelling has formed a core analytical backbone of a plethora of geographical studies. However, in its standard form it is not capable of integrating angular quantities. Circular regression methods (see the above text for a good entry point to this literature) offer the possibility to extend the conventional regression framework, permitting the investigation of new questions. Drawing on our commuting example, one could examine how the mean directions of travel across a region (as the dependent) are explained by a set of socio-economic variables such as age, income and industry sector of employment.

Jona-Lasinio, G., Gelfand, A. and Jona-Lasinio, M. (2012) Spatial analysis of wave direction data using wrapped Gaussian processes. *Annals of Applied Statistics*, 6: 1478–1498.

A final area of interest to geographers is the inclusion of a notion of spatial smoothing into the model. Geographers regularly employ models for real-valued data that incorporate space such as geographically weighted regression or the spatial autocorrelation model, all of which rely on the properties of the multivariate normal distribution. An adaption to circular or angular quantities requires a similarly useful probabilistic model. While there has been little work on multivariate circular distributions, the development of the wrapped Gaussian process outlined above is a notable exception that explicitly takes into account spatial aspects of the model, and Jona-Lasinio et al. (2012) demonstrate its ability in modelling wave direction.

PART III

MAKING GEOGRAPHICAL DECISIONS

8

GEODEMOGRAPHIC ANALYSIS

Alexandros Alexiou and Alex Singleton

Introduction

Geodemographic classification has been defined as 'the analysis of people by where they live' (Sleight, 1997: 16); it involves categorical summary measures that aim to capture the multidimensional characteristics of both built and socio-economic characteristics of small geographical areas. This chapter outlines the origins of geodemographic classifications, how they are typically constructed, and their application through an illustrative case study of Liverpool, UK.

Within sociology and geography there is a legacy of identifying aggregate socio-spatial patterns within urban areas through a variety of empirical methods. From the early 1900s onwards, researchers tried to systematically document spatial segregation and establish a series of general principles about the internal spatial and social structure of cities, commonly motivated by the ill effects of residential segregation of the poor and ethnic minorities (van Kempen, 2002). Within the UK, Charles Booth's poverty maps were one of the first attempts to map the socio-spatial structure of London in the early 1900s, although it was not until the late 1920s that the Chicago School formulated a comprehensive model of urban ecology, such as the concentric zone model of Burgess and Park (Burgess, 1925). Their research was largely based on the then recently introduced census data, alongside extensive field-work and map-making (Burgess, 1964: 11–13).

The analysis of detailed demographic, social and economic census data was further developed through the work of Shevky and Bell (1955). Their work introduced 'social area analysis', a methodology focused on a three-factor hypothesis that aimed to assert a typology of urban places measured in terms of urbanisation, segregation and 'social rank' (Brindley and Raine, 1979). This analytic framework inspired the adoption of a set of tools and techniques encapsulating a broader range of socio-economic census variables (Tryon, 1955; Rees, 1972), and such theoretical approaches were later collectively known as 'factorial ecologies', due to a widening of those aspects used to explain urban structure (Janson, 1980). Factor analysis (and, similarly, principal component analysis) dominated such quantitative geography in the 1970s, and was largely used to identify major underlying attributes of spatial structure, albeit with debatable results. Factorial studies were criticised not only because of their lack of theoretical context (Berry and Kasarda, 1977), but also because of their methodological weaknesses, for example their lack of extendability that constrained them to being city-specific (Batey and Brown, 1995).

During this period, much scholarly concern was also focused on the interpretation and categorisation of the fundamental processes by which cities operate. In spite of the numerous attempts to classify cities *per se*, studies failed to find a unified theory of city typology – if such a functional typology ever existed. Classifications started to focus alternatively on smaller-area geography, and on the 'methods flowing from identification of variations of cities and following from the selection of dimensions relevant to a specific purpose' (Berry, 1972: 2). There was a common belief that typologies aid in generalisation and prediction, and urban classification was much more comprehensive when applied with a narrow scope, in terms of both area and purpose.

Within such context, geodemographics emerged in both the United States and United Kingdom during the late 1970s as an extension of these earlier empirically driven models of urban socio-spatial structure. Geodemographic classifications organise areas, typically referred to as neighbourhoods, into categories or clusters that share similarities across multiple socio-economic and built environment attributes (Singleton and Longley, 2009).

Despite a lineage of use, geodemographic classifications lack a solid theory. In nomothetic terms, many view geodemographics as methodologically unsatisfactory since the underlying theory can be considered as 'simplistic' and 'ambiguous' (Harris et al., 2005). The conceptual framework is based on a fundamental notion in social structures, homophily – the principle that people tend to be similar to their friends. This manifests spatially as a general tendency for people live in places with similar people, much like the 'birds of a feather flock together' adage suggests; and it is consistent with Tobler's first law of geography, that 'everything is related to everything else, but near things are more related than distant things' (Tobler, 1970: 236). However, one paradox is that despite geodemographic representations showing spatial autocorrelation between taxonomic groups, the methods for building geodemographics as currently construed can be considered contradictory to Tobler's statement. The aggregations of zones into categorical measures based on attributes sweeps away contextual differences between proximal zones; and as such, the final classifications assume that areas within the same cluster have the same underlying characteristics. Standard geodemographic techniques have failed to incorporate near geography in a sophisticated way, and despite the term, geodemographics are in fact aspatial. Thus far, there have been very few attempts to build a unified framework, at least within which the relative benefits of both spatial interaction and geodemographic approaches can be maximised (see, for example, Singleton et al., 2010). For many applications, the issue of geographic sensitivity is usually experienced when normalising input variables globally and without taking into account local variation extents, thus obscuring potentially interesting local patterns. For instance, some argue that the relationship between areal typology and behaviour might not be spatially constant (Twigg et al., 2000). This type of ecological fallacy raises a series of methodological questions regarding the success of geoclassifications, given the high within-cluster variation that is already smoothed away (Voas and Williamson, 2001).

Geodemographic classification systems

Geodemographic analysis was initially developed as a 'strategy' that can be used to identify patterns from multidimensional census data (Webber, 1978). However, current geodemographics may use a variety of public and private data to generate profiles (Birkin, 1995). Some of the pioneering studies were applied in the UK to identify neighbourhoods suffering from deprivation (Webber, 1975). However, in the USA, geodemographics were first utilised in the private sector, as the macro-economic conditions, alongside the freedom-of-information tradition, created an environment that quickly enabled the exploitation of census data commercially (Flowerdew and Goldstein, 1989), and the first commercial applications started appearing during the early 1980s. In the following years, geodemographic classifications gained large popularity as their utility was demonstrated across a variety of applications – from strategic marketing and retail analysis to public sector planning (Birkin, 1995; Brown et al., 2000).

Despite a common starting point, there are arguably critical differences between the UK and the USA, as geodemographics evolved through different paths. While the US classifications have typically been commercial, in the UK context there is a long history of free and more recently open classifications, and they have seen greater application in public policy and academia (for a detailed review, see Singleton and Spielman, 2014). More generally, in the UK there has been a recent renaissance of interest in geodemographics from the public sector, mainly driven by government pressure to demonstrate value for money and the advent of new application areas (Longley, 2005).

For instance, Batey and Brown (2007) developed a method of evaluating the success of area-based initiatives by using a geodemographic classification to produce spatially targeted socio-economic profiles. In this way, they assessed the efficiency of spatially targeted urban policies by examining how many of the people these contained are in fact not those for whom the initiative is intended, in which case it is defined as inefficient or incomplete. Singleton (2010) and Singleton et al. (2010) explored patterns of access to higher education by linking summary measures of local neighbourhood characteristics with individual-level educational data; and through a spatial interaction framework, demonstrated the size of spatial flows between socio-economically stratified areas and institutions, with the aim that such a tool could be used by key stakeholders to examine potential policy scenarios.

Geodemographics have also been recently used in health screening epidemiology, where detailed geographical information is often unavailable. In these studies, finer geographic granularity is necessary in order to produce targeted ecological estimates and infer interaction effects between health and demographics (Aveyard et al., 2002). Small-area aggregates can also be used to increase statistical power, as small-area ecological data can alleviate bias due to measurement errors in individual-level data (Jackson et al., 2006). Other notable examples include the application of geodemographics in policing (Ashby and Longley, 2005). Geodemographic analyses of local policing environments, crime profiles and police performance can provide a neighbourhood classification that is produced explicitly to reflect differing policing environments and help allocate policing resources accordingly.

The composition of geodemographic classification differs quite radically depending on the scope and probable usage by the intended stakeholders; as a result, available geodemographic products include a variety of classification systems. Among the conventional general purpose classification systems are some privately developed classifications such as Mosaic (Experian), Acorn (CACI), P2 People and Places (Beacon Dodsworth), MyBestSegments (Nielsen) and CAMEO (EuroDirect). Such commercial geodemographic systems produce discrete classes primarily designed to describe consumption patterns. As such, their respective databases are not only populated with census data but compiled from large sets of consumer dynamics such as credit checking histories, product registrations and private surveys (Singleton and Spielman, 2014). Open classifications, on the other hand, are those that can be accessed by the public without cost, have transparent published methodologies and comprise freely available input data. One of the most popular open classifications available in the UK is the Output Area Classification (OAC) provided by the Office of National Statistics (see Vickers and Rees, 2007).

Building a geodemographic classification

Building a successful classification may seem fairly straightforward but it can be a difficult and very time-consuming process. It is important that a classification addresses end-user needs, but is also impacted by data availability, coverage and potential weighting (Webber, 1977). Harris et al. (2005) provide a good basis for the methodologies typically used to build geodemographic classifications, and also provide some examples in the UK context. Vickers and Rees (2007) also provide a detailed step-by-step analysis of the process of creating the OAC geodemographic classification, which was built upon previous work on clustering methodologies by Milligan (1996) and Everitt et al. (2001). Less is known about how geodemographic classifications are built within the private sector, beyond those details usefully presented in Harris et al. (2005). Commercial geodemographic classifications have an inherent commercial confidentiality, and as such, most of their methodologies remain a 'black box', which some have argued impairs not only reproduction, but also scientific questioning of the ways in which the clusters emerged from the underlying data (Longley, 2007; Singleton and Longley, 2009).

Scale, variable selection and evaluation

The first stage in building a geodemographic classification is to assemble a database of inputs that are deemed important for differentiating areas. The geographical unit of reference used to collate such data will depend on the purposes of the classification, and also pragmatically on those data available to the classification builder at different scales (including licensing constraints). For example, most open (and some commercial) geodemographic systems in the UK are based on data aggregated at the Output Area level, where zones represent an average population of approximately 300 people, and is the smallest scale at which public census data are provided. However, different sets of variables can have different scales and there are various ways in which these are managed, ranging from simple apportionment from aggregate to disaggregate scales, small-area estimation or microsimulation (Birkin and Clarke, 2012).

From the outset, geodemographic methods have typically employed a pragmatic variable selection strategy, combining the experience of the classification builder (what is deemed to work) with the overarching purpose of a classification (what is required), alongside some degree of empirical evaluation. Attributes can be collected and compiled with a variety of measurement types including percentages, index scores, ratios or composite measures (e.g. principal components). When standardising values it is important to remember that sometimes variables have varying propensities among different groups of people, typically by age or sex (Table 8.1). For instance, long-term illness indices frequently have higher values between groups of older people. An area that has a higher ratio of older to younger people will, *ceteris paribus*, tend to have higher rates of illnesses as well. In these cases, age standardisation is recommended since it can scale values in accordance with age structure; scaled ratios are calculated as the sum of the age-specific rates multiplied by the area population per age group. If area specific rates are not provided, they could be obtained from the national or regional average.

TABLE 8.1 Data formatting per aerial unit

Obtaining ratios per areal unit

Percentages	$x'_{a,i} = \dfrac{x_{a,i}}{P_a}$	where $x_{a,i}$ is the attribute value i of area a and P_a is the population of reference (denominator) of area a, i.e. total population, number of households, etc.
Standardised by group	$x'_{a,i} = \dfrac{x_{a,i}}{\sum_g r_{N,g} P_{a,g}}$	where $x_{a,i}$ is the attribute value i of area a, $r_{N,g}$ is the observed national ratio N for group g and $P_{a,g}$ is the population of group g in area a.

When managing quantitative data, in many cases variables will not seem appropriate to use in their raw format. Available data can have skewed distributions, contain a high rate of missing values or originate from sample sizes smaller than desired, thus generating uncertainty. In general, a detailed assessment of each variable is typical prior to the clustering process in order to identify 'unfit' data. Evaluation typically includes mapping, distribution plots (such as histograms) and correlation analysis.

A particular issue for effective cluster formation is non-normality of attributes or skew. Common techniques used to address this issue include normalisation of the variables when applicable, or weighting to adjust their influence on the final classification when normalisation is deemed by the classification builder not to be appropriate. Normalisation is the process of transforming the variable values to approximate normal distributions, usually through various power transformations. Other treatments include weighting or using principal component analysis to identify common vectors of variables that help reduce data complexity and noise (Harris et al., 2005). Table 8.2 summarises those common transformations used in geodemographics.

TABLE 8.2 Variable transformations used for normalisation

Normalisation transformations

Box – Cox	$x'_i = \begin{cases} \dfrac{x_i^\lambda - 1}{\lambda}, & \text{if } \lambda \neq 0 \\ \log x_i, & \text{if } \lambda = 0 \end{cases}$	The power λ achieves the best normalisation and can be estimated algorithmically
Square root transformation	$x'_i = \sqrt{x_i}$	
Log transformation* *(holds the place of zero)	$x'_i = \log x_i$	
Inverse hyperbolic sine	$x'_i = \sinh^{-1} x_i$	
Square transformation	$x'_i = x_i^2$	

Finally, a universal scale of measurement should be applied to every observation prior to clustering, such as range standardisation or standardised z-scores (Table 8.3), given that disproportionate measurements will frequently affect the dissimilarity function of the clustering technique towards variables with higher values. Techniques such as interquartile and interdecile range standardisation are useful when data contain outliers.

TABLE 8.3   Variable transformations used for scaling

Variable scaling

z-scores	$z_i = \dfrac{x_i - \mu}{\sigma}$
Range standardisation	$x'_i = \dfrac{x_i - x_{min}}{x_{max} - x_{min}}$
Interquartile and interdecile range standardisation	$x'_i = \dfrac{x_i - x_{Q2}}{x_{Q3} - x_{Q1}}$ $x'_i = \dfrac{x_i - x_{Q2}}{x_{90} - x_{10}}$

Clustering approaches and techniques

Clustering approaches and techniques can differ quite radically, depending not only on the purpose, but also on the nature of the data to be clustered (for a more in-depth analysis of clustering techniques, see Everitt et al., 2001; Hastie et al., 2009). A geodemographic typology is usually presented as a hierarchy; with different clusters produced for varying tiers of aggregated areas (Table 8.4). Such a hierarchy can be created from the top or the bottom. A top-down approach includes the creation of larger groups of cases that are subsequently divided into smaller subgroups. This method is typically implemented with the

TABLE 8.4   An example of a nested hierarchy for the 'blue collar communities' supergroup cluster from the 2001 Output Area Classification

Supergroup	Group	Subgroup
	1a: Terraced blue collar	1a1: Terraced blue collar (1)
		1a2: Terraced blue collar (2)
		1a3: Terraced blue collar (3)
1: Blue collar communities	1b: Younger blue collar	1b1: Younger blue collar (1)
		1b2: Younger blue collar (2)
	1c: Older blue collar	1c1: Older blue collar (1)
		1c2: Older blue collar (2)
		1c3: Older blue collar (3)

K-means clustering algorithm, and was used to produce the 2001 OAC, which included seven supergroups, which were respectively split into 21 groups and further into 52 subgroups.

A bottom-up approach is, however, more prevalent within the commercial sector, and includes the creation of numerous smaller groups (using *K*-means), which are then aggregated based on their similarities into larger groups (typically with hierarchical algorithms such as Ward's clustering).

K-means clustering uses squared Euclidean distance as a dissimilarity function, and so can be used only when variables are of a continuous measurement type. Essentially, *K*-means clustering assigns *N* observations into *K* clusters in such a way that, within each cluster, the average distance of the variable values from the cluster mean is minimised. Taking into account that for any set of observations *S* there is an argument that describes the minimum squared distance defined as

$$\bar{x}_s = \arg\min_m \sum_{i \in S} \|x_i - m\|^2$$

then for the aggregate of the total clusters there is a set of arguments that minimise the total within cluster variation of the multidimensional data points:

$$\text{WCSS} = \min_c \sum_{k=1}^{K} N_k \sum_{C(i)=k} \|x_i - \bar{x}_k\|^2$$

where WCSS is the within-cluster sum of squares for a cluster distribution *C* with *K* seeds, $x_i \in \mathbb{N}$ is the data observations and $\bar{x}_k$ is the *k*-cluster mean.

K-means is typically initiated with a random set of initial seeds, and then the algorithm assigns every observation to a seed based on the least squared distance. New means based on the assignments and then calculated, and observations reassigned to their new nearest cluster mean, again based on the least squared distances. The algorithm 'converges' when the within-cluster sum of squares is minimised, i.e. when the cluster assignments no longer change. This technique is straightforward to implement and perhaps explains the popularity in the classification of multidimensional inputs; however, the *K*-means algorithm needs a specific predetermined number of clusters (*K*), and furthermore, results can differ based on the initial *k* centres that are selected. As such, it is typical to run *K*-means multiple times for an analysis, extracting the results for each converged cluster set, and evaluating them on the basis of some metric – most commonly, an effort to minimise the within sum of squares (i.e. more compact, and therefore homogeneous clusters).

In hierarchical cluster analysis, Ward's algorithm can be applied to merge clusters with the least amount of between-cluster variance, thus producing the minimum increase in total within–cluster variance after merging (Everitt et al., 2001). Ward's clustering criterion is typically used in those geodemographics created from the bottom up to produce the more aggregate hierarchy of clustering typologies.

Although more prevalent in research rather than commercial applications, there are multiple other clustering techniques that have been implemented within the context of area classification. A self-organising map (SOM) is an unsupervised classifier that uses a type of artificial neural network to classify space (see Spielman and Folch, Chapter 9), based on the configuration of attributes that 'fit' each neuron (Skupin and Hagelman, 2005). Typically the SOM mapping process employs a lattice of squares or hexagons as the output layer, and the results are therefore easily mapped. SOMs have been tested as an alternative classifier of census data in the UK (Openshaw and Wymer, 1995) and the USA (Spielman and Thil, 2008) where they seem to perform well for socio-economic data at the census tract scale. They also have the advantage of not assuming any hypotheses regarding the nature or distribution of the data, and respond well to geographic sensitivity.

Another methodology to classify areal units is based on fuzzy logic algorithms or 'soft' classifiers. Fuzzy classifications have the inherent ability to assign spatial units to more than one cluster with varying membership values (i.e. probabilities). The degree of membership reflects the similarities or dissimilarities between groups and therefore is often addressed as a soft classifier (in contrast to hard classifiers such as K-means). Most studies regarding geodemographic analysis that use fuzzy classification employ the Fuzzy C-means algorithm or the Gustafson–Kessel algorithm (Feng and Flowerdew, 1998; Grekousis and Hatzichristos, 2012).

Other probabilistic classifiers that have been used less prevalently are multinomial logistic regression models, also known as m-logit models. A logit model has the advantages of using continuous, binary or categorical data to generate clusters, and these can also be considered as a soft classifier as they output the probability of areas belonging to each cluster category. Such models have been used in health geodemographics and epidemiology, where detailed geographical information is often unavailable (Jackson et al., 2006).

Cluster analysis and interpretation

The final step in building a geodemographic classification includes the review and testing of the cluster results, alongside description of the typology. For example, checking the size of clusters is one of the basic steps in the optimisation procedure. Clusters with relatively low representation of cases should generally be avoided, by either adjusting the number of clusters or by the re-evaluation of the data input. Furthermore, if, measured in terms of variance, two or more of the output clusters may look very similar, and merging might be considered, and inversely split if the clusters are too large. Harris et al. (2005) provides a 'rule of thumb' for merging similar clusters, if the loss of variance within the dataset is less that 0.22%. Other ways to test an output classification is to correlate it with existing classification systems, or via sampling, such as cross-tabulation with geocoded survey data.

If the classification appears successful, a final stage is typically naming and describing the resulting clusters with written 'pen portraits' that best fit the profile of areas represented by the clusters. The process of creating such descriptions can be quite difficult, especially in lower hierarchies, where the cluster dissimilarities are more subtle (Vickers and Rees, 2007). Here is an extract of the profile for the 'affluent achievers' cluster from the Acorn commercial classification by CACI:

> These are some of the most financially successful people in the UK. They live in wealthy, high status rural, semi-rural and suburban areas of the country. Middle aged or older people, the 'baby-boomer' generation, predominate with many empty nesters and wealthy retired. Some neighbourhoods contain large numbers of well-off families with school age children, particularly the more suburban locations. These people live in large houses, which are usually detached with four or more bedrooms. (CACI, 2013)

Classification systems also commonly augment such descriptions with other visual materials such as photographs, maps and bar graphs or radar charts. Depending on the intended end-users, labelling and description must be selected appropriately in order to expand the user's understanding of the group, while taking into account that the end user might not be accustomed to geodemographic classifications.

Liverpool Case Study

In this final section, a practical example of creating a geodemographic classification will be presented. For this purpose, the Local Authority of Liverpool will define the extent of the classification, which includes 1584 Output Areas. The analysis uses the R statistical programming language, and the dataset is assembled in its entirety with 2011 census variables, provided by the Office for National Statistics and aggregated at the Output Area level.

Methodologically, the cluster analysis follows a similar approach to that of the 2001 OAC, although it only aims to capture broad socio-economic categories for illustrative purposes. This analysis utilises the K-means clustering algorithm and produces a single aggregate typological level. As a first step, consideration was required to identify those variables that would form useful inputs to the classification. Although the census includes a very wide variety of potential candidate variables, a large number of them are homogeneous across space or highly correlated.

For variables to be effective in a classification they should ideally show variation over space. For instance, any variation by sex is considered to be of lower importance, since the majority of Output Areas have the same overall ratio of males to females. Furthermore, given the urban location of this case study area, variables that captured dichotomies between urban and rural space might also be considered as less useful for any resulting classification.

Three elements were initially selected to guide the classification process and included demographic, housing and economic activity indicators. In total, 29 preliminary attributes were selected over the three taxonomical elements (Table 8.5).

TABLE 8.5 Initial dataset used for the Liverpool classification

Variables	Variable Definition
Demographic	
V1: Age 0–4	Percentage of resident population aged 0–4 years
V2: Age 5–14	Percentage of resident population aged 5–14 years
V3: Age 15-24	Percentage of resident population aged 15-24 years
V4: Age 25–44	Percentage of resident population aged 25–44 years
V5: Age 45–64	Percentage of resident population aged 45–64 years
V6: Age 65+	Percentage of resident population aged 65 or more years
V7: Ethnic Group, White	Percentage of people identifying as white
V8: Ethnic Group, Black	Percentage of people identifying as black African, black Caribbean or other black
V9: Ethnic Group, Asian	Percentage of people identifying as Indian, Pakistani, Bangladeshi, Chinese or other Asian
V10: Population Density	Number of people per hectare
Housing	
V11: Privately owned	Percentages of households that are privately owned
V12: Rent (private):	Percentage of households that are private sector rented accommodation
V13: Rent (public):	Percentage of households that are public sector rented accommodation
V14: Detached	Percentage of all household spaces that are detached
V15: Semi-detached	Percentage of all household spaces that are semi-detached
V16: Terraced	Percentage of all household spaces that are terraced
V17: Flats	Percentage of households that are flats
V18: Central heating	Percentage of occupied household spaces with central heating
V19: No central heating	Percentage of occupied household spaces without central heating

(Continued)

TABLE 8.5   (Continued)

Economic Activity

V20: Working full-time	Percentage of household representatives who are working full-time
V21: Working part-time	Percentage of household representatives who are working part-time
V22: Unemployed	Percentage of household representatives who are unemployed
V23: Retired	Percentage of household representatives who are retired
V24: Student	Percentage of household representatives who are full-time students
V25: No qualifications	Percentage of people over 16 years without further education qualifications
V26: Higher education	Percentage of people over 16 years for which the highest level of qualification is level 4 qualifications and above
V27: No car household	Percentage of households with no cars
V28: 1 car household	Percentage of households with 1 car
V29: 2+ car household	Percentage of households with 2 or more cars

The variables were each transformed into percentages, taking into account their respective denominator, with the exception of density, which was the only non-percentage variable. The next stage was to check how the variables were distributed and correlated, and assess for any that might negatively affect the clustering process. On the basis of variables with problematic distributions, these were removed from the initial dataset. Following the 2001 OAC methodology (Vickers and Rees, 2007), a log transformation was fitted to the variables to create more

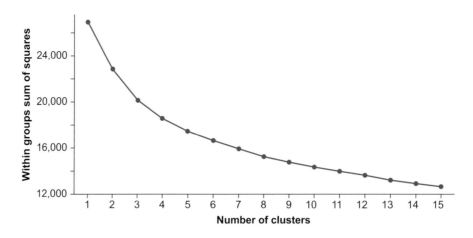

FIGURE 8.1   *K*-means: distance from mean by cluster frequency

normal distributions. A cross-correlation table was then generated to show those variable pairs with high correlation, and a number of further attributes were selected for removal on this basis, thus aiming to reduce redundancy within the input data, and also limit bias towards any particular dimension being measured.

The variable selection process returned 17 variables that would form the input data to the K-means clustering, and were then scaled uniformly with z-scores. In order to address the question of how many clusters might be suitable, a within-cluster sum of squares distance graph (scree plot) was used to help identify a point at which the total distance only marginally improves the cluster homogeneity (also known as the elbow or knee criterion). However, in this case (Figure 8.1) there is no significant elbow visible, and as such, for these illustrative purposes we select $K = 5$ as a number of clusters that would be useful when mapping urban areas – increasing the classes would create a more detailed, but potentially less easily interpretable representation.

The K-means algorithm was subsequently run 10,000 times, and the result returning the least within-cluster total distance through these multiple iterations was extracted as the optimal result. The cluster sizes were then checked, and these varied between 72 and 522 output areas. This size variation is within acceptable limits, taking into account the limited extent of the analysis area. A useful way of

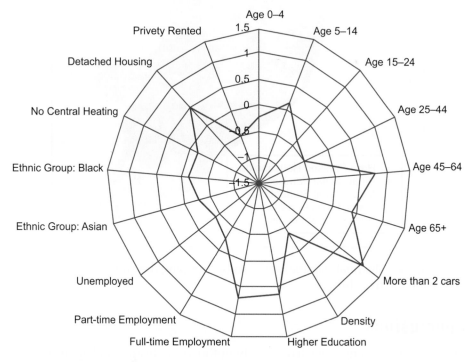

FIGURE 8.2 Within-cluster variable analysis of Cluster 2

obtaining information about how variables load onto each cluster is through a radar plot. Figure 8.2 shows a summary of the distribution of values within Cluster 2 (note that the Liverpool mean is 0). Cluster 2 consists mainly of neighbourhoods of middle-aged families, the majority of which are full-time workers with higher education degrees. Families are more prevalently living in low–density, detached houses, while the high ratio of car ownership indicates these areas may be more affluent. This cluster was named 'white collar families'. A map of the other clusters and their attributed names can be seen in Figure 8.3. As discussed earlier, patterns exhibit a degree of spatial autocorrelation, despite locational proximity being absent from the classification.

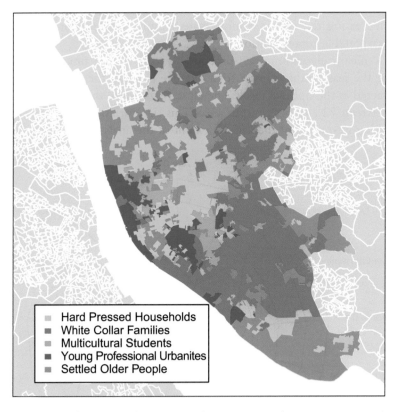

FIGURE 8.3 The final classification results, grouping the output areas of Liverpool into five clusters

Conclusions

In the previous sections we have briefly outlined the history and application of geodemographic classifications, concluding the chapter with an overview of the basic process of building a geodemographic using a case study of Liverpool. While it is true that such applications can produce useful results, geodemographic research may face substantial challenges in the near future. Many geodemographics have historically relied on the analysis of the decennial census of the population, but institutional shifts in both the USA and UK are already changing the nature and availability of such data, given the growing costs associated with their collection (Singleton and Spielman, 2014). As such, the granularity currently offered by census data might not be readily available in the future; and as such, more research is needed into how the linkage of non-census attributes (both commercial and non-commercial) can be both validated and made more accessible for these purposes.

Secondly, geographic classifications, as currently construed, do not account for spatial relations between proximal zones. This traditional 'aspatial' approach has a number of implications when generating profiles. For marketing-related applications of geodemographics, a lack of local sensitivity may have fiscal implications, such as a reduced uptake of a product or service. However, in public sector uses, the consequences may be more severe, with mistargeting having potential implications on life chances, health and well-being. Hitherto, methods used to take into account near geography are typically geographically crude, accounting for spatial context through either an arbitrary zonal distance, or by division of areas into administrative units that may not correspond with the organisation of actual communities. Future research is needed to produce measures of near geography that can capture such associations and evaluate these vis-à-vis traditional geodemographic models.

FURTHER READING

For an excellent introduction to geodemographics, we would highly recommend Harris et al. (2005). More generally, the majority of research articles utilising geodemographic models can be found online at https://www.zotero.org/groups/geodemographics.

9

SOCIAL AREA ANALYSIS AND SELF-ORGANIZING MAPS

Seth Spielman and David C. Folch

Introduction

Maps are flat. Flat in the sense that they are viewed on two-dimensional sheets of paper, or more often computer screens. However, the world described by maps is anything but flat. Here we are not concerned with physical topography of mountains and valleys, but rather this chapter develops a method to describe the social topography of a place. Unlike physical terrain where landscape features are defined by a single variable, elevation, the social landscape is multivariate (see Alexiou and Singleton, Chapter 8). Meaningful social variation occurs at the intersection of variables – race, class, wealth, and the built environment. The classical geographic approach to the visualization of social terrain is the creation of an atlas, a book of maps where each map shows the spatial pattern in an individual variable (see Cheshire and Lovelace, Chapter 1). To discern complex multivariate patterns in an atlas readers must compare map images and remember spatial associations between variables and places. The associations between social variables in geographic space not only structure the geographic world, but also create textures and patterns in a multidimensional abstraction of the real world called 'attribute space'.

The concept of an attribute space is central to this chapter. An attribute space is a p-dimensional region defined by the p variables selected for a particular analysis. The attribute space contains all of the variation (variance) in the variables selected for an analysis; it is bounded by the maxima and minima of these variables. Throughout the chapter we maintain a distinction between geographic space, i.e. the 'physical' world, and attribute space, a world defined by user selected variables. While analogies can be drawn between attribute and physical space (e.g. a univariate attribute space is like a line, bivariate a plane, and three variables a cube), attribute spaces are often bounded by many more than three variables and can thus be difficult to visualize. Nonetheless, the contours, peaks, and valleys of this p-dimensional space represent meaningful social variation such as the cores and edges of neighborhoods.

One can exploit the properties of attribute space to aid in the understanding physical geographic spaces. An alternative to the variable-by-variable approach taken in an atlas is to identify regions in attribute space and project these regions back into geographic space. A region in attribute space is a co-occurrence of observed values within the set of variables defining the attribute space. Such a co-occurrence is commonly called a 'cluster'. Algorithms that identify these clusters of co-occurring attributes are called clustering algorithms; there are entire books written on the subject of cluster analysis (Everitt, 2011; Kaufman and Rousseeuw, 1990). In a high-dimensional dataset it is rare for observations to be exact duplicates of each other, thus one of the fundamental challenges of cluster analysis is determining how similar observations must be to be considered members of the same group (cluster). In the social sciences cluster analysis applied to geographic space is referred to as 'geodemographic analysis'.

However, even when applied to zonal geographies, such as US census tracts or UK output areas, these clustering algorithms are not especially spatial in character. That is, cluster identification exploits the spatial properties of attribute space but it does not allow effective visualization of it. Traditional cluster analysis assigns observations to categories, and these categories can be viewed on a map. The enduring appeal of atlases is, at least in part, due to their visual nature. The maps contained within an atlas structure variables visually in space – the map reader's eye is drawn to regions of wealth (or poverty), ethnic concentrations, or areas with little population. One significant disadvantage of cluster analysis is that the visual appeal of thematic maps is lost since the outputs from cluster analysis are typically more cognitively demanding than a thematic map of a single variable. One of the reasons for this is that while the spatial properties of attribute space are exploited in cluster analysis, attribute space is not directly visualized. Figure 9.1 is an example of the

FIGURE 9.1 US block groups (eight types), created using a SOM and kaggle.com data (gray areas indicate missing data)

typical outputs of cluster analysis: a map that classifies over 200,000 small geographic zones (US block groups) into 22 categories on the basis of 44 variables.

In this chapter we provide a tutorial on the Kohonen Self-Organizing Map (SOM) algorithm, originally introduced in Kohonen (2001). We apply the general approach to social area analysis, but these methods are easily transferable to other domains. Figure 9.1 is a SOM visualization based on roughly 175,000 US Census block groups and seven variables; this chapter introduces the methods and the code necessary to produce this figure. In many ways SOM-based analysis is identical to other forms of cluster analysis. The principal advantage of the Kohonen SOM algorithm is that it identifies clusters in attribute space and provides a visualization of attribute space. That is, a SOM provides classification of input data *and* maps of attribute space. From a user's perspective, another advantage of the SOM is that it uses a fairly intuitive computational strategy to identify these patterns. This chapter focuses on building an intuitive understanding of the SOM algorithm, beginning from basic principles, and then demonstrates an application of the algorithm to a real-world problem: patterns of non-response to federal surveys. This particular problem was part of a data science competition hosted by kaggle. com and sponsored by the US Census Bureau. This chapter describes some of the methods used in the authors' winning entry into that competition; in particular, we illustrate the use of the SOM to build typologies of input geographies and visualize attribute space.

Before diving into the substance of the chapter a few technical caveats are necessary. All of the examples in this chapter use the R programming language exclusively. R is flexible, open source, free, and widely used, and thus ideal for the pedagogic purposes of this chapter; however, it is not the best available toolset for the Kohonen SOM algorithm. The SOM_PAK software library has superior model fitting tools, but it is distributed in C and can be difficult to compile. While Windows executables are available online, compiling this library on modern OSX and Linux operating systems requires small modifications to source code and is beyond the scope of this chapter. We assume some familiarity with R, but not prior exposure to the SOM algorithm.

Primer on cluster analysis

When thinking about the SOM algorithm, or really any clustering algorithm, similarity and distance are analogous and fundamental concepts. The analogy is especially useful for geographers who tend to think spatially. Observations near each other in attribute space are, by definition, similar across many attributes. Take the simple example shown in Figure 9.2, which contains three observations in a simple two-dimensional attribute space bounded by 0 and 3 in both dimensions. Using the distance analogy, we would say that observations *A* and *C* are more

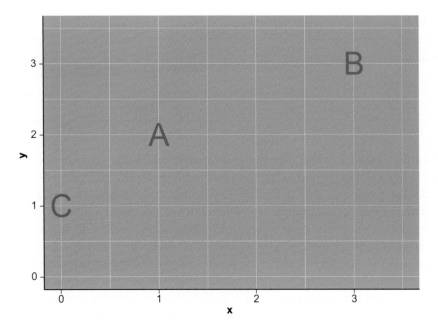

FIGURE 9.2 Two-dimensional attribute space

similar than observations A and B. One way to measure the similarity between these three observations is to calculate the Euclidean distance between them, $d(i, j) = \sqrt{(i_1 - j_1)^2 + (i_2 - j_2)^2 + \ldots + (i_p - j_p)^2}$, where i and j are observations described by $1, 2, \ldots, p$ variables. In the simple example in Figure 9.2 the distance between A and C would be $d(A,C) = \sqrt{(1-0)^2 + (2-1)^2} = 1.41$ and the distance between A and B would be $d(A,B) = \sqrt{(1-3)^2 + (2-3)^2} = 2.23$. Strictly speaking, Euclidean distance measures dissimilarity, i.e. the greater the distance the more dissimilar the observations are. Geographers often presume a relationship between geographic proximity and attribute similarity, an assumption sometimes called the 'first law' of geography. However, what is a contestable 'law' in geographic space is a simple fact in attribute space.

Many of the measures of similarity used in cluster analysis are analogous to measures of Euclidean distance. Thinking of observed data as a set of locations in attribute space, we can measure the similarity of observations by computing the distance between them. The pairwise dissimilarities can be stored in a distance matrix, which in this case would simply be the 3×3 matrix shown in Table 9.1.

Extending this example, we can imagine that observations A, B, and C were not single observations but were prototypes or exemplars of broader sets of observations. If we use these three observations as 'seeds' to generate data we end up with

TABLE 9.1 Simple distance matrix

	A	*B*	*C*
A	0.00	2.23	1.41
B	2.23	0.00	3.61
C	1.41	3.61	0.00

a plot like Figure 9.3. This plot was generated by taking three sets of 100 samples from a normal distribution, using the observed X and Y locations for observations A, B, C as means and 0.25 as the standard deviation. Each of the lowercase letters is a child of the corresponding capital letter. If we calculated a 300×300 matrix of the distance between all of the lowercase letters, we would find that in general the as and the bs are more similar to each other than the as and the cs.

Assume that we did not know the parent and we wanted to assign each lowercase letter to an uppercase letter. We could simply do this on the basis of distance to the original uppercase letters. It is plausible that some of the children of C are closer to A than to their true parent. Simply using distance could result in some misassignment because the division between the as and cs is not as crisp as those between the bs and the other classes. This is one of the central problems in cluster

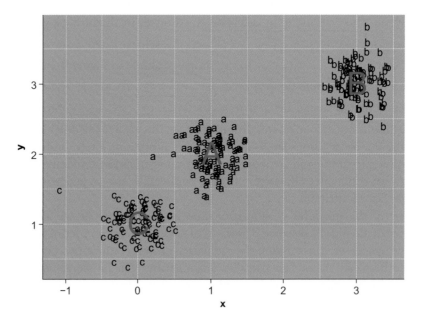

FIGURE 9.3 Three types

analysis; the 'parents' of each observation are usually unknown and the problem is to correctly assign each observation to the type it represents. In practice, however, not only are the prototypes (parents) of each observation unknown but so also is the correct number of types. A naive analyst might assume that both the a's and c's in Figure 9.3 were instances of the same type. More precisely, the central problem of cluster analysis is determining the correct number of types and correctly assigning each observation to a type.

In practice, once the number of types has been determined through cluster or other types of geodemographic analysis (see Chapter 8, this volume) the most common way to describe a type is to use the groupwise mean. In this case the groupwise means are contained in a 3×2 matrix consisting of the locations of the three uppercase letters used to generate the lowercase letters in Figure 9.3; in a more complex analysis with many variables the groupwise means would be a $k \times p$ matrix where k is the number of groups/classes and p is the number of variables. However, determining the similarity among the classes would be more complicated with a high-dimensional dataset; observations might be similar on some characteristics but dissimilar on others.

One of the weaknesses of cluster analysis is that it is difficult to describe the relationship among the classes. This is where the Kohonen SOM algorithm excels. It functions like a cluster analysis, in that observations are assigned to 'parents' or exemplars on the basis of similarity. However, the procedure used to assign observations to groups also creates a spatial relationship among the groups. This spatial relationship can be visualized, which makes it possible to map attribute space.

The Kohonen Self-Organizing Map algorithm

A self-organizing map is a network of 'neurons'. Each neuron is best conceptualized as a list of p values, where p is the number of variables in the analysis. Neurons are typically arranged in a lattice of some type, such that each neuron has some (non-zero) number of neighboring neurons. Figure 9.4 shows an example of a 10×6 hexagonal SOM; SOMs can also consist of rectangular neurons. While SOMs are typically organized as a flat two-dimensional network, they have been extended to toroidal and spherical forms (Schmidt et al., 2011). The advantage of these more complex three-dimensional shapes is that they do not suffer from edge effects; however, they add significant complexity to visualization of results.

Initially neurons are assigned arbitrary values for each variable, where the arbitrary values are bounded by the attribute space. The distance (or similarity) between an observation and any given neuron can then be computed using Euclidean distance as described in the previous section. The SOM algorithm iteratively uses observations (data) to adjust the neurons' values. One way of stating

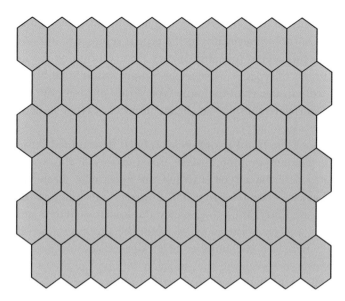

FIGURE 9.4 A SOM network of neurons

this is that neurons are 'trained' using the observed data. The key innovation of the SOM is that when a single neuron's values are adjusted the values of neighboring neurons are adjusted as well. This creates a spatial structure on the SOM, where observations are assigned to neurons and similar observations occupy proximal neurons. Through creating a SOM one builds a network of neurons; within this network the relationship between proximity and similarity is engineered, i.e. a pattern of spatial association is created through the process of training the SOM. The network of trained neurons that results from training a SOM can be seen as a 'feature map' or an atlas of attribute space.

Maps preserve topological relationships among objects in space. In the cartographic context, entities and features that are close to each other in the real world are represented close to each other on a map. The concept of a map can also be applied to non-geographic objects; or it can be used to visualize geographic objects (census tracts) in a spatial but non-geographic context. That is, census tracts can be organized in space based upon the similarity of their characteristics rather than their geographic proximity. This is the basic idea behind the Kohonen algorithm that creates SOMs. In the remainder of this section we explain how the algorithm works.

Training a SOM

Training a SOM is like shooting data at a honeycomb. Admittedly, it is hard to shoot data, but imagine it had some physical form. When a piece of data hits the honeycomb it would deform the cell it landed in and those surrounding it. It would be possible to control the extent of the deformation by adjusting the speed at which the data are launched. In training a SOM data are launched at a honeycomb of neurons; each observation (i.e. row of data) is 'aimed' at the neuron to which it is the most similar (the shooter never misses the targeted neuron). Once the data hits this specific neuron, which we call the best matching unit (BMU), the deformation in the honeycomb causes nearby neurons to adjust their values to be similar to the BMU that was hit by the data. Because neurons are randomly initialized, where a given observation hits the honeycomb is initially random, but the deformations quickly cause 'regions' to emerge on the honeycomb. Since each row of data is repeatedly shot at the honeycomb, observations that are similar to each other will eventually hit BMUs in the same region of the lattice. However, the speed at which data are shot at the honeycomb decreases over time, thus the spatial extent of the deformations, and hence the effect each hit BMU has on its neighbors, is reduced.

In practice, the 'honeycomb' is a lattice of square or hexagonal neurons. When the algorithm terminates, each observation is assigned to a final BMU. It is useful to think of these final BMUs as buckets for data; collectively they constitute the feature map. The feature map is a mapping of attribute space. Observations that are similar are placed either in the same bucket or in buckets that are close to one another on the feature map. For example, places with many wealthy householders, with high levels of education, high homeownership rates, and low poverty rates would end up in buckets that are near each other and clustered in a region of the feature map. On the other hand, census tracts where poverty is abundant and residents typically have low levels of education would end up clustered in buckets in a different region of the SOM; probably quite far away from the well-educated and wealthy people because distance on the feature map is analogous to similarity.

Creating a simple SOM in R

Figure 9.3 shows three types of observations, each of which contains 100 instances. In this simple example, the data are two dimensional (defined by an x and a y variable). We could order these types based upon their similarity to each other resulting in a simple list: $[C, A, B]$. C is more similar to A than B; the distance between A and B is less than the distance between C and B. The simplest type of SOM is a linear one, and we use this form for the first example by creating a three-neuron arrangement with the hope that it reproduces the C, A, B ordering. The goal is then that each c-type observation should be assigned to a single BMU (or bucket), and this BMU should be close to one containing all the a-type observations.

Listing 9.1 contains all of the R code necessary to run this example and will reproduce Figure 9.5. The code depends upon the kohonen R library. Unfortunately the plotting functions in the kohonen library use circles to represent neurons, regardless of their actual geometry.

LISTING 9.1   A simple SOM in R

```
1   ##LOAD THE KOHONEN LIBRARY
2
3   library(kohonen)
4
5   ##SET SEED TO REPRODUCE RESULTS IN CHAPTER
6
7   set.seed(1223)
8
9   #CREATE A SIMPLE DATASET
10
11  simpdata <- data.frame(x=c(1,3,0), y=c(2,3,1), z=c(1,3,4))
12  rownames(simpdata) <- c('A','B','C')
13
14  #CREATE A SIMPLE DATA SET CONTAINING CHILDREN OF A, B, AND C.
15  extSimpData <- data.frame(x=rnorm(n=100, mean=simpdata[1,1], sd=0.25),
      y=rnorm(n
      =100, mean=simpdata[1,2], sd=0.25), type='a')
16  extSimepData <- rbind(extSimpData, data.frame(x=rnorm(n=100,
      mean=simpdata[2,1], sd=0.25), y=rnorm(n=100, mean=simpdata[2,2],
      sd=0.25),
      type='b'))
17  extSimpData <- rbind(extSimpData, data.frame(x=rnorm(n=100,
      mean=simpdata[3,1], sd=0.25), y=rnorm(n=100, mean=simpdata[3,2],
      sd=0.25), type='c'))
18
19  #CREATE A RECTANGULAR 1 by 3 SOM
20  aGrid <- somgrid(xdim=3, ydim=1, topo="rectangular")
21
22  #TRAIN THE SOM
23  aSom <- som(data=as.matrix(extSimpData[,1:2]), grid=aGrid, rlen=100,
      alpha=c(0.05,0.01), radius=c(3,0))
24
25  #PLOT RESULTS
26  plot(aSom, type='mapping', labels=extSimpData$type)
```

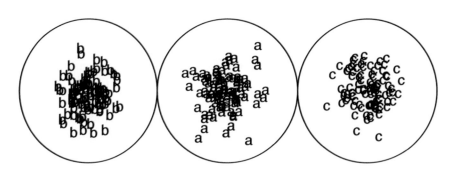

FIGURE 9.5   A simple linear SOM

In this simple analysis there are three key user specified parameters, which are the inputs to the `som()` function in the `kohonen` R library:

1. The size and shape of the SOM grid. This is specified in the `somgrid()` function, which allows the specification of X (`xdim`) and Y (`ydim`) dimensions. One can also specify a shape for the neurons (`hexagonal` or `rectangular`). Choosing these dimensions is one of the hardest aspects of using a SOM since there is no consistent rule of thumb. In general, if the goal is a fine distinction between categories, then increase the size of the SOM. If more general categories are needed, then use a smaller size. In this case we chose a 1 × 3 SOM because we knew the simple synthetic data had three categories. In practice, we prefer to build a large SOM and then post-process the output via cluster analysis to combine neurons into some desired number of groups. The choice of shape affects the connectivity within the SOM. In a rectangular SOM each bucket (neuron) has four neighbors. It is also possible to specify `hexagonal`, which gives each neuron six neighbors. It is also possible to create a toroidal SOM, in which the edges are connected to each other.

2. The run length (`rlen`). This specifies the number of times the data are presented to the SOM. In the example, where `rlen=100` is specified, each observation is presented to the SOM 100 times. In training a SOM, data are recycled because each observation affects the overall configuration of the SOM; reusing the data allows refinement of the feature map.

3. The learning rate (`alpha`). The learning rate controls the velocity at which data are 'launched' at the SOM. More precisely, the learning rate controls the influence an observation has on both the neuron it hits and its neighbors. One specifies a range because over the `rlen` runs of the SOM the learning rate declines. In the example the learning rate declines from 0.05 to 0.01 in a sequence of 100 steps.

4. The fourth and final parameter is the `radius`. This is the size of the neighborhood, specified as a range of values. This has an intelligent default, so specifying `radius` is not necessary. The size of the neighborhood declines over the `rlen` runs of the SOM. Near the end of the training the neighborhood is zero, meaning that data, when shot at the SOM, affect only the BMU.

These are the essential elements and logic of any SOM analysis. Typically, SOM-based analysis involves two training stages and multiple iterations; these have been omitted for simplicity. In the example, we generated data based on three known exemplars. We computed the distance/similarity between these exemplars to establish the relationships between them. We trained a SOM using data generated from the exemplars. The SOM correctly assigned observations to neurons that corresponded to the exemplars, and the spatial structure of the SOM captured the similarity of the exemplars.

The visualization of the output is a more complex problem and will be discussed in the next section, which outlines a 'real-world' application.

Using the SOM algorithm

In 2012 the US Census Bureau hosted a data analysis competition through `kaggle.com`. The aim of the competition was to predict survey non-response at the block group level in the USA. Survey non-response occurs when the US Census Bureau attempts to collect information from people but they do not respond to the government's inquiries. In theory, response to government surveys is legally required; in practice, for the largest annual survey (the American Community Survey) only two-thirds of the sampled households respond. This non-response triggers a very expensive follow-up procedure, and the cost of this process restricts the number of surveys the US Census Bureau is able to collect in a given year. In an effort to understand what causes non-response the US Census Bureau released data on 211,000 block groups and their response rate to the American Community Survey. The goal of the competition was to develop a predictive model of non-response.

Our approach, which won the visualization prize, exploited a SOM. There is significant regional and demographic variation in response rates that makes traditional statistical approaches ineffective. Thus we constructed a SOM, based on 44 variables for over 200,000 zonal geographies. Our entry into the competition was an interactive website. Here we present a simplified version of the analytical work that was necessary to construct the website. The original analysis was done using both R and SOM_PAK, an open source C library available from `http://www.cis.hut.fi/research/som_pak/`. To simplify the analysis, the example below is restricted to R.

In the competition we trained a large 160 × 100 neuron SOM and then ran a cluster analysis on the 16,000 vectors describing the resulting neurons. The cluster analysis reduced the 16,000 neurons down to a set of 22 clusters using k-means. Applying k-means or a hierarchical cluster analysis to SOM neurons creates regions on the feature map and provides a way to build hierarchical representations using a SOM. The goal of this chapter is not to highlight our specific results, although we do believe that it provides meaningful insights into the US population, but to present a tutorial on the use of the SOM algorithm for social area analysis. To simplify the tutorial we have reduced the original set of 44 variables down to seven variables measuring population density, poverty, percent African-Americans, percent Hispanic, percent Asians, percent of housing units that are owner occupied, and the percent of households that are married couples. The R programming language is not well suited to interactive graphics, so this chapter uses only static visualizations of the SOM results. Moreover, we assume some basic knowledge of R; code related to SOMs is explained in detail but the visualization code is not explained. We have bundled the data necessary for this tutorial into an RData file that is available from the companion website.

The first step in creating a SOM is deciding on the dimensions of the map; these are specified in the `somgrid()` function. In general, we find that making a large SOM allows more flexibility in the reporting of results. One can reduce complexity on the SOM via a cluster analysis. Unfortunately, training a SOM is computationally intensive, and a large SOM can take some time to train. The code in Listing 9.2 may take many hours to run, depending upon the computer. Reducing the size of the `somgrid` and the `rlen` will speed things up substantially. However, if the `rlen` is too small the resulting SOM may not be well trained.

LISTING 9.2 Fitting a SOM

```
1    ##LOAD THE DATA
2    load("rdata")
3
4    ##BUILD SOM GRID
5    aGrid <- somgrid(xdim=160, ydim=100, topo="hexagonal")
6
7    ##TRAIN SOM USING SCALED DATA
8    aSom <-
9    som(data=as.matrix(scale(na.omit(usa.bg.som[,1:7]))),
        grid=aGrid, rlen=1000, alpha=c(0.05, 0.01),
        keep.data=FALSE)
```

It is important to note that in Listing 9.2 the data input to the SOM has no incomplete cases. All missing data are expunged from the file using `na.omit`. Incomplete cases can be classified later using the `map.kohonen` function, which matches observations to BMUs based on available data. The BMU matching code included with this chapter will account for missing data. While incomplete rows cannot be used to train the SOM, they can be assigned to a BMU on the basis of available information. Data are also standardized to have a mean of zero and a unit variance using the scale function.

Once the SOM has been trained, the challenge is what to do with the results. How does one know if they are good? How should one communicate the results to an audience? Standardized answers to these two questions do not exist, and this limits the algorithm's wider use.

First, it is our opinion that a 'good' result cannot be determined without some manual intervention. However, some objective criteria can be helpful. For geographic applications of the SOM, investigation of the unified distance matrix, or *u*-matrix, is critical. The *u*-matrix measures the degree of similarity between a neuron and its neighbors. Because geographers often use distance on a SOM as an analogy for similarity, discontinuities in the *u*-matrix present a problem. Imagine reading a topographic map where different parts of the map sheet had a different scale. One inch on some parts of the map might represent 5 miles and on other parts of the map it might represent 0.5 miles (or 50 miles). Having an irregular scale would make a map of geographic space difficult to use. Having an irregular feature map similarly complicates interpretation of the SOM.

There are two ways to visualize the *u*-matrix: a Sammon mapping and a simple visualization of the *u*-matrix. A Sammon mapping projects links between neurons such that the length of the link represents the distance between neurons. One hopes that a Sammon mapping looks like a net when the aim is to use distance on the SOM as an analogy. A net–like Sammon map means that 'scale' or similarity as a function of distance is constant on the SOM. Similarly, a *u*-matrix with relatively constant values implies that a unit of distance on the feature map has the same meaning throughout the map. In practice, these SOMs are never perfect in that distance is never uniform throughout the grid. Discontinuities in the data usually prevent a 'perfect' solution; consider the simple 1 × 3 SOM presented earlier. The best analytical solution had discontinuities in the distance between the neurons. Figure 9.6 shows the *u*-matrix for the aSom object created above. In this case, the lack of variation is a positive indication of the relatively uniform scale on the feature map. There is one notable problem area near the top of the *u*-matrix, but it involves only a few neurons.

One of the persistent challenges of using a SOM is understanding the patterns on the feature map and how these correspond to geographic patterns and/or locations. One can examine individual component planes to gain an understanding of patterns on the SOM. A component plane is the value of the *i*th element of the 1 × *p* vector defining each neuron.

FIGURE 9.6 *u*-matrix

In this case if $i = 1$ we are examining the neuron values for population density, if $i = 2$ we are examining the neuron values for percent poverty. The relative differences in these values show the affinity of the various regions of the SOM for observations with various values of each 'component'. In reading these component planes it is important to remember that the process of training a SOM is equivalent to adjusting the $1 \times p$ vector for each neuron, following the procedure described previously. Once trained, observations are assigned to their best fitting neuron (BMU). By standardizing the data prior to training the SOM we ensure that each variable has a zero mean and unit variance. This means that the values on the component planes are all on the

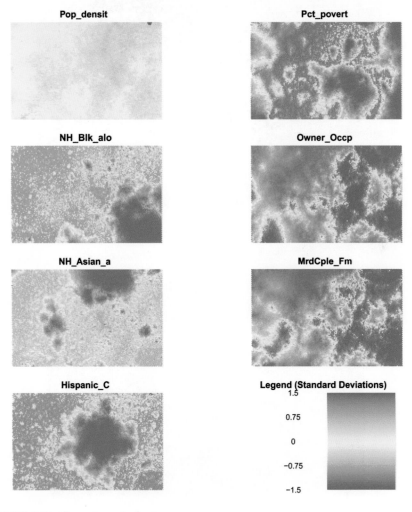

FIGURE 9.7   Component planes

same scale and can be interpreted with a single legend. Viewing the component planes in conjunction with places with known attributes is key to understating the geography of the feature map.

Examining Figure 9.7, it is clear that places with a high Hispanic population do not, in general, have a high Asian population. One also notices that areas with a high African-American population and a high Hispanic population overlap with regions of high poverty. However, regions with high poverty do not generally overlap with areas with a high proportion of married couple families.

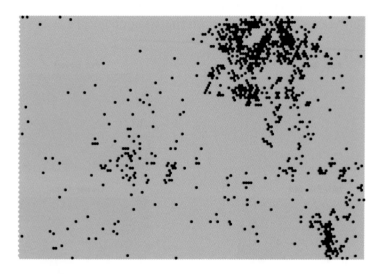

FIGURE 9.8   Geographic activation of neurons; Queens County, New York

FIGURE 9.9   Geographic activation of neurons; Audubon County, Iowa

It is also notable how uninteresting population density is as a variable. This is surprising and may be because, relative to other variables, the variance in population density is low.

Another way of examining a SOM is to look for regions with known properties and see how they map onto to feature space. For example, Figure 9.8 shows how the block groups in Queens County, NY, perhaps the most diverse county in the United States, maps onto the SOM. One sees that observations in Queens cover a large portion of the SOM, at least when compared to Audubon County, Iowa, shown in Figure 9.9. The demographic diversity translates into scattered BMUs on the SOM feature map.

Reducing SOM complexity through cluster analysis

One commonly used strategy to reduce the complexity of the SOM is to conduct a cluster analysis on the neurons. When the number of neurons is large, clustering neurons can vastly simplify cartographic display. The clustering can be done by applying *k*-means, for example, to the vectors describing each neuron. The result is a hierarchical classification system that facilitates interpretation of the results. Classifying regions on the SOM is simple, as seen in Listing 9.3. Figure 9.1 shows all block groups in the US (for which data were available) classified into one of eight types, where color represents a type. Figure 9.10 shows how these eight types map onto the attribute space.

LISTING 9.3 Clustering neurons and visualizing results on a feature map

```
1   ##CLUSTER NEURONS
2   somClusters  <-  kmeans(x=aSom$codes,  centers=8,  nstart=1000,  iter.
    max=10000)
3
4   #Plot Centers
5   plot(0, 0, type = "n", axes = FALSE, xlim=c(0, 100),
6     ylim=c(0, 160), xlab="", ylab= "", asp=1)
7   ColRamp <- rev(designer.colors(n=8, col=brewer.pal(8, "Set1")))
8   ColorCode <- rep("#FFFFFF", length(somClusters$cluster))
9   for (i in 1:length(somClusters$cluster))
10    ColorCode[i] <- ColRamp[somClusters$cluster[i]]
11
12  offset <- 0.5
13  ind <- 1
14  for (row in 1:aSom$grid$ydim) {
15  for (column in 0:(aSom$grid$xdim - 1)){
16    Hexagon(column + offset, row - 1, col = ColorCode[ind])
17    ind <- ind +1}
18    offset <- ifelse(offset, 0, 0.5)
19  }
```

FIGURE 9.10   SOM neurons grouped into eight classes

Conclusion

The Self-Organizing Map algorithm is a useful computational tool for the analysis and visualization of high-dimensional patterns. The SOM is interesting from a geocomputaitonal perspective because it 'spatializes' data via the creation of a feature map. One can examine these feature maps and draw connections between feature space and geographic space. The algorithm has seen extensive use in the geographic literature but its use is somewhat complicated by the lack of good tools in the software commonly used by geographers. In this chapter we have explained the logic of the algorithm in plain English and illustrated its use via the R kohonen library and an analysis of demographic patterns in a relatively large dataset of over 200,000 geographic zones. We have augmented this library via some visualization functions that are available online from the companion website; these functions are used to produce the figures in this chapter. Readers familiar with R's spatial and cartographic functions should be able to easily augment the figures in this chapter for their own purposes.

FURTHER READING

Kohonen (2001) is the central reference work on self-organizing maps. The text provides the details of the algorithm and discusses the use of SOM-PAK, an open source library for computing self-organizing maps.

Skupin and Hagelman (2005) present an interesting application of self-organizing maps to the temporal domain. They explore how neighborhoods move through attribute space over time.

Agarwal and Skupin (2008) provided a detailed discussion of the application of SOMs to geographic problems.

10

KERNEL DENSITY ESTIMATION AND PERCENT VOLUME CONTOURS

Daniel Lewis

Introduction

This chapter presents a method for making geographical decisions about services based on the spatial distribution of their users, particularly in terms of analysing existing systems of service provision. Many commercial companies and public services operate in a location-based way, and anyone who wants to use such services has to physically visit them. Young people have to go to school to learn; ill people need to visit a doctor's surgery or hospital (Chapters 11 and 13) to receive treatment; and hungry people might choose to visit a supermarket to buy food. Each of these services is effectively in competition with each other for pupils, patients, customers and consumers. Making sure that services are successful requires an understanding of the community of people being served, which is a product of both the individual characteristics of service users and their respective geography.

The community of people served by a given facility, like a doctors' surgery, might ideally be defined as everyone who lives within distance x, or travel time y, according to the capacity of that facility. This, we could claim, is a *normative* assessment (i.e. a standard or normal expectation) of how people ought to be arranged geographically with respect to any particular service. Unfortunately, reality does not usually reflect a *normative* distribution. People will often choose (or be constrained) to use particular services for reasons other than simple proximity to their residence or place of work. In the past, researchers often had to rely on models such as the gravity model, or its more sophisticated counterpart, the spatial interaction model, to get a sense of the community of people who used a particular service. Recently, though, the increasing collection and availability of administrative data from local and national government concerning the use of public services, and research cooperation from the commercial sector, has meant that we can look directly at patterns of service use.

A straightforward example might be hospital records. In the UK, for any publically funded (NHS) hospital, administrative records are kept detailing patient visits. These records tell us something about the patient (their demographics), something about the hospital visit (any diagnoses, procedures, operations or outcomes), and also something about the patient's geography (where they are domiciled). This information can tell us a number of things about how that hospital is used, all of which are very important to delivering efficient and high-quality hospital services. It is more complex than this, though; hospitals are in competition with one another, so we do not just want to know about any one hospital, we also want to know about competing hospitals. Soon enough we are dealing with very large amounts of complex data about patients in multiple hospitals that are difficult to meaningfully assess.

This chapter will discuss the geocomputation of 'service areas', also often called 'catchment areas' or 'market areas'. Delineating service areas is an effective way of spatially summarising complex or voluminous data on service use in a way that can help service managers, administrators or policy-makers better understand and make geographic decisions about location-based services. Spatial methods can help us simplify complexity.

Service area analysis methodology

This section will discuss the creation of service areas. The methodology is a three-step process: first a density surface is computed, then it is thresholded according to the percentage volume of service users we wish to enclose, and finally the thresholded surface is contoured to create the service area. In order to make service area analysis accessible, an ArcGIS 10.x Python script tool has been provided on the companion website. The service area tool allows users to input a set of data points and output service areas, whilst the percent volume contour tool allows more advanced users to input a pre-computed density surface and output the service area based upon that surface. The R package adehabitat (http://cran.r-project.org/web/packages/adehabitat/) also implements the method with the kernelUD function (Calenge, 2013).

Service area analysis has existed for many years in the spatial sciences, both conceptually and practically (see Wilson, 2000). It emerged from the work of Von Thunen (1783–1850) and Weber (1868–1958), who theorised the locations of farms and industries, respectively. Applebaum (1965) traced the beginnings of contemporary service area analysis to commercial definitions of 'market areas' in store location research in 1930s America. Early research into service areas depended on Reilly's law of retail gravitation (Reilly, 1931), which constructs service areas in terms of 'break points' between cities subject to the relative size of their populations. These kinds of insights from a gravity modelling perspective were important in the development of different attitudes and approaches to service area delineation.

Huff (1964) is one of the first outside of the 'gravitationalists' to consider the geographical delineation of retail trading areas, arguing that existing break point and buffer approaches may be subject to errors, including differences with 'transportation facilities, topographical features, population density, and the locations of competing firms' (p. 35). Huff (1964) instead focuses on the consumer, not the firm, creating probability surfaces for firms based on the behaviours of consumers, with contours delineating equi-probability of customers visiting a given firm. Huff (1964) and Huff and Batsell's (1977) service area delineations allow for the possibility of overlapping, or 'congruent' areas (Huff and Rust, 1980).

The method implemented in this chapter and used to delineate service areas is also analogous to an approach in ecology designed to tackle a long-standing problem called 'home range estimation'. A home range is the 'area traversed by the individual [animal] in its normal activities of food gathering, mating, and caring for young' (Burt, 1943: 351). Home ranges are also commonly referred to in ecology as 'utilisation distributions', which Burt (1943: 351) tells us are 'rarely, if ever, in convenient geometric designs'.

A key component in this analytical method is that data are required about the locations of service users. The simplest form of data that we can work with requires two variables: the location of the service users; and an identifier for the service location used (assuming data relate to a system of services). Using these two identifiers, we can both locate individuals spatially and group individuals by the service that they use.

In an ideal world we would always work with population data, that is, data that accounts for every user of a given service; however, in reality it is often too expensive, difficult or simply impossible to obtain data on a complete population. If we do not have population data, we instead have sample data, that is, data about a subset of that population. If we are working with sample data it is important to think about the caveats of that sample – the conditions underlying its creation. If, for example, the sample of people has been randomly selected from a known population of all users, we could assume that it is likely to be representative of the population, as the random selection is unlikely to have introduced significant bias. However, if a particular selection strategy has meant that a specific subset of the population has been sampled, we would need to be careful about how we address that sample.

In addition to a basic dataset of locations and service enrolment, it would be nice to have a range of other variables to allow us to create stratified service areas (i.e. service areas for particular subsets of the data, for instance men and women) and also to test hypotheses about the types of people who use the service. Knowing something about individuals and their habits is important to understanding the conditions of service. We might also want to know about characteristics of the service visited, such as the range or type of services provided. In addition to these features of services and service users, we might also want to have some geospatial reference data available, that is, data about the wider environment that the service and its users are located within. Geospatial reference data are important from a visualisation perspective. Finally, we might employ secondary data about the places that are covered by service areas, for instance data from the UK Census of Population, allowing us to compare the service users within a given service area with the whole population of users and non-users.

Creating a service area

Following the procedure set out in ecology by Worton (1989) and subsequently detailed by Gibin et al. (2007) and Sofianopoulou et al. (2012) with reference to public health applications, this section details the three steps required to generate service areas: firstly, the creation of a density surface; secondly, the thresholding of the density surface; and finally, its delineation to form a bounded service area.

Our goal is to create a service area that bounds a particular percentage of the distribution of service users for a given service. We use a density surface to do this as we want to bound the smallest possible area that satisfies a given percentage; this will be defined by the areas of highest density. Density is a commonly used meas-ure of the count of observations per area; you will be aware of it in maps of population density, which measure the number of people per area. One way of creating a continuous density surface is to use a method called kernel density estimation (KDE).

Kernel density estimation is, in essence, a smoothing technique that trans-forms a discrete distribution of points (such as the location of service users) to a continuous surface. The important property of KDE is that the transformation preserves the density characteristics of the discrete point distribution – areas that have lots of points receive higher density values than areas that have fewer points. When we look at point distributions that relate to service use, we usually find lots of points near to the service in question, and fewer points further away; this is typical of the kind of 'distance decay' relationships we expect. Figure 10.1 shows the transformation from a point distribution of users of a service to a continuous density surface using KDE.

Density surfaces like the one shown in Figure 10.1 are often used to detect the clustering of points, in which areas of high density are referred to as 'hotspots'. If we look at the location of the service, there is a clearly a high density of service

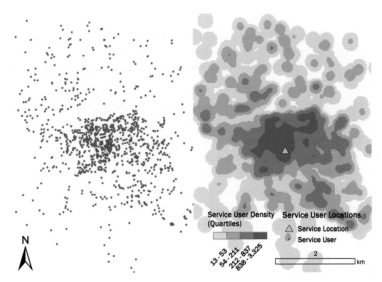

FIGURE 10.1   A point distribution of service users (left), and its density surface (right) computed using KDE (25 m cell size, 200 m bandwidth, Epanechnikov kernel)

users near to the service location, and a tendency towards lower densities further away from the service location.

Creating the density surface using the KDE method requires the user to make several choices, notably the cell size, bandwidth and the kernel function. Let us consider the significance of each of these in turn.

Cells size is the easiest to deal with: the density of observations is estimated on a continuous surface; this continuous surface is a 'lattice', that is, a regular grid of cells. Thus, cell size specifies the resolution of the density surface – the length of the side of each square cell. Large cell sizes produce a grainier image, and smaller cell sizes a sharper image. In Figure 10.1, the density surface would appear more blocky if we increased the cell size from 25 m and sharper if we decreased it; however, each time we halve the cell size, we quadruple the number of cells we need to hold the density values for the whole surface. More cells mean larger file sizes and slower processing, therefore we may need to find a compromise between smaller cells sizes and processing time. Generally speaking, though, as computing performance has increased, the need to restrict cell sizes for most applications has lessened.

Bandwidth and kernel function decisions are associated, and determine the process of 'spreading' (smoothing) the point distribution. KDE works by positioning a 'kernel' over each point in the point distribution. A kernel is a window of a given 'bandwidth' that spreads the point over the area of the window according to a mathematical function. Usually we use a function that is peaked over the point itself, and gradually decays as it gets to the edge of the window. It is easy to understand this in one dimension, as shown in Figure 10.2; in both cases the discrete point is spread along the number line (the *x*-axis) according to the shape of the kernel. The Gaussian kernel (also called the normal kernel) is one of the most used kernels: the bandwidth is the standard deviation of the function. This particular function is 'unbounded', meaning that the curve never quite decays to 0, thus a density value is calculated for all cells in your area of interest, for each point. The Epanechnikov kernel is the kernel implemented in ESRI's ArcGIS KDE tool, and is the kernel used in Figure 10.1. As you can see from Figure 10.2, unlike the Gaussian kernel, the Epanechnikov kernel is 'bounded': it is only defined for the bandwidth (in this case from −1 to +1), which means that density values are only calculated within the bandwidth. Once you have positioned a kernel over each point in a point distribution and smoothed the point according the function of the kernel and its bandwidth, you simply add all of the windows together to get your density surface; this is what has happened in Figure 10.1.

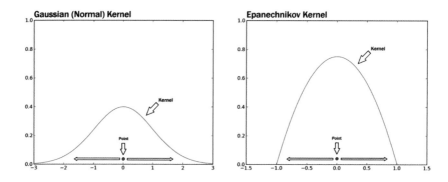

FIGURE 10.2   Two kernel functions smoothing a point

In practical terms, the choice of kernel function will not make a significant difference to our application of calculating service areas, so you may as well use whichever has been implemented in your software (usually either normal or Epanechnikov). However, de Smith et al. (2009) give a good run-down of other possible kernels. The same is not true of the bandwidth, however, the appropriate setting of which is very important to the resultant density surface.

The size of the bandwidth defines how big the window is that we use to smooth the point distribution we are working with. A small bandwidth creates a small window, a large bandwidth a large window, and this has the effect of giving low or high amounts of smoothing, respectively. Unfortunately if we mis-specify the bandwidth we can end up with a density surface that is over- or under-smoothed. An over-smoothed surface will have indistinct features, perhaps looking a bit blurred, whereas an under-smoothed surface will look too rough or spiky.

The goal is to find a bandwidth that strikes a balance between over-generalisation and over-discretisation of a point distribution, Figure 10.3 gives an example of what this might look like using a three-dimensional visualisation.

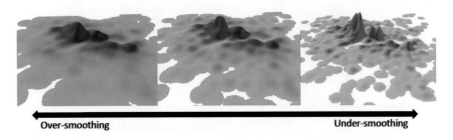

Over-smoothing   Under-smoothing

FIGURE 10.3 Over- and under-smoothing of a point distribution

Bowman and Azzalini (1997) describe the 'normal optimal smoothing' approach to heuristically specifying a bandwidth (h) for KDE as

$$h = \sigma n^{-1/6} \qquad (10.1)$$

Here σ is the standard deviation of the point distribution, usually the mean of the standard deviation in the x and y directions; n is the number of points in the point distribution. Silverman (1986: 86) notes that equation (10.1) is an appropriate heuristic for Epanechnikov kernels if h is multiplied by 1.77.

In point distributions that represent the locations of service users we often find that a minority of users live a long way away from the service, giving a 'long-tailed' distribution. Such distributions can have the effect of inflating the standard deviation of a point distribution, and thereby over-smoothing. Therefore, following Bowman and Azzalini (1997), the average absolute deviation (AAD) or median absolute deviation (MAD) can be substituted for σ in order to counter the inflation of h due to outlying data points:

$$\text{AAD} = \text{mean}(|\, x_i - \text{mean}(x)\,|) \qquad (10.2)$$

$$\text{MAD} = \text{median}(|\, x_i - \text{median}(x)\,|) \qquad (10.3)$$

Figure 10.4 demonstrates the differential effect of bandwidth selection for a general practioner (GP) surgery in Southwark, London. Ultimately, the selection of bandwidth will depend on subjective preference. Although completely automated bandwidth selection procedures do exist, they can be unreliable for the purpose of creating service areas, tending to over-smooth.

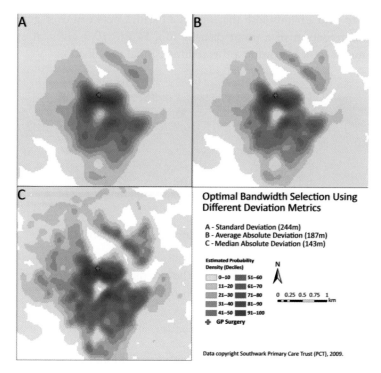

FIGURE 10.4 KDE surfaces using an Epanechnikov kernel, resulting from bandwidth selection using equations (10.1), (10.2) and (10.3) as A, B and C, respectively

Thus far we have considered using a single 'fixed' bandwidth. However, a more advanced approach may be undertaken in which 'adaptive' bandwidths are used. In this situation bandwidth is dependent upon the local distribution of points, not their global distribution. In adaptive KDE, the bandwidth is allowed to vary so that local areas are neither under- nor over-smoothed as might be the case using a fixed bandwidth. Figure 10.5 gives an example of adaptive KDE using the sparr package in R (Davies et al., 2011) for the same GP surgery as in Figure 10.4: Figure 10.5(a) is the density surface using an adaptive Gaussian kernel, and Figure 10.5(b) gives the hypothetical bandwidths over the study area. It is clear from Figure 10.5(b) that the adaptive bandwidth is much smaller for areas near to the GP surgery, and tends to increase with distance. As might be expected, computing an adaptive KDE for a large dataset, over a large area, on a fine grid is more resource-intensive than the equivalent fixed bandwidth approach.

Having defined an appropriate density surface for a point distribution, a service area can be defined by working out the smallest area within which a predefined percentage of the surface lies, and drawing around that area. This is called a percent volume contour; the term 'volume' is used because, as in Figure 10.3, the density

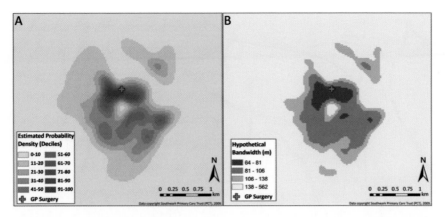

FIGURE 10.5   Kernel density estimate using (a) Gaussian kernel and (b) hypothetical bandwidths

surface can be seen as having a volume (edge length squared × cell density value) and hence can be represented arbitrarily in three dimensions. The procedure for thresholding the density surface is as follows:

1. Compute the percentage of the total volume of the density surface that each cell contributes.
2. Sort the cells from high volume to low volume.
3. Recode the cells as 1 until a given cumulative percentage is reached; thereafter recode cells as 0. This creates the thresholded density surface.

All that remains is to convert the thresholded density surface into a vector geometry. Converting to a line geometry is the percent volume contour (PVC), whilst converting to polygon geometry is the service area. Figure 10.6 demonstrates the 50%, 75% and 95% service areas for a GP surgery in Southwark, London.

Aside from the adaptive KDE methods suggested above, there exist several possible advanced refinements to the service area creation method. Here we will briefly discuss smoothing techniques for zonal data and the use of network KDE. So far we have assumed that data relate to the locations of individuals; however, data will often be aggregated to an administrative geography. When data are reported at an aggregate level of geography, a technique such as KDE, which is applicable to point and line data, is unsuitable; this is because the presence of an aggregate geography implies boundary conditions that are ignored by the kernel approach. The implication of smoothing an areal aggregation using KDE is that values in one area could affect the estimated density in neighbouring areas, despite the fact that we know empirically that those values lie within bounded, neighbouring but non-overlapping areal units. Inferring a surface from an areal

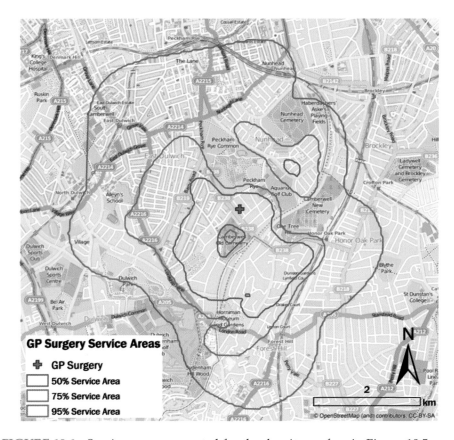

FIGURE 10.6   Service areas computed for the density surface in Figure 10.5

aggregation implies a 'change in support' – from a larger, irregular areal aggregation, to a finer, regular grid (or set of irregular point locations). Tobler's (1979) pycnophylactic surface model is a good candidate for effecting a change in support from aggregate administrative units to a fine, regular gridded surface model. An application of Tobler's pycnophylactic smoothing – pycno – is available for R (Brunsdon, 2011). Kyriakidis (2004) also identifies a subset of the geostatistical technique of 'kriging', known as 'area-to-point kriging', which would also be applicable. The appropriate model in the case of service area delineation would be area-to-point Poisson kriging, owing to the fact that the areal aggregation containerises counts of service users visiting a particular service; such models are detailed by Goovaerts (2006).

Another assumption we have made is that it is appropriate to conceptualise space as isotropic, that is, uniform in all directions. However, services and their customers are usually constrained to a network of roads and paths, so that most of the time we cannot travel between places in a straight line. Standard KDE implementations use a straight-line distance, but it might be more relevant to use a network-based KDE instead. Network KDE calculates the density of a given set of observations on a network, and thus may be a better representation of the density of customers who utilise a service. Okabe and Satoh (2009) have created the SANET toolkit (Spatial Analysis on NETworks) for ESRI ArcGIS, which supports network KDE. However, network KDE is more computationally expensive than traditional KDE, there is less certainty about how to specify appropriate bandwidths, and creating PVCs is untested. Bounding a subset of a network in order to create a PVC requires thought, although initially implementing a method similar to how network buffers are drawn (using triangulated irregular networks) might be a viable solution.

We recommend using the KDE and PVC approach detailed above for estimating service areas; for the most part it is a fast and efficient procedure that is well integrated and understood in GIS. There is support for choosing the appropriate kernel, cell size and, most importantly, bandwidth. The procedure produces good outputs for visualisation purposes, and is generally acceptable analytically, even in situations where there is a change in support (although there may be some biases or uncertainties introduced here). The primary role of the researcher is to decide what you would consider an appropriate percentage at which to delineate a service area. There are no easy answers to this, although most researchers point towards enclosing a group of 'core' users which can range from 50% to around 80% in service area terms, although others, such as Sofianopoulou et al. (2012), use 95%, 98% and 99% service areas owing to the nature of their research question. In the following section we provide a short overview of analytical options using service areas.

Examples of service area analysis

There are a number of different analytical contexts in which service areas can offer new and novel understandings of how existing systems of service work. In many, if not the majority of cases, service area analysis will be implemented simply in order to effectively visualise the distribution of service users for a particular service or system of services. Maps are easily understood forms of communication, and enable the effective transmission of information in a number of areas; this might include operational or managerial situations within a company, as well as customer-facing contexts.

Singleton et al. (2011) define service areas for secondary schools in the UK, presenting a web-based decision support system in order to aid parental choice. Not only is a school's service area depicted, using data on school and pupil locations from the pupil-level annual schools census, but a whole range of additional data is also presented, at both the school level and the neighbourhood level. The visualisation of service areas, and the association of a wide range of additional data, allow for a comprehensive insight into local secondary school enrolment and attainment.

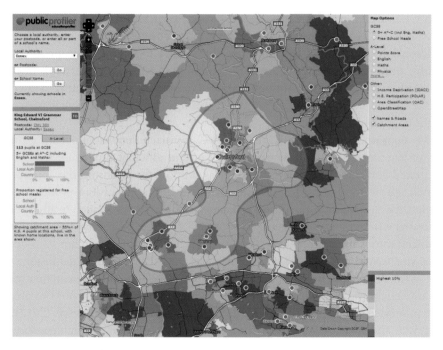

FIGURE 10.7 A school service area in Singleton et al.'s (2011) education profiler (source: http://www.educationprofiler.org/)

Whilst Singleton et al.'s (2011) web-based implementation of service areas is a powerful geovisualisation-based spatial decision support system, it is strictly communicative. We can only make visual comparisons between service areas and their statistics. The next logical step is to begin to compare the spatial congruence, or overlap, between different services. Figure 10.8 demonstrates the overlap for a system of hospital trust service areas in London *c*. 2009 according to their 50% and 80% service area polygons.

Initially, we might simply enumerate the number of overlaps experienced by hospital trust service areas; this will provide crude estimates for the presence of locally competing markets. However, we can go further by estimating the areal extent of the overlap between hospital trusts; this gives us an idea of the importance of the overlap. Of course, whilst it might be enlightening to consider service area overlaps in terms of shared area, it is actually much more pertinent to think in terms of people, rather than geometry; it is, after all, people who are being provided care.

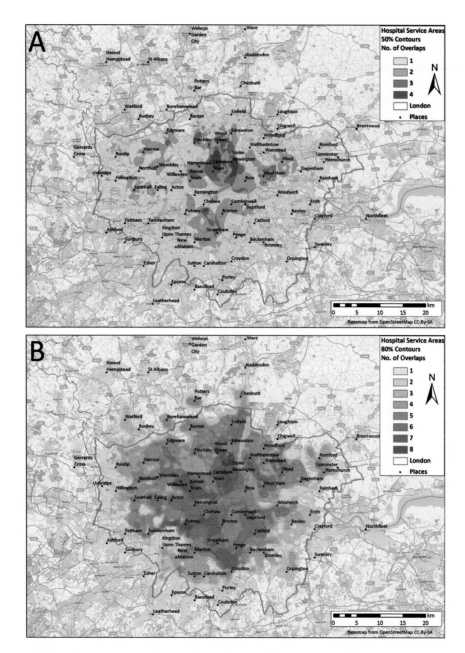

FIGURE 10.8 Overlapping NHS trust service areas for 25 NHS London trusts, for outpatients in 2008/2009

We can use the set of service areas as bespoke geographies, and estimate any number of demographic or socio-economic variables based on the service area geometry. This can give us a richer idea of the kinds of places covered by each service area, and also provide real context for areas of congruence between locally competing services.

A service area can also be used as a sampling frame or geographical filter through which questions about social and spatial processes related to a given service, or system of services, can be answered. Generally speaking, a service area encloses a group of people, some of whom will be users of that service, and some of whom will not. The difference between the community of people that are served, and the community of people that are within a service area but not served, can help answer questions important to particular services. In the case of publically provided schools or health-care services it might provide monitoring information about the spatial equity of provision, whereas commercial companies might use the analysis for marketing purposes to help them target local groups of non-users according to their demographic or socio-economic characteristics.

Harris and Johnston (2008) use service areas to look at segregation in primary schools in Birmingham, UK, demonstrating that there is evidence for the concentration of children of particular ethnicities in particular schools in a way we would not expect given a sampling of eligible children based upon primary school service areas. This raises questions about the reasons for these patterns, and the potential for academic disadvantages to be brought about as a spatial process. Harris and Johnston (2008) use their service areas to select individuals who either fall inside or outside each service boundary, and can use that information to create an index. We can, for example, create a new randomly sampled population for a service based upon its service area, and compare that sample with the observed population. Comparing an observed population to a hypothetical population allows us to say to what extent an observed population might deviate from a hypothetical population.

A similar analysis for GP surgeries in Southwark, London was interested in ethnic drivers of patient choice. Figure 10.9 suggests that there is variation in uptake of local primary care GP services by ethnicity. Indeed, when individual GP surgeries are considered, evidence from Southwark suggests that patient ethnicity is a major factor in the sorting of population by GP surgeries, with some groups, notably African, Muslim and East Asian patients, being over twice as likely to use particular surgeries than might be expected given their proportion in the normal resident population.

As with any research, we have to be careful in our interpretations: whilst using the service area as a sampling frame can help us uncover patterns, and stress the need for better spatial understandings of how people use services, it would be difficult to draw conclusions based upon this analysis alone.

Service areas do, however, also offer the potential to contribute to statistical models as the dependent or explanatory variable. Lewis and Longley (2012) consider a range of demographic and spatial explanatory variables in a study of patient

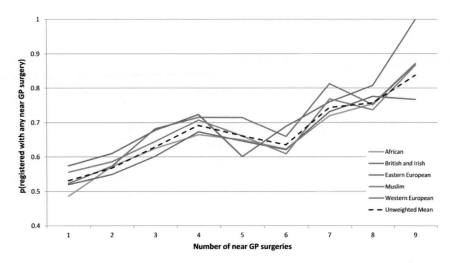

FIGURE 10.9 Registration with GP surgeries in Southwark, London (2009), based upon the number of 50% service areas a patient falls within, stratified by patient ethnicity

choice of GP surgeries; a control variable is defined according to the number of service areas that each patient resides within. This variable is important in accounting for the possible choice set of a given patient, which varies depending on where that patient lives. This measure of accessible GP surgeries by service area is found to be almost as significant an explanatory variable as distance to a GP surgery, or the number of doctors of a GP surgery has, is in explaining why patients might not use their most proximal GP surgery. These kind of insights are significant, as Soufianopoulou et al. (2012) suggest: service areas are unlikely to be coterminous with other administrative boundaries, or in the case of GP surgeries, with doctor-defined service boundaries. There is clearly a spatial component to service provision that is not currently well understood or accounted for.

In the future, there may be scope to use the geometric properties of the service areas of a system of services to actually define the analysis. In analyses of the competition between services it might make sense to create a 'spatial weights matrix' that represents the service area overlaps; this would allow for statistical models to adjust for the spatial dependencies created by locally competing services and mitigate the effect of bias on estimates of competition.

Conclusion

This chapter has outlined and illustrated a methodology for defining the service areas of a given service or system of services, and provided an overview of the analysis options available, with examples. In some respects service area analysis is one of the older spatial analytical traditions, with notable examples from some of the most significant spatial thinkers. However, new insights from external disciplines, the increasing availability of appropriate data and modern high-powered desktop computing have the potential to yield new insights in a topic area about which little has been published in the last ten years within the social science literature.

As we continue along an era of growth in spatial analysis and applied geocomputation in both the public and private sectors, detailed and sophisticated ways of dealing with increasingly large datasets will increasingly be sought after. Policies in the public sector that advocate more competition and greater cost savings in services such as health care and education can benefit greatly from spatial approaches, provided a solid data infrastructure is in place. Commercial companies, meanwhile, will seek to understand how location-based analyses of customer demographics can mesh with the increasingly online world of business.

FURTHER READING

Many of the statistical and geostatistical procedures mentioned in this chapter, particularly kernel density estimation, can found in an accessible form in de Smith et al. (2009) or online at http://www.spatialanalysisonline.com. This is an excellent practical resource.

Longley et al. (2011) is an excellent resource for providing a wider context for the application of geographical information science, whilst Cromley and McLafferty (2012) provide an excellent context for the use of GIS in health research, which has been the main focus of this chapter.

Wilson (2000) provides a useful overview of neoclassical models, in a text that is also relevant to spatial interaction modelling. Some of these models have been referenced in the chapter as shaping the early thought on market areas, and should help the reader get a sense of the wider context and roots of geocomputation.

Finally, for applied examples, I recommend any of the literature cited in the 'Examples' section above.

11

LOCATION-ALLOCATION MODELS

Melanie Tomintz, Graham Clarke and Nawaf Alfadhli

Introduction

Location–allocation models are a form of optimisation models designed to find the optimal location for service provision across a city or region given the spatial distribution of demand for that service. That demand is normally represented as the number of persons likely to use or need that service and is often located in a set of points or polygons across the region. Those polygons are most likely to be census zones or tracts or perhaps zones based on postal geography. There are a number of alternative versions of the models which use different rules to find those optimal locations. The most common model is probably the *p*-median, which aims to find a set of service locations that will minimise the total distance or time travelled across the city or region to visit those service points. In most cases the planner (or user of the system) has to stipulate the number of facilities that need to be located (based on resources available). Then, the model will locate that number of facilities and allocate demand zones to each facility. Thus demand in each polygon is allocated (only once) to its nearest facility.

The aim of this chapter is to introduce the technique of location–allocation modelling and to give some illustrations of their use. We look at different forms of the model and how they can be operationalised in a GIS package such as ESRI's software ArcGIS. Then we give two examples from our own work – one model to find the optimal locations for stop-smoking clinics in Leeds, UK, and another to find the optimal locations for police stations in Kuwait.

Location-allocation models and their application

Location–allocation modelling is a technique used to determine the set of optimal locations for service provision under a certain set of criteria or constraints.

With such models it is possible to understand the relationships between access and facility location and variations over space and time (Nemet and Bailey, 2000), for both existing patterns of service location and for potential future patterns. Most models will allocate demand to the nearest facility available. Then the objective is to minimise total travel costs, time or distance travelled for either all persons or a selected geodemographic group. The earliest model produced is known as the *p-median problem*. This was first defined by Hakimi (1964) and is the most commonly used method (Church and Sorensen, 1996). To solve the *p*-median problem the following data are required (Cromley and McLafferty, 2012: 354):

- number of demand sites;
- number of possible supply sites;
- distance, time or cost of travel from each demand site to each potential supply site;
- number of facilities to open.

The objective function for the *p*-median problem can be written as

$$\text{minimise } Z = \sum_{i \in I} \sum_{j \in J} a_i d_{ij} x_{ij}$$

subject to the following constraints:

- An individual demand site must be assigned to a facility $\sum_{j \in J} x_{ij} = 1$ for all i.
- Demand must be assigned to an open facility $x_{ij} \leq \sum x_{ij}$ for all (i, j).
- Exactly p facilities must be located: $\sum_{j \in J} x_{jj} = p$.
- All demand from an individual demand site is assigned to only one facility.
- $x_{ij} = (0, 1)$ for all (i, j).

Here Z is the objective function; I is the set of demand areas and the subscript i is an index denoting a particular demand area; J is the set of candidate facility sites and the subscript j is an index denoting a particular facility site; a_i is the number of people at demand site i; d_{ij} is the distance or time (travel cost) separating place i from candidate facility site j; x_{ij} is 1 if demand at place i is assigned to a facility opened at site j or 0 if demand at place i is not assigned to that site; and p is the number of facilities to be located.

There are a number of alternative formulations that can be used, depending upon the aims of the project. In these cases the constraints can be modified and new ones added as necessary. For example, the *location set covering problem* involves locating facilities to cover a set of demand sites within a pre-defined travel distance or time (Toregas et al., 1971). The *maximal covering model* estimates the distance or time that the user most distant from a facility would need to travel to reach the set of facilities located by the model – then the aim is to ensure that a set of facility locations are chosen that ensures no person will be farther than some set maximum service distance from a facility (Church and Revelle, 1974). These models are most commonly used to locate 'outreach' services, such as ambulance or fire stations, where service planners are keen to ensure that no one is more than a given distance (or time) from such a station.

Location-allocation models are often used in combination with geographical information systems (GIS, see O'Brien, Chapter 17) which are important for storing, manipulating, retrieving and mapping large spatial datasets. Nevertheless, most early GIS lacked a sophisticated modelling capability to address the needs of location analysts (Church and Sorensen, 1996). However, location-allocation models are now integrated into some GIS, such as ArcGIS. Church and Sorensen (1996) point out that there are two basic approaches to solving the p-median model: optimal and heuristic techniques. Optimal techniques take a long computation time for larger datasets, and hence mostly heuristic processes are used to obtain quick but reasonable results. Heuristics are algorithms that work faster when working with large datasets by providing a result close to optimal, but do not necessarily guarantee that the best result will be found. The first heuristic for the p-median problem was developed by Teitz and Bart (1968), and other heuristics followed such as genetic algorithms, simulated annealing, tabu search, GRASP (greedy randomised adaptive search procedures), hybrids and GRIA (global–regional interchange approach). Teitz–Bart heuristics are also embedded into ESRI's software which was a former network module developed for ArcInfo (for more detail, see Church and Sorensen, 1996) and is now integrated into ArcGIS Desktop Advanced. ArcInfo workstation was largely command-based software; ArcGIS Desktop Advanced is more user-friendly and no programming skills are needed. The latest version to date is ArcGIS 10.2.

There are various ways to calculate distance travelled. The easiest and computationally fastest is the *Euclidean distance* which calculates the route from one point to another as a straight line. The disadvantage is that it is less accurate as travel on a road network is more common. A further possibility would thus be to calculate the distance using a transportation network where it is possible to find the shortest distance or lowest-cost path from one point to another. A set of solutions to this were introduced in the 1950s by Dijkstra, called the shortest-path algorithm or Dijkstra algorithm (Dijkstra, 1959). When the interest lies in serving multiple locations, then planners are more interested in finding the order of stops to minimise the total travel distance; this is known as the travelling salesman problem and is more complex than the standard location-allocation model (Lawler et al., 1985).

Performing location–allocation analysis within ArcGIS requires the Network Analyst extension. Seven problem types to answer specific location analysis problems are supported: minimise impedance, maximise coverage, maximise capacitated coverage, minimise facilities, maximise attendance, maximise market share and target market share. But how does it work and which input data do we need? In general, three types of data are used: *candidates*, *demand* and a *network*. The 'location–allocation analysis layer' stores the input, parameters and results of the defined problem. The 'location–allocation analysis classes' consist of six analysis classes that are stored within the analysis layer. The six classes are 'facilities class', 'facilities properties', 'demand points class', 'demand point properties', 'lines class' and 'line properties'. The facilities class represents the candidates (e.g. facilities) to allocate the demand according to the specified problem type to the most appropriate facility location. The facility properties class specifies the facility type (*candidate 0* means the candidate facility might be part of the solution; *required 1* means this candidate facility must be part of the solution; *competitor 2* includes all the rival locations and removes demand from the problem (used for the problem types 'maximise market share' and 'target market share'); *chosen (3)* means that the candidates change to 'chosen' as they are part of the solution now. Also a weight can be added (e.g. if one candidate is more attractive than another, it can be given a higher weight), and the capacity (e.g. how many persons a facility can handle) can be included. The demand point class is typically a location that stores demand (e.g. business customers). Often areas are represented as centroids to work with point layers. Within the demand point properties, weights and impedance parameters can be set up. The line class is responsible for the visualisation that connects demand points allocated to facility points (candidates). This type of visualisation is typically called 'spider maps'. The line properties are the results including weight and impedance on the lines. The 'location–allocation analysis layer properties' consist of several tabs where settings are added by the modeller (e.g. the impedance to specify the cost attribute, possible U–turns, restrictions and barriers of the network). In the end, the results can be viewed in tables and visualised in the form of spider maps, where the centre of the spider is the location of the facility and the legs point to the zones that will be allocated to that centre. This is useful as visualisation helps people who are not familiar with these techniques (often the policy-makers) to understand and interpret the results more easily.

There have been many examples of location–allocation models in the literature. To give an example, we simply list some key papers in relation to two key application areas: the location of emergency facilities (fire stations, police and ambulance centres) and health care (GPs, hospitals, clinics, etc.). For emergency services, Fujiwara et al. (1987) use the model to find the optimal location for ambulance stations in Bangkok, Thailand. Other examples include Rahman and Smith (2000) who were concerned that there is considerable evidence that because of poor geographical accessibility, basic health care does not reach the majority of the population in developing nations, and hence a complete reassessment of health-care location planning was required.

Goldberg et al. (1990) used a location-allocation model to determine the best location for ambulance centres in Montreal, Canada, based on two criteria (seven-minute and 14-minute response time). In Louisville, Kentucky, emergency managers have applied an optimisation model to identify the optimal location for all types of emergency vehicles. The result is that the response time of ambulances, for example, was reduced by 36% (Rosero-Bixby, 2004).

For health, Tanser et al. (2010) report that the first applications of location-allocation planning were used by Gould and Leinbach (1966) to determine the capacity for hospitals in Guatemala based on the population and the road network. Hodgson (1988) aimed to locate primary health-care facilities in Goa, India, by using a location-allocation model based on hierarchies. Logan (1985) contrasted the distribution of medical centres according to the government of Sierra Leone with his own model derived using the location-allocation model. Ayeni et al. (1987) utilised the model to determine the ideal location of maternity and other health-care facilities in Nigeria, and managed to reduce the distance to these facilities considerably. Ross et al. (1994) explored locations of breast cancer screening services in Eastern Ontario, Canada, and found that women from many catchment areas had to travel very long distances to reach a service. Møller-Jensen and Kofie (2001) used a location-allocation model for health-service planning in rural Ghana by evaluating different scenarios for future accessibility issues. Tomintz et al. (2008, 2009, 2013) applied location-allocation models to optimise service provision for stop-smoking services and antenatal classes in Leeds, UK, analysing travel distances and service provision based on small geographical areas.

It is noticeable that most applications relate to public sector location problems. The models are more problematic to use for private sector applications where competition is a significant factor in terms of facility location. As Hodgson (1978, 1981) argues, minimising distance travelled is less appropriate when consumers do not routinely visit their nearest centres. So for sectors such as retailing, location-allocation models may only be appropriate for convenience style outlets, chemists and newsagents, and so on. For other types of retailing some choice model needs to be added. Models that incorporate choice and facility attractiveness, as well as distance travelled, such as spatial interaction models or utility-based choice models, are a more appropriate location technique for the private sector. That said, there are some interesting discussions of location-allocation models for private sector applications (see Hodgson, 1978, 1981; Goodchild, 1984; Ghosh and Craig, 1984; Farhan and Murray, 2006).

Modelling optimal locations for stop-smoking services given various policy alternatives

The first example we present here is related to health care and the location of stop-smoking clinics in Leeds, UK. One aim of the UK government is to tailor health services to community needs, and to provide easier and more equal access to those services for all population groups (often called ambulatory or walk-in centres). More disadvantaged population groups should especially benefit from such policy as they often have less income and less access to cars. Between October 2006 and September 2007 (the time period considered), 66 stop-smoking locations were used at different days and times throughout Leeds. Services could be relocated, added or removed on a quarterly basis, based on demand, budget and availability of locations. Thus the first model run shows the 'optimal' locations for the 66 stop-smoking services using a location-allocation model, with the model allocating smokers to their nearest stop-smoking service based on travel distance along the road network. To model the 'optimal' locations for 66 stop-smoking services, the following data are required:

- estimated number of smokers for a Census geography, in this case 2001 UK output areas (OA) (demand site);
- number of possible supply sites (66 stop-smoking services);
- road network (distance element).

First, the number of smokers at an OA level was estimated using a spatial microsimulation approach (the technique and results are shown in detail in Tomintz et al., 2008, 2009). As the location-allocation model was set up using a road network, the location of the smokers were converted into a point file rather than a polygon file as the smokers needed to be allocated to their nearest node on the road network. To create a point file of smokers, the centroid of each OA was calculated, and the estimated smokers assigned to the centroid of their respective zone. Second, the number of supply sites was entered into the location-allocation model, and the model estimated the 'optimal' locations for 66 services so that it was possible to compare the current versus 'optimal' service locations. Third, the distance element used here was a road network, as this is more representative of true distances travelled than straight-line distance. As the target is to provide easier access to services for all people, the model was set up to locate stop-smoking services by minimising total distance travelled and all road nodes could be possible locations for a stop-smoking service.

Figure 11.1 shows the result for the 'optimal' locations for 66 stop-smoking services (red spiders) given the objective to minimise the total travel distance. The centre of the red spider is the location of the stop-smoking service and the red lines show the area this stop-smoking location would cover (allocation). The blue stars show the existing 66 stop-smoking services for the time period from October 2006 to September 2007. Not surprisingly, the location-allocation model distributes facilities much more widely to cover all geographical areas of the city (especially for 66 centres). Thus, the model allocates stop-smoking services in the north and east of Leeds where no services currently operate. The majority of services remain in the centre of Leeds and this is reasonable as there is a high population and high numbers of smokers.

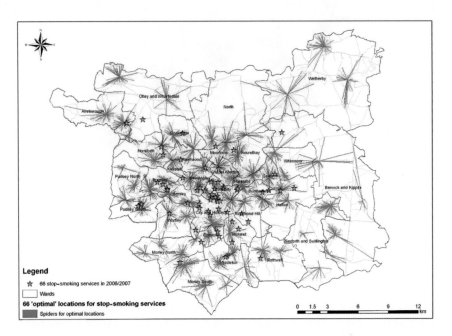

FIGURE 11.1   Current versus 'optimal' locations for 66 stop-smoking services

Tomintz et al. (2008, 2009) explored data relating to attendance at stop-smoking clinics, and it was found that smokers aged 65–74 were the least likely to attend a stop-smoking service. If this group could be targeted by optimising services in close proximity to them then perhaps attendance within this target group could be improved. Thus, the spatial microsimulation model was used to estimate the location of smokers aged 65–74 only. Although 66 locations have been used for smoking clinics in Leeds, on a typical day, between nine and 19 are actually open for use. Thus the second model allocates just nine clinics in relation to demand from the 65–74 smokers only. The results are shown spatially in Figure 11.2. Table 11.1 summarises the numbers of smokers a stop-smoking service would cover, the average distance smokers would need to travel to their nearest stop-smoking centre, and the furthest distance a smoker lives away from the nearest stop-smoking centre. Location 9 covers most smokers (1380), whereas location 4 covers fewest smokers (303). The average distance a smoker needs to travel to the nearest centre is 3.0 km. The furthest distance a smoker would need to travel is 11.8 km.

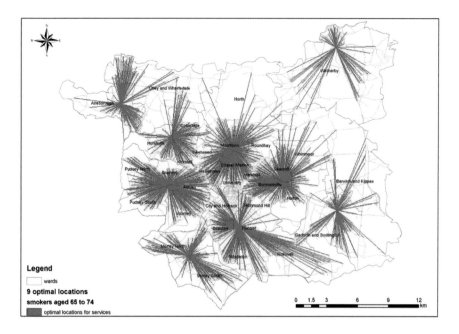

FIGURE 11.2 Nine optimal locations for stop-smoking services to target smokers aged 65–74 most effectively by using a location-allocation model

TABLE 11.1 Summarised information for nine optimal locations for stop-smoking services based on the location-allocation model to target estimated smokers aged 65–74

	Smokers covered	Average distance (km)	Furthest distance (km)
Location 1	984	2.8	8.1
Location 2	481	3.2	7.8
Location 3	665	2.3	7.0
Location 4	303	3.5	8.4
Location 5	587	2.7	6.8
Location 6	1,181	3.6	11.8
Location 7	1,305	2.8	8.6
Location 8	466	3.2	8.7
Location 9	1,380	2.9	6.2

One of the UK government targets is also to reduce smoking rates in more disadvantaged areas. Heavy smokers (people who smoke 20 or more cigarettes a day) are mainly distributed in more disadvantaged areas where there are also high rates of morbidity and mortality from lung cancer. The third example of a model run uses the location–allocation model to find optimal locations for four additional stop-smoking services (where the number of services is chosen randomly for demonstration purposes) if additional money became available to target heavy smokers (Figure 11.3). To run the location-allocation model, the demand, supply and distance elements have to be set up as before. Demand here relates to a simulated population who smoke 20 or more cigarettes a day, and the centroids for the location of these smokers are calculated for all output areas in Leeds using ArcGIS. The supply items are the 19 stop-smoking services most often available on a typical day. These existing 19 clinics are set up in the way that the location-allocation model has to select these locations even if better locations could exist. All other locations are set up to be a possible location, which means that the model will find four optimal locations in addition to the 19 existing ones (by, as usual, minimising the distance travelled). The distance element remains a road network as it is more accurate than straight-line distance and the results are shown in terms of travel length. The blue dots on the map represent the existing 19 frequently open clinics and the red lines show the areas that the 23 stop-smoking services would now cover (19 existing plus four new ones). The four red spiders without blue dots are the best new locations. The model wants to locate these additional stop-smoking services in the north-west, west, north-east and east of Leeds. This shows the 'untapped' demand from heavy smoker residents in these areas. When adding four additional services, the average distance a smoker would need to travel is 2.2 km, whereas the furthest distance is 9.4 km.

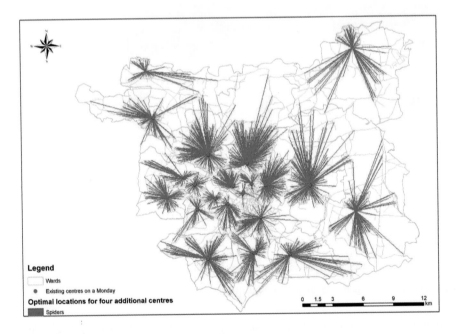

FIGURE 11.3 'Optimal' locations for four additional stop-smoking services based on the number of simulated heavy smokers

A location-allocation model for police stations in Kuwait

Our second example concerns the location of police stations in Kuwait. This time the maximum coverage model will be used. In addition, the models are used to reduce the number of existing police stations, in order to save resources and concentrate services at fewer, larger centres. In many parts of the world, governments are attempting to save money by reducing expenditure on public services. In the UK, for example, the number of police stations was reduced by 630 between 2002 and 2012.

To set up a model for Kuwait the following datasets are required:

- location of police stations: this file also includes police station attributes (for 2011);
- districts file: this includes 83 census polygons, alongside the total population and crime rates within each district;
- roads network file: these roads are attributed as uni- and bidirectional.

FIGURE 11.4   Distribution of existing police stations in Kuwait

Source: Kuwait Institute for Scientific Research 2011

Figure 11.4 shows the existing locations of police stations in Kuwait. As can be seen, all government municipalities or governorates in Kuwait are very well

covered by existing police stations. However, the number varies from one governorate to another based on the size of population as well as the nature of each
governorate (administratively and economically). For example, the Capital governorate (the capital of Kuwait located in the north-east of Figure 11.4), is the most
important governorate in Kuwait in terms of administrative importance and
hence has a high number of stations per head of population. There are 21 police
stations located in this governorate, although the felony offence rate in 2005 was
relatively low; there were 192 crimes in a population of 244,453 in 2011. In
comparison, Mubarak Al-Khabeer (centre east in Figure 11.4), which is the new-
est governorate, has just two police stations. Similarly, Jleeb Al-Shiyoukharea has
one police station to serve about 180,000 people. The crime rate here is particularly high because of the presence of many poor, marginal and manual labourers
of non-Kuwaiti origin located together in a very overcrowded area. This area is
coloured red in the centre of Figure 11.4.

Given limited resources, the location–allocation model can be used to reduce the
number of police stations, as well as locating them more optimally in terms of the
population and the number of crimes committed. The number of police stations
needed is chosen as 15, despite the fact that the current number of the existing
police stations is more than 50, as can be seen in Figure 11.4. Although this is an
arbitrary number, it would reduce the provision by two-thirds and save considerable

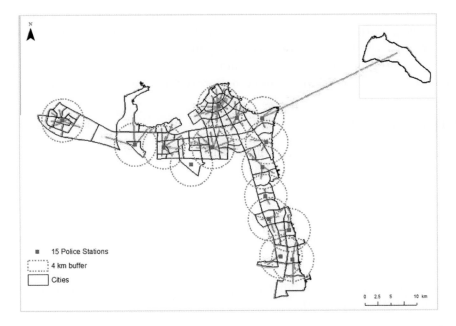

FIGURE 11.5 The proposed locations of police stations in the governorates of
Kuwait, produced by ArcInfo Workstation

resources (whilst at the same time still providing a good geographical spread of stations across Kuwait). By applying the maximum demand model, most areas in the governorates of Kuwait could be covered by the new pattern of police stations (Figure 11.5). Between two and five minutes' travel time is considered an appropriate response time for an emergency services facility, especially a police station (Tong and Murray, 2009). Therefore, the model worked on the constraint that approximately three minutes by car with a speed limit of 80 kilometres/hour is an ideal time response, with a maximum coverage of 4 km serving as a buffer. Figure 11.5 indicates that, using three minutes by car, the majority of major districts can be covered by police stations.

Conclusion

This chapter has described the use of location–allocation models for facility location. The models find the optimal location of X facilities (where X is determined by the planner) to minimise the distance travelled by users of those facilities. Models typically minimise total distance travelled by all users or ensure that no user is further from a predefined distance to their nearest facility. Given that competition between facilities is more difficult to incorporate into these models, they are typically used more prevalently in public service facility location than in commercial applications. The examples given in the chapter show how they can be used for important application areas such as health care and emergency planning. They are ideally suited to be an integral part of modern GIS, where the spatial data for the models can be pre-prepared and the results visualised in easy to interpret forms. There is no doubt that they are an important element within applied geocomputation and will continue to be refined in the future for even more sophisticated application areas.

FURTHER READING

It is always good to read original sources. One of the earliest and most cited references concerning location–allocation models is Cooper (1963, 1964):

- Cooper, L. (1963) Location-allocation problems. *Operations Research*, 11(3): 331–343.
- Cooper, L. (1964) Heuristic methods for location-allocation problems. *SIAM Review*, 6(1): 37–53.

A useful single collection of different applications of location–allocation models is Ghosh and Rushton (1987):

- Ghosh, A. and Rushton, G. (1987) *Spatial Analysis and Location-Allocation Models*. New York: Van Nostrand Reinhold.

A complementary article dealing with public sector applications, especially in relation to health planning, is Tanser et al. (2010):

- Tanser, F., Gething, P. and Atkinson, P. (2010) Location–allocation planning. In T. Brown, S. McLafferty and G. Moon (eds), *A Companion to Health and Medical Geography* (pp. 540–566). Oxford: Blackwell.

Richard Church and Alan Murray have made many innovative contributions to location-allocation modelling over the years. Church and Murray (2009) is a nice summary of these many advances, but also shows how the technique can be embedded more broadly in GIS and spatial analysis, especially for business site location:

- Church, R.L. and Murray, A.T. (2009) *Business Site Selection, Location Analysis, and GIS*. Hoboken, NJ: John Wiley.

PART IV

EXPLAINING HOW THE WORLD WORKS

12

GEOGRAPHICALLY WEIGHTED GENERALISED LINEAR MODELLING

Tomoki Nakaya

Introduction

One of the most popular approaches to statistical model building is regression analysis. Here, we would associate a response variable we wish to predict with explanatory variables, using an assumed relationship such as a linear function. We fit the models to the observed data statistically, and then use the model to infer the properties of underlying processes hidden within the dataset. This chapter focuses on a specific type of regression modelling for spatial analyses, namely geographically weighted regression (GWR), and its extended form, semi-parametric geographically weighted generalised linear modelling (S-GWGLM). Such models replicate the geographically varying aspects of the association between the response and explanatory variables. The S-GWGLM modelling framework is implemented in GWR4, a free Microsoft Windows application, which we introduce in the case study part of this chapter.

GWR captures smoothed geographical variations in the relationships between variables using kernel-based non-parametric regression. This technique is based on the premise that 'processes in different places are likely to be more similar as the places become closer together'. In particular, mapping the estimated coefficients of GWR models enables us to see the geographical structure of the spatial heterogeneity in the processes. In this sense, GWR is considered a tool of exploratory spatial data analysis (ESDA). ESDA lets us visually explore and infer the localised nature of geographical processes. Figure 12.1 shows a conceptual image of GWR.

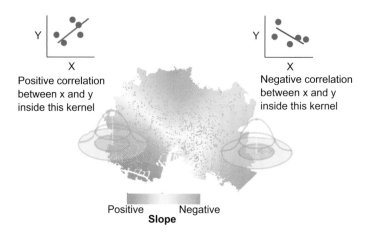

FIGURE 12.1   The concept of the GWR approach

Since the first seminal papers of GWR (Brunsdon et al., 1996; Fotheringham et al., 1996), many theoretical and applied studies of GWR have used the technique in fields such as geography, economics, ecology, epidemiology, and applied statistics; and Fotheringham et al. (2002) provides the most comprehensive guide to GWR. This chapter focuses on two important extensions of GWR (Nakaya et al., 2005, 2009):

1. Geographically weighted generalised linear modelling (GWGLM): GWR was proposed originally as an extension to the ordinary regression model to predict a continuous variable with a Gaussian (normal) error. However, when predicting a non-continuous variable, it is more appropriate to use different regression models, such as a logistic regression model for binary data, and a Poisson regression model for count data. Using a linear predictor, these regression models are integrated as generalised linear models (GLM: McCullagh and Nelder, 1989). Then, by integrating GLM and GWR, GWGLM models estimate binary and count data with coefficients that vary geographically.
2. Semi-parametric framework of GWR/GWGLM (S-GWR/S-GWGLM): A semi-parametric model is a regression model that combines non-parametric and parametric terms. In the case of GWR/GWGLM, a semi-parametric model mixes terms of geographically varying/local coefficients and fixed/global coefficients. This technique is more flexible than GWR/GWGLM. By fixing some effects as global rather than local, we can reduce the complexities of local relationships. This, in turn, may enhance the readability of geographically varying relationships, as well as the predictive performance of

the model. Importantly, it also enables model comparisons, which we can use to determine which explanatory effects on the response variable are fixed globally and which vary geographically in GLM.

The following section explains S-GWGLM models in more detail; it then presents an implementation of the modelling framework using GWR4 by applying semiparametric geographically weighted logistic regression (S-GWLR) to a place-attachment dataset in a citywide geographical setting.

Outline of the method

Generalised linear models are a framework that covers a wide range of regression models that use a linear predictor and probability distributions from the exponential family. Suppose we have an ordinary regression model:

$$y_i = \beta_0 + \beta_1 x_{1,i} + \beta_2 x_{2,i} + \ldots + \varepsilon_i = \sum_k \beta_k x_{k,i} + \varepsilon_i, \tag{12.1}$$

where y_i and x_{ki} are the response/dependent and kth explanatory/independent variable for sample/area i, respectively. β_k is the kth coefficient to be estimated; β_0 is the intercept, and we assume $x_{0,i} = 1$ for all i. In addition, ε_i is the error term that describes fluctuations not explained by the linear relationship of the explanatory variables. The error term is assumed to follow a normal distribution, with a mean of zero and variance of σ^2. The error distribution is represented by $N(0, \sigma^2)$.

In the GLM framework, a model has three components: the stochastic component of the prediction, the linear predictor using explanatory variables, and the link function that monotonically and continuously transforms the expectation of the response to the linear predictor. We can rewrite the ordinary regression model to represent these components as follows: the stochastic component is given by

$$y_i \sim N\left(\mu_i, \sigma^2\right) \tag{12.2}$$

for the systematic component, we have the linear predictor

$$\eta_i = \beta_0 + \beta_1 x_{1,i} + \beta_2 x_{2,i} + \ldots = \sum_k \beta_k x_{k,i} \tag{12.3}$$

and a link function

$$\mu_i = \eta_i \tag{12.4}$$

Equation (12.2) means that the prediction of the model follows a normal (Gaussian) distribution, with mean μ_i and variance σ^2. Equation (12.3) is the linear predictor that reflects the systematic component of the model prediction. Equation (12.4) tells us that the link function is the so-called 'identity function', which means that the expectation of the model's prediction on the response is the same as that of the linear predictor.

All GLM use linear predictors, but there are variations in the stochastic components and link functions. The most popular modes of GLM include Poisson regression for count data modelling and logistic regression for binary (dichotomous) data modelling. Table 12.1 summarises the three most popular GLM models.

With regard to the stochastic component, the model prediction is assumed to follow an exponential family distribution. The generic form of the distribution is as follows:

$$\log f(y_i \mid \eta_i, \phi) = \frac{y_i \eta_i - b(\eta_i)}{a(\phi)} + c(y_i, \phi) \tag{12.5}$$

where $f(y_i)$ is the probability distribution function that evaluates the likelihood of observation y_i, given its model-based estimate using the two parameters η_i and ϕ. Therefore, we refer to the distribution function, f, and its log transformation, $\log f$, as the likelihood and log-likelihood function, respectively. The function has three functional components, a, b and c, which are defined for each type of probability distribution.

TABLE 12.1 Three popular GLM models

Model	Gaussian regression (ordinary regression)	Poisson regression	Logistic regression
Type of response (range)	Continuous $(-\infty, \infty)$	Count $(0, 1,2,3, \dots)$	Dichotomous (0 or 1)
Probability function	Gaussian	Poisson	Bernoulli
$\log f(y_i)$	$-\frac{1}{2}\log(2\pi\sigma^2)$ $-\frac{1}{\sigma^2}(y_i - \mu_i)^2$	$y_i \log \mu_i - \mu_i$ $-\log(y_i!)$	$y_i \log(\mu_i)$ $+(1-y_i)\log(1-\mu_i)$
Deviance	$\Sigma_i(y_i - \mu_i)^2$	$2\Sigma_i\left(y_i \log \frac{y_i}{\mu_i} -(y_i -\mu_i)\right)$	$2\Sigma_i\left(\begin{array}{l} y_i \log \mu_i \\ +(1- y_i)\log(1-\mu_i)\end{array}\right)$
Link function	Identity $\mu_i = \eta_i$	Log $\log \mu_i = \eta_i$	Logit $\log \dfrac{\mu_i}{1 - \mu_i} = \eta_i$
Dispersion parameter ϕ	σ^2	1	1

Specification of GWGLM and S-GWGLM

GWGLM is a variant of GLM that includes coefficients that vary geographically. In this case, we replace equation (12.3), representing the linear predictor, with

$$\eta_i(u_i, v_i) = \beta_0(u_i, v_i) + \beta_1(u_i, v_i)x_{1,i} + \beta_2(u_i, v_i)x_{2,i} + \ldots = \sum_k \beta_k(u_i, v_i)x_{k,i} \quad (12.6)$$

In this model, the coefficients vary depending on the geographical coordinate of the sample geographical position i, (u_i, v_i). In the ordinary regression model, all coefficients model fixed/*global effects*, which remain spatially unchanged. However, these geographically varying coefficients model *local effects*.

There are three major modes of GWGLM that can be implemented in GWR4. Firstly, the Gaussian regression model is equivalent to conventional GWR. Here, equation (12.4) is replaced by $\mu_i = \eta_i(u_i, v_i)$.

Secondly, the geographically weighted Poisson regression (GWPR) is

$$y_i \sim \text{Poisson}[\mu_i] = \text{Poisson}[\exp(\eta_i(u_i, v_i))] \quad (12.7)$$

$$\log(\mu_i) = \eta_i(u_i, v_i) \quad \text{or} \quad \mu_i = \exp(\eta_i(u_i, v_i)) \quad (12.8)$$

Poisson$[\mu_i]$ denotes the Poisson distribution with mean μ_i. In the Poisson regression model, a zero or positive integer represents a response count. It is worth noting that the log link transformation ensures that the expected value is non-negative. The Poisson regression model often includes one additional term, called the offset, ρ_i:

$$\log(\mu_i) = \log \rho_i + \sum_k \beta_k(u_i, v_i)x_{k,i} \quad \text{or} \quad \mu_i = \rho_i \exp\left(\sum_k \beta_k(u_i, v_i)x_{k,i}\right) \quad (12.9)$$

The response count is usually examined by the ratio of the count to the size of the collected samples. For example, suppose the count is the number of deaths in a population in an epidemiology study. In this case, the offset term, ρ_i, can represent the expected size of the outcome or the size of the population at risk at position i. Therefore, we can interpret $\exp(\eta_i)$ as the rate of the outcome occurrence. For more information, see the application of GWPR/S-GWPR to spatial epidemiology in Nakaya et al. (2005).

Thirdly, geographically weighted logistic regression (GWLR) is given by

$$y_i \sim \text{Bernoulli}[\mu_i] \quad (12.10)$$

$$\log\left(\frac{\mu_i}{1 - \mu_i}\right) = \sum_k \beta(u_i, v_i)x_{k,i} \quad \text{or} \quad \mu_i = \frac{1}{1 + \exp\left(-\sum_k \beta(u_i, v_i)x_{k,i}\right)} \quad (12.11)$$

In this model, the response is either 0 or 1, so μ_i should be between 0 and 1. We obtain this condition using a logit function. Bernoulli$[\mu_i]$ denotes the Bernoulli distribution with parameter μ_i. This simply means that y_i is expected to be 1 with a probability of μ_i, and 0 with a probability of $1 - \mu_i$. For an application study, see Atkinson et al.'s (1993) application of GWGLM to geomorphology.

We refer to the semi-parametric variant of the models, S-GWGLM, as mixed GWGLM models, because both local and global effects are occur in one model:

$$\eta_i(u_i, v_i) = \sum_k \beta_k(u_i, v_i) x_{i,k} + \sum_l \gamma_l z_{i,l} \qquad (12.12)$$

where $z_{i,l}$ is the lth explanatory variable for area i, and its coefficient, γ_l, is assumed to be fixed. GWGLM is equivalent to the so-called partially linear models in local regression. One reason for using this kind of model is that we occasionally want to analyse varying effects by adjusting other effects. If we set all explanatory effects as global terms, except for the intercept term, we can consider the geographical distribution of local intercept estimates as spatially smoothed areal effects, adjusted by the other explanatory variables. It is also useful to judge if a term should be a local or global term, as we will discuss later.

Estimating coefficients of GWGLM and S-GWGLM

GWR estimates the local coefficients by repeatedly fitting the regression model to a geographical subset of the data using a geographical kernel weighting. Imagine a circle of a certain radius measured from the regression point at which the coefficients are to be estimated. Fitting the original regression model to the subset of the data within the circle provides estimates of the local coefficients. By repeating the local fitting of the regression model for the regression points, we obtain the set of local coefficients for the points. Instead of using a simple circular window, GWR uses a geographical kernel weighting, which yields a fuzzy local subset to generate a smoother surface for the local coefficients. In essence, the kernel weighting is used to evaluate the goodness of fit of the predicted response to the observed response, giving more importance to locations that are nearer to the regression point (cf. Figure 12.1).

GWGLM uses geographically weighted maximum likelihood estimation (GWMLE) to estimate the local coefficients of the ith regression point. To do so, it solves the following maximisation problem of the geographically weighted log-likelihood model:

$$\{\hat{\beta}_k(u_i, v_i)\} = \arg\max_j \sum_j \left\{ \log f(y_j \mid \eta_j(\{\hat{\beta}_k(u_i, v_i)\}), \hat{\varphi}) w_{ij}(d_{ij}) \right\} \qquad (12.13)$$

where the symbol $^\wedge$ refers to the estimate, $\{\hat{\beta}_k(u_i.v_i)\}$ is the set of estimated coefficients focusing on regression point i, and d_{ij} is the distance between locations i and j. The dispersion parameter of distribution ϕ is σ^2 in the case of the Gaussian model, while the Poisson and logistic regression models usually assume $\phi = 1$ a priori. The numerical process used to estimate the local coefficients is the geographically weighted variant of the scoring algorithm (Fotheringham et al., 2002).

The geographical weight of the jth observation at the ith regression point, w_{ij}, is introduced here as a non-negative and monotonically decreasing function of the distance between ith and the jth location. A popular choice is the fixed Gaussian kernel function

$$w_{ij} = \exp\left\{-\frac{1}{2}\left(\frac{d_{ij}}{G}\right)^2\right\}$$

(12.14)

where G is the bandwidth parameter that controls the local geographical extent of the weighting. Here, the fixed kernel means that a fixed G is applied for all local model fitting.

A popular alternative is the following adaptive bi-square function:

$$w_{ij} = \begin{cases} \left[1-\left(d_{ij}/B_i\right)^2\right]^2 & \text{if } d_{ij} < B_i \\ 0 & \text{otherwise} \end{cases}$$

(12.15)

Given that M is the bandwidth parameter applied for all local model fitting, the bandwidth, B_i, which is dependent on regression point i, is defined as the distance to the Mth nearest observation point from the regression point. Adaptive kernels are useful in preventing unreliable coefficient estimates because of the number of observations in local subsets. This is particularly true when the distribution of the regression points is highly uneven. We can also use adaptive Gaussian or fixed bi-square functions.

Estimation using S-GWGLM is more complicated than GWGLM. In essence, the method employs a 'back-fitting procedure' (Hastie and Tibshirani, 1990), which repeats the following steps until the estimation converges: (i) estimate the local coefficients given the global coefficients, based on GWMLE; and (ii) estimate the global coefficients given the local coefficients, based on the usual maximum likelihood estimation (Nakaya et al., 2005). It is also possible to derive indicators for model diagnostics, including standard errors for the coefficients and the degrees of freedom of the model based on the estimation principle.

The deviance is an indicator of the goodness of fit of the fitted model, based on the log-likelihood in GLM. The (residual) deviance of a fitted model, D_{fitted}, is defined as

$$D_{\text{fitted}} = -2\sum_i \left\{ \log f_{\text{fitted}}(y_i) - \log f_{\text{full}}(y_i) \right\} \tag{12.16}$$

This is the difference in the log–likelihood between the fitted model and the full/saturated model. The full model has a parameter for each observation, so it fits the data perfectly, but has no degrees of freedom. Here, a better fit between model and data means a smaller deviance. A related statistic, with a limited range from 0 to 1, is the pseudo R-squared or 'percentage of deviance explained':

$$PD = 1 - \frac{D_{\text{fitted}}}{D_{\text{null}}} \tag{12.17}$$

Here, D_{null} is the deviance of the null model, which only has a constant term, and no explanatory variables. In the case of the Gaussian model, this is equivalent to R-squared. If the model better fits the data, the value of PD increases.

Bandwidth selection

The bandwidth size regulates the GWGLM model complexity because the bandwidth controls the degree of variability of the estimated coefficients. An extremely small bandwidth means we repeat the local model fitting to a very small geographical subset of the data. In this case, the estimates of the coefficients are likely to fit the data well, but might be unreliable because the estimates will show large variances due to the lack of degrees of freedom (DOF) in the local model fitting. On the other hand, using a large bandwidth may ignore meaningful spatial variations in the coefficients when the true distribution varies spatially. In this case, a model using an excessively large bandwidth yields strongly biased estimates. Therefore, selecting a bandwidth involves a trade-off between the goodness of fit and DOF, or between the bias and variance of the model estimates.

One solution is to find the optimal bandwidth size using a model selection criterion, such as cross-validation (CV), Akaike's information criterion (AIC), or AIC corrected for a small sample size (AICc). This concept is shown in Figure 12.2. In particular, AICc is useful when modelling relatively small DOF, as often encountered in non-parametric regression, even for non-Gaussian models (Burnham and Anderson, 2002). AICc is defined as follows:

$$AICc = -2\log\left[\sum_i f\left(y_i \mid \eta_i(\{\hat{\beta}_k(u_i, v_i)\}), \hat{\varphi} \right) \right] + 2q + 2\frac{q(q+1)}{N-q-1} \tag{12.18}$$

where $\log\left(\sum_i f(y_i)\right)$ denotes the log–likelihood of the fitted model, representing its goodness of fit. The better the model fits the observed data, the smaller the log–likelihood. Here, q represents the number of parameters in the model. Given p as

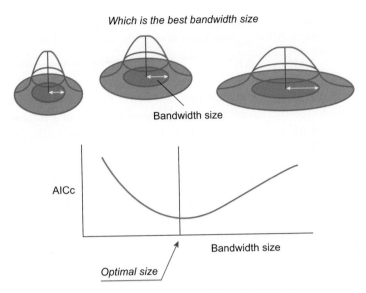

FIGURE 12.2   How to decide on the optimal bandwidth size

the total number of coefficients in the linear predictor, $q = p$ for Poisson and logistic regression models, and $q = p + 1$ for Gaussian models. This is because the dispersion parameter, σ^2, is estimated as $\hat{\phi}$. A smaller value of q means that the model is simpler, and hence has smaller DOF. The smaller the AICc of the model, the better the predictive performance of the model will be.

However, it is not straightforward to apply the criteria to GWGLM, because the geographic variability of coefficients is not specified as an explicit function with independent parameters. Therefore, effectively equivalent numbers of parameters are derived from local regression theories (Loader, 1999), and use the values for model selection (for details, see Nakaya et al., 2005).

Geographical variability test of coefficients

An advantage of S-GWGLM is that we can incorporate a fixed/global effect of a subset of the explanatory variables on the response variable based on prior knowledge. However, it is not always obvious which coefficients should be assumed as fixed or variable. A natural way to overcome this difficulty is to conduct empirical model comparisons of different S-GWGLM models using different combinations of fixed and varying coefficients.

To assess the variability of the kth coefficient in a GWGLM model, we compare two models; the fitted GWGLM model (pivot model) and the model in which only the kth coefficient is constant, while the other coefficients vary spatially (k-fixed model). If the pivot model is better than the k-fixed model according to a model comparison criterion such as AICc, we consider the variability of the kth coefficient to be statistically supported. If we have the AICc of the pivot model, $\text{AICc}_{\text{pivot: } k\text{-varying}}$, and the k-fixed model, $\text{AICc}_{k\text{-fixed}}$, we can calculate the difference between the two as

$$\Delta\text{AICc}_k = \text{AICc}_{\text{pivot: } k\text{-varying}} - \text{AICc}_{k\text{-fixed}} \qquad (12.19)$$

If this quantity is positive, the k-fixed model is better than the pivot model. In this case, it is better to use the kth term as a global term rather than a local term. When the absolute size of AICc difference is within 1 or 2, there is essentially no difference between the two. A rough rule of thumb is that, when the absolute difference is larger than 4, the judgement is more clearly supported (Burnham and Anderson, 2002).

When comparing the nested models, hypothetical testing is also possible. When the dispersion parameter for the GWPR and GWLR is set to 1, the difference in deviance between the two models follows an approximately chi-square distribution under the null hypothesis that there is no difference in the performance of the two models. The DOF of the chi-square distribution is the difference between the DOF of the two models:

$$\Delta D(k) = D_{k\text{-fixed}} - D_{\text{pivot}} \sim \chi^2_{\text{DOF}_{\text{pivot}} - \text{DOF}_{k\text{-fixed}}}. \qquad (12.20)$$

GWR4 computes the equivalent statistics for (S-)GWGLM. In the case of the Gaussian model, it is better to evaluate the stochastic error for the inference of the dispersion parameter. In this case, the F-statistics approximately follow an F distribution under the null hypothesis that there is no difference in the performance of the two models (Mei et al, 2006):

$$F = \frac{\left(D_{k\text{-fixed}} - D_{\text{pivot}}\right) \Big/ \left(\text{DOF}_{\text{pivot}} - \text{DOF}_{k\text{-fixed}}\right)}{D_{\text{pivot}} \big/ \text{DOF}_{\text{pivot}}} \qquad (12.21)$$

$$\sim F_{\text{DOF}_{\text{pivot}} - \text{DOF}_{k\text{-fixed}}, \, \text{DOF}_{\text{pivot}}}$$

In both hypothesis tests, when the null hypothesis is rejected, we can say that the term shows significantly geographical variation at the significance level used for the test.

Automated model building

GWR4 contains two separate fitting techniques for automated variable selection in the S-GWGLM models. One is the LtoG (from local to global term) variable selection routine, which executes a series of model comparisons to search for the optimal combination of varying and fixed terms, given the explanatory and response variables. The concept is similar to that of stepwise variable selection:

Step 1: Begin with an (S-)GWGLM model with local terms and varying coefficients as a pivot model.

Step 2: Try to fit a series of model by switching the varying local terms to global, one by one, and establish which switch gives the best performance in terms of the model selection criterion.

Step 3: If the best term from step 2 improves the model performance, the pivot model is updated by changing the relevant local term to a global term.

Step 4: Repeat steps 2 and 3 until no further improvement is possible by switching any terms from being local to global.

An alternative model selection procedure, also implemented in GWR4, is GtoL (from global to local term). This is the reverse procedure to that described above. In this case, the default model is the global model (or an S-GWGLM model), and the first round of model comparisons allows each parameter in turn to be spatially varying. Here, model selection occurs in the same way. In other words, we select an optimal model based on the model selection criteria, and then repeat this until no further improvement is possible.

A practical S-GWGLM analysis of place attachment using GWR4

To demonstrate the S-GWGLM framework, this section uses a case study to analyse place attachment using S-GWGLM modelling. Place attachment refers to the affective bonds that link people to a certain place (Lewicka, 2011). The concept is closely related to 'topophilia' and 'sense of place' in human geography (Tuan, 1974), and is associated with people's self-identity, well-being, and social participation. The data are taken from an internet-based questionnaire survey conducted in 2009 by a social survey company. The respondents relate to residents of the central Tokyo metropolitan area, consisting of 23 wards of Tokyo and the surrounding cities ($n = 1393$).

Although place attachment is a multi-dimensional construct, the indicator we use here is based on a simple question in the survey: 'Do you feel attached to the neighbourhood in which you live?' Respondents could indicate one of five choices {1, yes; 2, yes, weakly; 3, hard to say; 4, no, weakly; 5, no}. In this analysis, we focus on the following recoded dichotomous variable representing the response: PA (place attachment) = {0 if choice 3, 4, or 5 is selected; 1 if either 1 or 2 is selected}. Therefore, the research question examines the association between the place-attachment response and the other explanatory factors, and how this association varies geographically.

There are three explanatory variables in this case study. LIVLEN represents the length of time the respondent has lived in their current residence, and is defined as {0, less than 10 years; 1, greater than or equal to 10, but less than 20 years; 2, greater than or equal to 20 years}. The longer a person has lived in a place, the more likely it is that place attachment will occur. OWNH indicates if the respondent owns his or her house {0, no; 1, yes}. House ownership may enhance place attachment because a person who owns a house expects to be there for a long time. Therefore, we expect OWNH to be positively associated with place attachment. Finally, UPRE indicates whether the respondent prefers living in the suburbs or city centre {0, prefers living in suburbs or hard to say; 1, prefers living in city centre}. UPRE is based on the question 'If you could choose between living near the city centre and in the suburbs of the city, which would you prefer?' Since the study area is the core region of the metropolitan area, the preference for a lifestyle that depends on an urbanised environment is thought to be positively correlated to the place-attachment indicator. The result of the global logistic regression model generally supports these expected effects, though the effect of LIVLEN is insignificant. This may be because LIVLEN is correlated with OWNH.

$$\log\left(\frac{PA_i}{1-PA_i}\right) = 0.158 + 0.129 \text{ LIVLEN} + 0.597 \text{ OWNH} + 0.282 \text{ UPRE} \atop {\scriptstyle(0.116)} {\scriptstyle(0.076)} \qquad\qquad {\scriptstyle(0.124)} \qquad\qquad {\scriptstyle(0.116)}} \qquad (12.22)$$

The numbers in parentheses below the coefficients are the standard errors of the estimates. Note that the data file must contain the geographical coordinates. Although the coordinates in the sample data set are not precise residential locations, being randomly perturbed with spatial constraints (the maximal size of the dislocation is about 5.5 km), the results are essentially the same as when we use the original coordinates.

Considering S-GWLR models

The simplest GWLR model is a constant model without any explanatory variables. In this case, the estimator takes the following simple form:

$$\exp(\hat{\beta}_0(u_i, v_i)) = \frac{\sum_j w_{ij} y_j}{\sum_j w_{ij}(1 - y_j)} = \frac{\sum_j w_{ij} PA_j}{\sum_j w_{ij}(1 - PA_j)} \qquad (12.23)$$

In the case study, this is a local ratio of 'the number of people feeling place attachment' to 'the number of people feeling weak or no place attachment' around the point (u_i, v_i). The general distributional pattern of the local ratio using this estimator is a good way to start exploring possible geographical effects on weak place attachment among residents. However, some analysts may consider adjusting LIVLEN to see this geographical variation in case the effect of the length of residence is established. In this case, we use the following S-GWLR (Model 1):

$$\log\left(\frac{PA_i}{1-PA_i}\right) = \beta_0(u_i, v_i) + \gamma_L \text{LIVEN} \qquad (12.24)$$

where the effect of length of residence is set as a global term so that the geographically varying intercept term, $\exp(\beta_U(u_i, v_i))$, is considered as the ratio described earlier, but adjusted by the sample's length of residence. We can also view the model as a special logistic regression model by considering spatially autocorrelated unknown factors, which are captured by the varying intercept term.

To further explore any geographic variability in the explanatory variables, we can compare the above models with the following GWLR (Model 2):

$$\log\left(\frac{PA_i}{1 - PA_i}\right) = \beta_0(u_i, v_i) + \beta_L(u_i, v_i)\text{LIVLEN} + \beta_H(u_i, v_i)\text{OWNH} +$$
$$\beta_U(u_i, v_i)\text{UPRE} \tag{12.25}$$

The final model used here (Model 3) is obtained from the LtoG automated model building process, based on Model 2:

$$\log\left(\frac{PA_i}{1 - PA_i}\right) = \beta_0(u_i, v_i) + \gamma_L\text{LIVLEN} + \gamma_H\text{OWNH} + \beta_U(u_i, v_i)\text{UPRE} \tag{12.26}$$

where the effects of LIVLEN and OWNH are considered to be global. To interpret the coefficient of logistic regression, it is common to use the exponential form of the coefficient (e.g. $\exp(\beta_U(u_i, v_i))$, which represents the *odds ratio* (OR) of the response, shown as $PA_i/(1 - PA_i)$ for an increase of +1 in the explanatory variable, x_k.

Using GWR4

GWR4 was developed to implement S-GWGLM modelling with a GUI-based interface. It uses tabbed sub-windows so that a modelling session proceeds intuitively in a step-by-step manner. The program also offers a wider range of options related to GWGLM, including geographical variability assessment and the automated variable selection routines explained earlier. The program operates within a Microsoft Windows environment and is based on the .Net Framework 4.0.

GWR4 consists of five steps when running a model: Data, Model, Kernel, Output, and Execute. Each step has a corresponding tab (see Figure 12.3). The user first opens the data file in the Data tab, and then moves to the Model tab to specify the model. Figure 12.4 demonstrates the various settings available in the Model tab for Model 4. The dataset opens in the File tab, and the list of fields/variables appears in the centre box. If you wish, you can move fields to other boxes to specify the response/dependent variable, the explanatory variables as geographically varying (local term) or fixed (global term), and at least two geographical coordinates. In the case of GWGLM, the model type should be changed from Gaussian to logistic. Geographical variability testing and automated model building are both optional.

Step1: Data>

Start your session by giving it a title then open your data file.

Step2: Model>

Specify one regression type and the variable settings needed for GWR modelling.

Step 3: Kernel>

Choose a geographic kernel type and a method to search for optimal bandwidth size.

Step 4: Output>

Specify filenames for the files storing the modelling results.

Step 5: Execute>

Execute the session to compare necessary calculations and read results.

FIGURE 12.3   The steps in a GWR modelling session

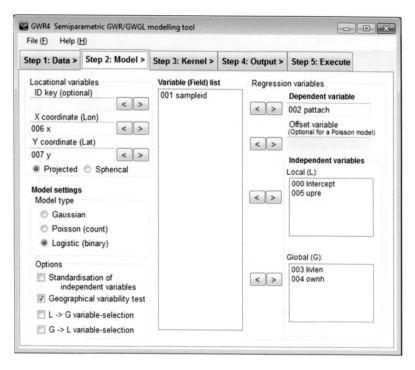

FIGURE 12.4   Screenshot of the specification in the Model tab

Since an S-GWGLM model requires a much longer computational time than the corresponding GWGLM model, these options may take a relatively long time to complete, particularly when implemented as part of automated model building.

In the Kernel tab, the default setting is to use an adaptive bi-square kernel and to find the optimal bandwidth using the 'golden section search' option, based on AICc. However, these settings can be changed. Although the golden section search is a useful way to automatically find the optimal bandwidth, it is not always straight-forward to find adequate initial bandwidths. Since binary data is less informative than continuous and count data, model fitting using a small bandwidth often fails to converge. In a small subset of data, the variation in the response and explanatory variables is likely to be too small for model fitting. GWR4 automatically adjusts the search range using trial and error. This might be time-consuming in the case of GWLR. Therefore, we recommend that a user specifies the minimum bandwidth for the search after several trials. In the sample data analysis, the nearest 400 samples for an adaptive bi-square kernel are used as the minimum search range for the bandwidth, as shown in Figure 12.5. Note that, when the optimal bandwidth in the routine is the minimum bandwidth size, the software generates a warning message in the summary output files.

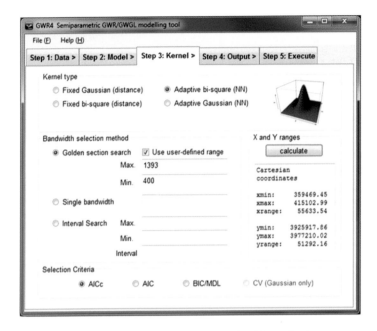

FIGURE 12.5 Screenshot of the specification in the Kernel tab

In the Output tab, the user specifies a file name for the *session control file*. This file saves the settings entered in the previous tabs, including the data and geographic kernel information. The file can be reused later by opening it from the File menu or running the model from the command line. Finally, to run the model, click the 'Execute' button in the Execute tab. The output is displayed on the screen, and is stored in the files specified in the Output tab.

As noted earlier, there are two output files that contain the modelling results. The summary file contains the general summary information of the fitted model, and comprises:

(i)   the settings for the modelling session;
(ii) the result of the global regression;
(iii) the result of the bandwidth selection;
(iv) the result of the GWR/GWGLM model;
(v)   the list of estimated fixed coefficients, if any global terms were included;
(vi) the summary table of estimated geographically varying coefficients, if any local terms were included;
(vii) an ANOVA table that compares the global and GWR/GWGLM models.

The supplementary material provides a sample of the summary file. The indicators used to evaluate the statistical performance of the fitted model are available in part (iv) of the summary file. Table 12.2 summarises the indicators for the fitted models in the case study.

If the user selected a variability test and automated model building, these results are added to the bottom of the summary. For example, Table 12.3 shows the

TABLE 12.2   Summary of fitted geographically weighted logistic regression models

Model	Global terms	Local terms	Optimal bandwidth	AICc	Effective number of parameters	Percent of deviance explained
Global model	Intercept LIVLEN RNET UPRE		Not applicable	1715.5	4	0.023
Model 1	LIVLEN	Intercept	403	1726.1	10.5	0.024
Model 2		Intercept LIVLEN OWNH UPRE	489	1706.6	29.8	0.058
Model 3	LIVLEN OWNH	Intercept UPRE	403	1694.5	20.4	0.054

reported result of the geographically varying test when applying Model 3. In this case, the coefficients all vary geographically. The difference in the AICc, shown as 'Difference of model selection criterion (AICc)' in Table 12.3, suggests that the coefficients of LIVLEN and OWNH are better as fixed, while the intercept and the coefficient of UPREF are better as geographically varying. Rounding up the fractions in the difference in the DOF, we can use the chi-square distribution with DOF = 7 for the hypothetical test. The rejection region of the chi-square test with DOF = 7 and a significance level of 0.05 is $\Delta D(k)>14.07$, according to chi-square distribution tables. Based on $\Delta D(k)$, shown as 'Diff of deviance, $\Delta D(k)$' in the table, we can conclude that the geographic variability of the intercept and the coefficient of UPREF are statistically significant at the 5% level.

The geographic listwise output file stores the estimated coefficients, their standard errors, and local model diagnostic indicators, including the local version of percentage

TABLE 12.3   Statistics for the geographical variability tests for local terms of the geographically weighted logistic regression model (Model 2)

Term (k)	Difference of deviance, $\Delta D(k)$	Difference of degree-of-freedom	Difference of model selection criterion ($\Delta AICc_k$)
Intercept	24.04	6.67	−10.16
LIVLEN	8.74	6.45	4.58
OWNH	6.33	6.50	7.19
UPRE	25.15	6.10	−12.46

deviance explained. Figure 12.6 shows the length-of-residence adjusted ratio for the place attachment rate as the distribution of $\beta_0(u_i, v_i)$, using Model 1. The figure shows a general trend that the values are high in the central and western part of the region. In other words, people living in these regions are more likely to have strong place attachment than those in other regions. The distribution of the z-value (pseudo t-value), defined as the local coefficient divided by its standard error, is shown using contour lines on the map. Broadly speaking, the values are mostly above 1.98 or below −1.98, which suggests that $\hat{\beta}_0(u_i, v_i)$ is not equal to zero at the 5% level.

Model 3 attains the lowest AICc and is considered the best model in this analysis. Using this result, Figure 12.7 shows the distribution of $\hat{\beta}_U(u_i, v_i)$ and the effects of

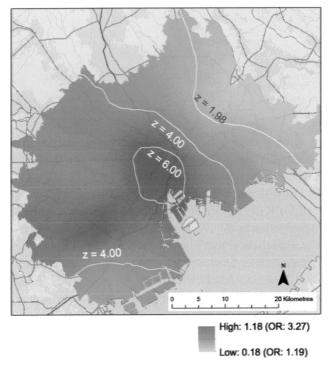

FIGURE 12.6 Distribution of the intercept parameter of Model 1 (OR: local odds ratio of people feeling place attachment adjusted by the length-of-residence effect)

UPREF, which exhibits significant geographical variability. The value in the central part of the study area is clearly positive, indicating that people who prefer living in a city centre are likely to exhibit place attachment. On the other hand, the peripheral parts of the study region show negative values. This may reflect that people living in peripheral regions perceive the neighbourhood environment as being less urbanised, and those who aspire to a city-centre lifestyle are therefore likely to be less satisfied.

This case study shows that the development of an affective feeling of bonding with a living area is inhibited if there is a disagreement between the expected and experienced neighbourhood environment. In addition, the case study also shows the utility of the S-GWLR, which enables us to infer the geographic context associated with such a psychological disagreement about a living place.

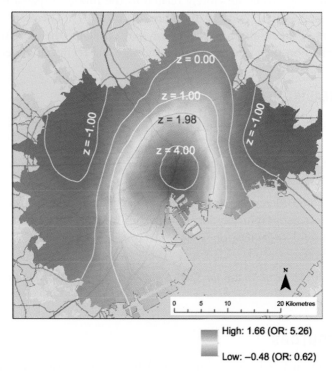

FIGURE 12.7 Distribution of local effects of UPRE on weak place attachment according to Model 3

Conclusion

This chapter has introduced the GLM-based semi-parametric geographically weighted regression (S-GWGLM) modelling framework. This framework allows us to mix geographically varying and fixed coefficients in a generalised linear model. It also allows us to explore which explanatory terms should be varying or fixed by comparing possible S-GWGLM models. GWR4 is a computer application that was developed as a platform to implement S-GWGLM modelling. The software includes the new methods of assessing geographical variability for estimated coefficients and automated model selection, enabling the user to search for an optimal combination of fixed and varying explanatory terms in a model. While GWR4 currently only fully supports S-GWGLM functions, other recent software applications are equipped with other types of advanced GWR functions. For example, ArcGIS provides a function to integrate GWR model fitting into its cartographic display, and the R packages spgwr and GWmodel provide a wide array of GWR variants and related tools to manage several problems in local modelling, such as local collinearity or outlier problems. Note that geographically local properties explored using these tools are closely related to the mode of spatial thinking that emphasises geographical context (Fotheringham, 1997). Therefore, meaningful interpretations of the results from these tools may challenge the user's spatial thinking ability, based on their geographical knowledge and technical understanding.

FURTHER READING

Fotheringham et al. (2002) provides a most comprehensive guide of GWR and its extensions:

Fotheringham, A.S., Brunsdon, C. and Charlton, M. (2002) *Geographically Weighted Regression*. Chichester: Wiley.

GWR can be considered as a type of 'local spatial analysis' which has been proposed to extract geographically local properties in geospatial information. Lloyd (2011) is a useful overview on a wide array of modern local spatial analyses including GWR:

Lloyd, C.D. (2011) *Local Models for Spatial Analysis*, 2nd edn. Boca Raton, FL: CRC Press.

Web resources

Information about GWR, including the GWR4 download sites, is available at the following GWR portal sites: https://geodatacenter.asu.edu/gwr_software (GeoDa Center for Geospatial Analysis and Computation, Arizona State University) and http://ncg.nuim.ie/ncg/GWR (National Centre for Geocomputation, National University of Ireland, Maynooth).

ACKNOWLEDGEMENTS

This work was supported by JSPS KAKENHI Grants 20298722 and 24300323. The author wishes to thank Stewart Fotheringham (Arizona State University), Martin Charlton (National University of Ireland), Chris Brunsdon (National University of Ireland), Paul Lewis (National University of Ireland), and Jing Yao (University of St Andrews) for their kind support and collaboration on the development of GWR4.

13

SPATIAL INTERACTION MODELS

Karyn Morrissey

Introduction

Both conceptual and analytical methods for evaluating spatial flows have traditionally been based on how distance and transportation options impede an individual's travel decision-making. This chapter outlines the method of spatial interaction modelling, which allows a full set of complexities involved in travel decisions to be encompassed within a single framework. Such complexities might include the demographic and socio-economic profile of the populations under study, the profile or 'attractiveness' of the end destination, and travel cost – in terms of distance travelled and any impedances of the transport network. This chapter introduces those steps and data required to model flows of individuals from one destination to the next using spatial interaction models. A worked example in the spreadsheet package Excel illustrates a spatial interaction model estimating trips to hypothetical medical services in Liverpool.

An important requirement of product and service success, in both the private and public sector, is an understanding of how, where and when consumers access goods and services. Thus, the spatial flows of individuals to and from particular locations are important for planning likely users of new facilities or the demand for a certain product such as a new community health centre or a new retail park. Initial studies on spatial flows (or, more broadly, accessibility) attempted to assess flows in terms of transport opportunities alone (Jones, 1981). However, how an individual chooses to move across space is a much broader concept than transport availability alone. The work of Moseley (1979) moved away from the idea of mobility as a proxy for spatial flows and defined trip choices in terms of accessibility. Moseley (1979) asserted that accessibility is more than just the transport modes or distance involved in reaching a particular service; accessibility is also about the attributes of the people wishing to access a service, the attributes of the destinations and the means by which these opportunities are reached. Thus, to fully understand spatial flows, one must identify three components: the people wishing to travel, the potential places/opportunities to be visited and the distance/separation of people from these places/opportunities.

Further work by Joseph and Phillips (1984) classified access according to two categories, potential and revealed access. Potential access refers to a person's ease of accessing these services based on existing conditions but does not warrant the utilisation of the service. Revealed accessibility, based on potential access, focuses on the actual use of services (Joseph and Phillips, 1984). Both types of access can be further classified into spatial and non-spatial access (Luo and Wang, 2003; Wan et al., 2012; Bissonnette et al., 2012) based on how the accessibility is influenced by spatial factors (e.g. spatial location and travel distance) and non-spatial factors (e.g. socio-economic status, or cultural background). Thus, to model spatial flows that represent a realistic pattern of human behaviour, how individuals chose to access services, both spatial and non-spatial considerations must be taken into account. Spatial interaction models (SIMs) involve determining through demand, supply and interaction information the attributes that promote flows of people and goods between different locations.

Modelling framework

Spatial interaction models have a long tradition of being used to estimate flows of people to service outlets (Birkin and Clarke, 1991; O'Kelly, 1986; Yano et al., 2000). For different consumers and destination choices, one can model the trade-off between spatial convenience (visiting an outlet close by) and the attractiveness of particular outlets (measured by proxies such as size, brand and quality of the service). Although there is a classical family of SIMs (Wilson, 1974), one model, the production-constrained SIM, has dominated the literature (Fotheringham et al., 2001; Clarke et al., 2002). Such models allow the user to estimate the trip end totals (revenue in shopping models, number of patients attending hospital or surgery in health-care models) and may be used by planners to locate or reconfigure new services. Thus, a production-constrained SIM may be used as a location model (Clarke and Wilson, 1994) From the outputs of these models, it is also possible to build a suite of accessibility indicators that measure how well served residents are for services under consideration (Clarke and Wilson, 1994; Clarke et al., 2002; Morrissey et al., 2008). Thus, such models can be used to quantify accessibility according to where individuals consume services as predicted by the SIM, and as such, provide a more realistic representation of access to services than, for example, simply taking the number of service outlets in a zone or estimating accessibility through a simple straight-line nearest facility type indicator.

The steps required to model an SIM and develop a suite of accessibility indicators may be broken down into six major processes:

- problem definition;
- model specification and data collection;
- travel impedance measurement;
- calculation of accessibility measures;
- interpretation and evaluation of the results;
- visualisation of accessibility values.

The first step is to define those spatial flows that are to be estimated. For example, do they relate to the number of potential trips to a new supermarket, or the demand for hospital services in a region? Within the model specification stage, this involves identifying what data you are going to use, both spatial and non-spatial. Spatial interaction analysis is typically a data-intensive process, and four pieces of information need to be included:

- definition of the spatial unit for analysis;
- definition of the socio-economic groups involved;
- the attributes of the destination;
- available modes of transport.

Sources usually include secondary data such as the census of population (as used in the example presented later in the chapter) and/or survey data. However, a key requirement of any dataset used for SIMs is that the data contain spatial referencing, such as an individual's address or attached nested zone (census tract, etc.). Examples of SIMs relate to a range of spatial scales from the local (e.g. facilities in the neighbourhood), to regional (e.g. cities and their hinterlands) and inter-regional levels (e.g. connectedness of a region or country).

In practice the selected scale needs to be linked to the planning and policy decisions required at each level: within communities, neighbourhoods and larger administrative areas. For example, an aim of using SIMs to improve accessibility plans at the regional level does not always lead to improved accessibility at the local level. This means that a policy that is to be implemented at the regional level needs to have its ramifications examined at the local level to ensure that it does not hinder local level accessibility.

With regard to aspatial factors, it is not possible to identify every need for every group since there are potentially hundreds of combinations of socio-economic groups. However, to maximise the usefulness of spatial interaction modelling in policy and planning, it is important to categorise people and places as thoroughly as possible. Spatially referenced demographic and socio-economic data are typically available through the census of population for a country, or modelled surrogates such as those derived from large-scale rolling surveys (e.g. the American Community Survey). The types of opportunities offered at the destination sites are assessed through a variety of mechanisms depending on application area and include the prevalently used square footage of a retail park or the availability of certain opportunities at the destination (e.g. a course of higher education, or type of retail). The goal of any transport system is not mobility *per se*, but access to facilities. Private transport by car will invariably offer the best means of accessibility in many areas. However, there are a multitude of transport methods, from private car to public transport, walking and cycling. These options should be considered in a well-executed spatial interaction analysis.

The worked example illustrated in this chapter involves creating an SIM to estimate flows of individuals to five hypothetical medical practices across 12 output areas (OAs) in Liverpool. An OA corresponds to around 15 households. Thus, data at the OA level are required, and the best source of information about the characteristics of these zones is found within the Census of Population. Data on the number of doctors in each medical practice may be taken from the medical practices' websites (attractiveness of the destination) or health-service databases. Travel impedance for the purpose of this worked example used Euclidean distance to calculate distance between the centroid of each OA and each of the medical practices, although road network distances could also be used, as these take travel conditions into account.

Travel impedance

Travel impedance represents the spatial separation between an origin, i, and a destination, j. Travel impedance can be measured in terms of travel distance, time or cost estimated by straight-line distance or network distance (Liu and Zhu, 2004). Due to the complexity of travel behaviour and data limitations, it is not always practical to have accessibility measures with a full range of travel options. Indeed, due to the complexities of measuring public transport, it is often assumed that an individual is travelling by private car (Liu and Zhu, 2004).

Factors that lead to travel impedance can be categorised under five headings: spatial, physical, temporal, financial and information. Spatial barriers relate to the distances involved in accessing required goods and services (Kwan and Weber, 2003). Time is an integral element of individual accessibility. This refers not only to the amount of time available to an individual for carrying out travel and activities, but also to the scheduling of activities throughout the day (Kwan and Weber, 2003). Thus, temporal barriers to accessibility arise within two contexts: firstly, when there is a mismatch between service times; and secondly, when the required travel times exceed some maximum threshold of practicability and acceptability. Temporal accessibility can be greatly improved by scheduling service delivery and transport provision jointly. In debates about the financial cost of travel as a factor affecting accessibility, the emphasis is often on affordability. As demand for public transport declines in rural areas, user costs increase, leading to increases in fares.

Travel costs are a more significant barrier to access for some groups than others. Access to low-wage employment will only be practical if fares are low enough to make employment viable. As a result, people on low incomes tend to work closer to home. As incomes are lower in rural areas in general, this can lead to issues such as rural deprivation and social exclusion. Physical accessibility barriers are often perceived as being the easiest to understand, and they are classified in terms of the assistance that an individual needs to make a journey.

Calculating spatial flows using spatial interaction models

SIMs involve determining through demand, supply and interaction information those attributes that promote flows of people and goods between different locations. For different consumers, one can model the trade-off between spatial convenience (visiting an outlet close by) and the attractiveness of particular outlets (measured by proxies such as size, brand and quality of the service). SIMs differ in form from other model-based approaches to accessibility analysis in that they may be derived via entropy-maximising techniques (Wilson, 1974) or contingency table theory (Willekens, 1983), rather than Newton's law of gravitational attraction. Entropy-maximising models are commonly used to find the most probable numbers of pairings x_{ij} between locations i and j given the number of origins, O_i, in location i and the number of destinations, D_j, in location j. Thus, the most probable macro-distribution is one that replicates the maximum number of micro-level events (Roy and Thill, 2004). Contingency table theory focuses on the pattern of association among variables and cross-classifies these interactions in a table of spatial interaction flows. Contingency tables are generally calculated via multivariate analysis such as log-linear models or logistic regression models.

There are four types of SIM commonly in use:

- Destination-constrained SIMs assume that the attributes of the supply point are known, that is, the locations of various service providers (supply points).
- Origin-constrained SIMs assume that the attributes of the demand point are known, that is, the location of households (demand points).
- Doubly constrained models assumes that the both the supply and demand point attributes are known.
- Unconstrained models assume that neither demand nor supply attributes are known.

Among these, the origin-constrained SIM is by far the most popular (Clarke et al., 2002; Morrissey et al., 2008). Models of this type allow the user to estimate the trip end totals (revenue in shopping models, the number of patients attending hospital in health-care models) and thus serve as location models. From the outputs of these models it is possible to build a suite of performance indicators to measure how well served residents are for the service under consideration (Clarke and Wilson, 1994; Clarke et al., 2002). Thus, these models quantify accessibility according to where they predict individuals consume services, and, as such, provide a more realistic representation of access to services than, for example, simply taking the number of service outlets in a zone or estimating accessibility through a simple straight-line nearest facility type indicator.

An example of an origin-constrained spatial interaction model is the following, which includes a worked example in Excel to estimate trips to a medical practice:

$$T_{ij} = O_i A_i W_j \exp(-\beta d_{ij}) \qquad (13.1)$$

where T_{ij} is the flow of individuals from residential zone i to each service centre j, O_i is the demand for the service in a predefined spatial area i, such as a postcode or a ward, W_j is the attractiveness of outlet j, d_{ij} is the distance from the origin i to the destination j, β is a distance decay parameter and A_i is a balancing factor that ensures that

$$\sum_j T_{ij} = O_i \qquad (13.2)$$

A_i is calculated as

$$A_i = \frac{1}{\sum_j W_j \exp(-\beta d_{ij})} \qquad (13.3)$$

The demand side (zone i) is usually represented as households or individuals aggregated into the smallest geographical output level available within the dataset. The supply element of the SIM represents the attractiveness of any given destination.

In this chapter the SIM described above is used to measure access scores from each of the OAs to their nearest medical centre using hypothetical data for Liverpool and a model developed in Excel. For the purpose of this chapter the attractiveness parameter for each health-care centre, W_j, is the number of practitioners in each health centre (a measure of how easy it is to be examined quickly). The demand variable, O_i, is the potential demand for medical services given the number of individuals with long-term illness in a particular OA. The distance variable, d_{ij}, is the distance from each OA centroid, i, to each medical practice, j. In the worked example in Excel, distance is calculated using the formula for Euclidean distance. However, it is important to note that the most accurate method of calculating D_{ij} is too use a network analysis tool in GIS. This allows road distance and increasingly temporal aspects to commuting such as congestion to be estimated for each origin–destination pairing. Figures 13.1–13.3 represent a worked example using our medical example within an SIM created using Excel. The spreadsheets used in this example are available as an online resource as part of this book. An additional spreadsheet presenting the formulas for each for each of the cells is also available, to aid students to develop their own SIMs.

It is possible to derive accessibility indicators from the predicted levels of inter-action as calculated by an SIM. Such indicators can quantify accessibility according

		G	H		J	K	L	M	N	O	P
1	Formula: Tij=Oi*Ai*Wj exp(-β dij)										
2	β= 0.2				*j* - destination						
3					L6	L2	L9	L7	LI2		
4			Number of GPs in each postcode	gps in L6	gps in L2	gps in L9	gps in L7	gps in L12			
5			Wj	9	10	23	12	27			
6				20	20	20	20	20			
7	*i* - residence										
8	postcode	Oi							SUM	DIFF	
9	L1	9		1.14	5.79	0.01	1.97	0.10	9.00	0.00	
10	L2	2		0.08	1.80	0.00	0.11	0.01	2.00	0.00	
11	L3	12		1.40	8.91	0.04	1.54	0.11	12.00	0.00	
12	L4	24		8.37	3.41	3.55	5.79	2.88	24.00	0.00	
13	L5	14		8.15	1.32	0.70	4.21	1.61	14.00	0.00	
14	L6	22		12.09	0.68	0.08	8.24	0.91	22.00	0.00	
15	L7	24		6.23	0.67	0.04	16.26	0.79	24.00	0.00	
16	L8	29		6.79	3.17	0.05	17.42	1.57	29.00	0.00	
17	L9	5		0.01	0.00	4.95	0.00	0.04	5.00	0.00	
18	L10	8		0.03	0.00	7.19	0.03	0.75	8.00	0.00	
19	L11	14		0.10	0.01	9.87	0.09	3.93	14.00	0.00	
20	L12	17		0.14	0.01	0.10	0.16	16.60	17.00	0.00	
21				42.53	25.78	26.58	55.82	29.29	180.00	0.00	
22	Oi = percentage of individuals with a long term illness in each postcode										

FIGURE 13.1   SIM output calculation in Excel

	A	B	C	D	E	F	G	H	I	J
1	exp (-βdij)		1							
3		L6	L2	L9	L7	LI2				
4	*i*									
5	L1	0.060326	0.275044	0.00022	0.07789	0.001685				
6	L2	0.050955	1	0.000381	0.049116	0.001278				
7	L3	0.082109	0.471738	0.000805	0.068101	0.002177				
8	L4	0.078447	0.028799	0.013026	0.040715	0.008997				
9	L5	0.198344	0.03836	0.008898	0.101888	0.017351				
10	L6	1	0.050955	0.002585	0.511121	0.024973				
11	L7	0.511121	0.049116	0.001408	1	0.021617				
12	L8	0.079614	0.033481	0.00022	0.153174	0.006156				
13	L9	0.002585	0.000381	1	0.001408	0.007012				
14	L10	0.002368	0.000219	0.198207	0.001416	0.01762				
15	L11	0.002252	0.000171	0.082924	0.001437	0.028111				
16	L12	0.024973	0.001278	0.007012	0.021617	1				
20	dij									
21	FROM									
22					L6	L2	L9	L7	LI2	
23				x	336906	334092	338020	337088	340300	
24				y	391352	390381	397205	390706	392800	
25		X	Y							
26	L1	334907	389380		2.8	1.3	8.4	2.6	6.4	
27	L2	334092	390381		3.0	0.0	7.9	3.0	6.7	
28	L3	334424	391055		2.5	0.8	7.1	2.7	6.1	
29	L4	335652	393567		2.5	3.5	4.3	3.2	4.7	
30	L5	336246	392829		1.6	3.3	4.7	2.3	4.1	
31	L6	336906	391352		0.0	3.0	6.0	0.7	3.7	
32	L7	337088	390706		0.7	3.0	6.6	0.0	3.8	
33	L8	337114	388830		2.5	3.4	8.4	1.9	5.1	
34	L9	338020	397205		6.0	7.9	0.0	6.6	5.0	
35	L10	339580	396774		6.0	8.4	1.6	6.6	4.0	
36	L11	340366	396371		6.1	8.7	2.5	6.5	3.6	
37	L12	340300	392800		3.7	6.7	5.0	3.8	0.0	

FIGURE 13.2   Calculation of distance using Euclidean distance (d_{ij}) and the distance decay parameter exp($-\beta d_{ij}$) for the SIM

Ai calc

1/[Wj1 exp(-Bdij)+Wj2 exp(-Bdij)...]

	L6	L2	L9	L7	LI2
Wj	9	10	23	12	27
	20	20	20	20	20

i below are the exp(-0.2*dij) values multiplited by the Wj factor at the head of each column.

PC sector:	L6	L2	L9	L7	LI2	SUM	1/SUM
L1	10.85874686	55.00882344	0.101240381	18.69361	0.909901114	85.57232	0.011686021
L2	9.171861482	200	0.175075575	11.78777	0.68999229	221.8247	0.004508064
L3	14.77963572	94.34754169	0.370486696	16.34424	1.175785125	127.0177	0.007872919
L4	14.12038844	5.75976302	5.992060347	9.7716	4.858404483	40.50222	0.024690007
L5	35.70185125	7.672006513	4.093148951	24.45323	9.369553248	81.28979	0.012301668
L6	180	10.1909572	1.189051479	122.6691	13.48517346	327.5343	0.003053115
L7	92.00184039	9.823144173	0.647743221	240	11.67300737	354.1457	0.002823696
L8	14.33055638	6.696209965	0.100999864	36.76178	3.324221646	61.21377	0.016336194
L9	0.465281013	0.076119815	460	0.337953	3.7866661	464.666	0.002152083
L10	0.426319117	0.04384233	91.17515928	0.339868	9.515022114	101.5002	0.009852194
L11	0.405309521	0.034185111	38.14492648	0.344935	15.17970409	54.10906	0.018481193
L12	4.495057821	0.2555527	3.225678529	5.188003	540	553.1643	0.001807781

exp (-βdij)

i	L6	L2	L9	L7	LI2
L1	0.060326371	0.275044117	0.000220088	0.07789	0.001685002
L2	0.050954786	1	0.000380599	0.049116	0.001277764
L3	0.082109087	0.471737708	0.000805406	0.088101	0.00217738
L4	0.078446602	0.028798815	0.013026218	0.040715	0.008997045
L5	0.198343618	0.038360033	0.00889815	0.101888	0.017351025
L6	1	0.050954786	0.002584895	0.511121	0.024972543
L7	0.511121336	0.049115721	0.001408137	1	0.02161668
L8	0.079614202	0.03348105	0.000219565	0.153174	0.006155966
L9	0.002584895	0.000380599	1	0.001408	0.007012345
L10	0.00236844	0.000219212	0.198206868	0.001416	0.017620411
L11	0.00225172	0.000170926	0.082923753	0.001437	0.028110563
L12	0.024972543	0.001277764	0.007012345	0.021617	1

FIGURE 13.3 Calculation of SIM attractiveness parameter (A_i)

to where individuals travel to (as predicted by the SIM) and, as such, provide a more realistic representation of access based on the movements of individuals rather than simply on the geographical distribution of service outlets within a zone. In this section, two model-based performance indicators are described, and were first introduced by Clarke and Wilson (1994). The first type of performance indicator relates to individuals and households. These indicators are based on residential location and relate to the ways in which the individual is served by facilities. As such, these indicators can be used to estimate the effectiveness of service provision to individuals. The second type of performance indicator relates to service providers and the specific services that they provide. Thus, these indicators measure the efficiency of provision by service providers.

By identifying spatial variations in effectiveness and efficiency of provision, such performance indicators allow the targeting of resources to increase the efficiency of a service facility or to increase the effectiveness of service provision to households. To analyse accessibility of rural services, one needs to use both the efficiency and effectiveness indicators together. Using both indicators to analyse spatial interaction provides an understanding of how households and facilities are interdependent. Thus, these indicators allow planners to examine whether the problem is in the effectiveness of delivery to residential locations or in the efficiency of provision at facility locations.

The following model-based indicators measure the effectiveness of service delivery to residential areas. As such, they measure the aggregate level of provision and the level of provision per household, respectively. These indicators are important because the effectiveness of a provider in delivering its services is based on its size and location. The aggregate level of provision for a particular origin zone i is given as

$$W_i = \sum_j \frac{T_{ij}}{T_{\star j}} W_j \tag{13.4}$$

This equation for estimating the aggregate level of provision for an area is calculated by dividing each SIM output (equation (13.1)) by the sum of all outputs for each zone j, where $\star$ indicates summation across all zones i. This is then multiplied by the attractiveness of zone j. The sum of all these values for residence zone i provides the aggregate provision for each zone i. This indicator ensures that even if an area does not have a service facility, the area will not have a zero accessibility score (unlike traditional indicators). In order to identify areas with low service provision it is necessary to relate this level of provision to the number of households in each area. For example, if service provision for a particular area is low, but population is also low, then the area may not be classified as a problem zone. On the other hand, if an area has relatively low provision and population is high, then the results of the model will identify it as a problem area. Also, because of the nature of this performance indicator, it is possible that an area with high provision and high population can still appear to be relatively poorly served because the indicator is a measure of the share of a facility that a residence area has (Clarke et al., 2002). Relating this aggregate provision indicator to population in an area will allow the identification of areas where a significant number of households suffer poor accessibility to a particular service. Figure 13.4 presents the calculation of the service provision indicator for our worked example. Using the outputs from the SIM, Figure 13.4 calculates the level of service provision for each of the 12 OAs in Liverpool to a number of hypothetical medical practices.

	A	B	C	D	E	F	G	H
1			L6	L2	L9	L7	LI2	
2		Number of GPs per practice	12	15	23	12	27	
3			20	20	20	20	20	
4	PC sector:							Aggregate Service Provision Accessibiliy Indicator
5	L1	Service Provision Accessibiliy Indicator	6.444301	67.32544	0.184281	8.453683	1.764266	84.17197
6	L2		0.466621	20.98397	0.027319	0.456978	0.11469	22.04958
7	L3		7.878938	103.7254	0.60577	6.639347	2.047886	120.8974
8	L4		47.21348	39.71685	61.45072	24.89869	53.07461	226.3524
9	L5		34.69523	15.37583	12.20023	18.10805	29.74895	110.1283
10	L6		68.22214	7.965601	1.382245	35.42784	16.69872	129.6965
11	L7		35.18136	7.746712	0.759715	69.9333	14.58387	128.205
12	L8		38.30883	36.91808	0.82811	74.88399	29.03351	179.9705
13	L9		0.028251	0.009532	85.86536	0.015836	0.751185	86.46997
14	L10		0.189603	0.040212	124.3706	0.115187	13.8259	138.5415
15	L11		0.591741	0.102928	170.8099	0.383742	72.40725	244.2956
16	L12		0.779502	0.091393	1.715672	0.885549	305.9492	309.2213
17								148.3333
18								

FIGURE 13.4 Service provision accessibility indicators for hypothetical medical practices in Liverpool

The level of provision per household is an indicator that divides the aggregate level of provision score by the number of households in the residence zone:

$$v_i^m = \frac{W_i^m}{H_i^m} \tag{13.5}$$

where H_m^i are households of different household types m. Similarly, a catchment population indicator can be calculated as

$$C_i = \sum_i \left(\frac{S_{ij}}{S_i} \right) \times P_i \tag{13.6}$$

where S_i is expenditure in area i, S_{ij} is expenditure from area i in area j, and P_i is the population in area i. In equation (13.5), a typical term on the right-hand side involves taking the proportion of provision at j which is used by residents of i and then summing to obtain a measure of total provision for residents of i. Clarke and Wilson (1994) state that, similarly, equation (13.6) represents partitions of the residential population which are combined to form a catchment population for the centre (or outlet) at j. One can also calculate a version of this in terms of demand, which we can call Δ_j:

$$\Delta_j = \sum_j \left(\frac{S_{ij}}{S_i} \right) \times W_j \tag{13.7}$$

Other typical indicators are W_j / P_j for effectiveness and W_j / Π_j for efficiency, where W_i is the aggregate level of provision in zone i (as calculated above), P_j is the population in zone j and Π_j is the catchment of zone j. These indicators may also be disaggregated by type m (social class, car owner, etc.) and provision is disaggregated by type of good, g. The performance indicators are thus derived directly from outputs from the model to assess the levels of provision, based on the model's predicted interaction set (Smith et al., 2006). These results may be combined with data on socio-economic status, age, ethnicity and so on. to identify areas/groups of individuals with poor access to a particular service. Thus, the key when assessing accessibility to services in rural areas is to ensure that the spatial distributions of provision and demand are such that both sets of effectiveness and efficiency indicators achieve appropriate targets.

From this example, using the service provision accessibility indicator, the access scores for residences in 12 OAs in relation to medical services were calculated for Liverpool (the highlighted column in Figure 13.4). The access scores ranged from 22.05 (indicating poor access) in L2 to 309.22 (indicating very good access), while the average access score across the 12 OAs was 148. The sixth part of accessibility analysis is the visualisation of accessibility values. Examining accessibility to services across space can be very difficult and GIS aids accessibility analysis by combining and analysing complex information from multiple sources, displaying the information in a map format. Modelling service provision in GIS can be broadly summarised under two general headings:

- the mapping of service provision in an area, region or country;
- modelling of accessibility to each service by using accessibility indicators and integrating these accessibility values into a GIS.

The development of more sophisticated GIS techniques and their widespread use among the social sciences has greatly aided accessibility analysis. Using visualisation tools, the researcher may present their results for policy analysis and evaluation. Maps can be used to display service location patterns, to provide information on where residents live in relation to service facilities, and to visualise the spatial match between service needs and resources. Indeed, previous research carried out by Halden et al. (2005) found that mapping accessibility to key services has significant advantages over analysis based on population data alone. This is because the impacts of policies taken by planners and policy-makers to increase accessibility may be seen explicitly.

Spatial interaction analysis is used in a variety of disciplines, the most prominent of which are transport, health care and retail studies. The policy emphasis on equity issues and rural accessibility has increased in recent years. Governments and regional administration centres across Europe are becoming increasing aware of the concept of *spatial equity* with regard to accessibility and services (European Commission, 1996, 1999). As such, policy-makers are beginning to use spatial interaction and accessibility analysis to enhance their decision-making for a number of different policy issues, including:

- assessing overall levels of expenditure and its distribution with regard to service provision (who gets what?);
- setting national and regional targets for improving equitable access to services;
- analysing local and sub-national need for services;
- examining the ex-post and forecasting the ex-ante effect of locating or relocating a service.

Conclusion

The goal of this chapter was to introduce the concept of individual-level spatial flows and how these flows can be modelled to analyse real-life events such as travel to and from medical practices. Spatial interaction analysis requires a considerable range of network and socio-economic data. Although a spatially constrained case study was presented in this chapter, calculating accessibility measures for full extents of cities, regions or countries requires a larger amount of computation, and indeed there may be trade-offs between resolution and extent of the models implemented. SIMs and the accessibility indices that may be calculated from their output provide an invaluable set of tools for describing and understanding the spatial pattern of accessibility to key services. Using information on spatial flows and accessibility indicators allows policy-makers to simulate what-if analyses using relevant transportation and socio-economic attributes. Furthermore, in conjunction with GIS, the important visual dimension to accessibility analysis can be highlighted.

FURTHER READING

Students may be interested in reading a number of follow-up books to aid their understanding of SIMs and their application. Suggested texts include: Clarke and Wilson (1994), an excellent overview of SIMs and how they are constructed; Birkin and Clarke (1991), a beginner's introduction to spatial interaction modelling from a retail perspective; and Joseph and Phillips (1984), an excellent overview of the concept of accessibility, and why modellers should include both spatial and aspatial parameters when modelling accessibility.

Birkin, M. and Clarke, G.P. (1991) Spatial interaction in geography. *Geography Review*, 4: 16–24.

Clarke G.P. and Wilson A.G. (1994) A new geography of performance indicators for urban planning. In C.S. Bertuglia (ed.), *Modelling the City: Performance, Policy and Planning*. London: Routledge.

Joseph, A. and Phillips, D. (1984) *Accessibility and Utilization: Geographical Perspectives on Health Care Delivery*. London: Harper & Row.

14

PYTHON SPATIAL ANALYSIS LIBRARY (PySAL): AN UPDATE AND ILLUSTRATION

Sergio J. Rey

Introduction

PySAL is a library for spatial analytical functions written in the open source object-oriented language Python. Since the first publication introducing PySAL (Rey and Anselin, 2007), much has transpired in the development of the library, and this chapter provides an update of this progress. PySAL was born of a collaboration between two earlier projects: PySpace and GeoDa, developed at the Spatial Analysis Laboratory at the University of Illinois at Urbana-Champaign, directed by Luc Anselin; and the STARS project, which I directed at San Diego State University (SDSU). The collaboration recognized that by pooling the efforts of the two labs, a good deal of duplication of effort could be avoided since the constituent projects were relying on a number of common core algorithms, data structures, and related modules. Rather than each group implementing the same algorithm, shared developer resources could be used to implement a single version of the algorithm for the library which each group could then leverage in their own projects. Additionally, by providing the code via a library, it now became open to a much wider user community beyond the two project groups.

In 2007, Anselin moved from Urbana to Arizona State University (ASU) to become director of the then School of Geographical Sciences and Urban Planning where he also established the GeoDa Center for Geospatial Analysis and Computation. One year later I moved from SDSU to join the faculty at ASU and become a core member of the Center. This ushered in a number of important changes in how the project was organized. First, we moved away from internal development of the code base to a more open structure by centralizing the code repository at Google Code under a BSD license. This was shortly followed by the adoption of a fixed, six-month release cycle, with the first formal release of PySAL 1.0 in July 2010. PySAL 1.8 is the current stable release, with version 1.9 set for release in January 2015.

Since the initial release of version 1.0, PySAL has been downloaded over 30,000 times, with 20,000 downloads in 2012 alone. Beyond 2012 tracking of downloads has become complicated by two developments. First, PySAL has been incorporated into the Anaconda Python Distribution,[1] which is a collection of specialized Python packages for high-performance scientific computing. Only one download of PySAL is required to build Anaconda, which in turn can be downloaded many times. The second change in PySAL's development infrastructure was the transition of our code repository from Google Code to GitHub. Users downloading PySAL source code from GitHub are not tracked as the repositories are available for cloning and downloading by any interested party. Nevertheless, it has been gratifying to witness the growing interest in PySAL and it is timely to provide an update on the library.

The remainder of this chapter is organized as follows. The key modules that comprise the library are first described. This is followed by an overview of the different delivery mechanisms for PySAL that include interactive prompts, toolkits for GIS packages, graphical user interface (GUI) based exploratory spatial and space-time packages, high-performance computational gateways, and web services. The focus then shifts to an illustration of one particular module in PySAL: the spatial dynamics module where a selection of the analytics is applied to a case study of four decades of homicide rates for 1412 counties in the southern United States. The chapter ends with some comments about future directions for PySAL.

PySAL components

PySAL is designed as a modular library with individual components focusing on suites of analytical methods, data structures and algorithms related to a particular type of spatial or space–time analysis. Figure 14.1 provides a high-level view of the key modules in PySAL.

Spatial weights

At the core of many spatial analytical techniques is formal representation of neighbor relations between observations embedded in geographical space. There are a wealth of approaches to defining these relations and the *weights* module implements

---

[1]http://docs.continuum.io/anaconda/pkgs.html.

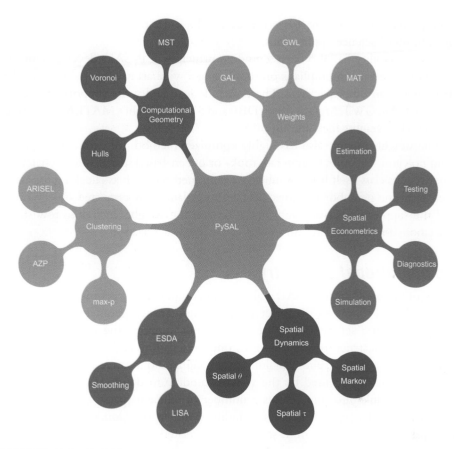

FIGURE 14.1 PySAL components

many of the most widely used, as well as lesser known, methods. The weights module organizes these into three different classes of spatial weights:

- contiguity-based weights;
- distance-based weights;
- kernel weights.

In general terms, the weights, defined as $w_{i,j}$, express the potential interaction between a pair of observations i and j. In most cases the number of pairs under consideration is substantially reduced due to the specific criterion adopted which gives rise to highly sparse representations of the spatial weights.

Given their centrality in analysis, efficiencies in memory footprint and in computations involving the weights have been a high priority of our development. These efficiencies derive from heavy use of sparse matrix methods, implemented in the package SciPy (Oliphant, 2007), and allow us to scale our analytics up to large problem sizes, where 'large' can be on the order of several million observations in social science applications.

Interoperability has also been a guiding principle in PySAL's development, and we have placed much emphasis on supporting a wide array of spatial weights data structures from other packages. These include read and write support for GeoDa weights, GAL, GWT, ArcGIS (Text, DBF and SWM) weights, MATLAB, GeoBugs, and STATA weights files, among others.

The weights module also has highly optimized methods for extracting topology from polygon shapefiles to generate rook- or queen-based spatial weight structures. The weights class itself has a number of useful methods such as checking for asymmetries in the neighbor relations, detection of islands (disconnected observations), and the support of various transformations on the weights including row standardization, double standardization and variance standardization.

Computational geometry

Many of the modules in PySAL make use of computational geometry algorithms, and we have centralized the latter in the computational geometry module. For example, given a set of points, the *CG* module supports the construction of a number of data structures, including Voronoi tessellations and minimum spanning trees, Gabriel graphs, sphere-of-influence graphs and relative-neighbor graphs. These in turn can be used by the weights module to define neighbor relations. The CG module also implements a number of efficient spatial data indices that support particular types of queries including point-in-polygon, segment intersection, projections of points onto segments and related operations.

Clustering

Methods for defining spatially constrained partitions of a set of areal units are implemented in the *region* module. The key approach is the max-*p* algorithm (Duque et al., 2012) which is a heuristic that attempts to find the maximum number of regions that satisfy some minimum floor constraint, such as the size of the population in a region or the number of areas combined, while maximizing intraregional homogeneity subject to a contiguity constraint. A key distinguishing feature of this algorithm is that the number of regions formed is endogenous, rather than having to be specified a priori by the user. In the *clustering* module are a set of methods to generate synthetic regions that respect the cardinality of solutions from the max-*p* algorithm. These provide a mechanism to evaluate the quality of the heuristic solution.

Exploratory spatial data analysis (ESDA)

Methods for global and local spatial autocorrelation analysis form the core of the *ESDA* module. The global methods include the analysis of binary outcomes via join count statistics with inference based on normal approximations as well as permutation-based approaches. For continuous variables, global version of Geary's C, Moran's *I* and the Getis–Ord G statistics are included, again with multiple approaches to inference. Local autocorrelation statistics include the local Moran and local indicators of spatial association (LISA) statistics (Anselin, 1995), and local versions of the Getis–Ord G statistics (Getis and Ord, 1992).

In addition to these standard measures for autocorrelation analysis, the ESDA module also includes bivariate Moran statistics as well as a suite of approaches for continuous variables that are rates, expressed as a ratio of the count of some event over a population at risk, where special care is needed due to variance instability of the attribute reflecting heterogeneity in the population at risk over the enumeration units.

Spatial dynamics

The *spatial dynamics* module was initially based on the space–time analytics from STARS (Rey and Janikas, 2006) but has grown with the addition of a number of newly developed methods. Three broad sets of space–time analytics are currently implemented. Markov chain-based methods, which depart from the classic discrete state Markov chain (DSMC), have been widely used in spatial analysis to model dynamics of many spatial processes including land-use change, migration, industrial structure and regional inequality dynamics among others. PySAL extends the DSMC in a number of directions to include a consideration of the role of space in shaping transition dynamics. Spatial Markov chains, first introduced by Rey (2001), allow for the influence of regional context which can introduce a form of spatial heterogeneity in the dynamics.

The spatial dynamic module also includes a LISA Markov chain (Rey and Janikas, 2006) which measures the transitions of observations across the four quadrants of the Moran scatterplot (from the ESDA module), thus extending the LISA to a dynamic context. A decomposition of the LISA Markov chain generates a pair of derived chains, one for the focal unit and one for the spatial lag, which in turn provide the basis for formal tests of the independence of the transitional dynamics for the lag and the focal chains (Rey et al., 2012). The combined use of the LISA Markov chain, together with the focal and lag marginal chains, supports the development of a rich taxonomy of space-time dynamics. Later in this chapter, I will illustrate the use of these two sets of methods.

In addition to the Markov chain-based space–time measures, this module includes several rank-based concordance measures (Rey, 2014) as well as various popular tests for space–time interaction (Knox, 1964; Jacquez, 1996).

Spatial econometrics

Modern methods of spatial econometrics are implemented in the *spreg* module. These include diagnostics for spatial autocorrelation in regression models based on Moran's *I* (Cliff and Ord, 1981), the classic and robust versions of Lagrange multiplier (LM) statistics (Anselin and Rey, 1991; Anselin et al., 1996), and LM-based statistics for use with two-stage least squares (2SLS) residuals (Anselin and Kelejian, 1997).

Estimation methods include non-spatial ordinary least squares and 2SLS, maximum likelihood estimation of the spatial lag and error models (Ord, 1975; Anselin, 1980, 1988; Smirnov and Anselin, 2001), spatial 2SLS of the spatial lag model (Anselin, 1980, 1988), generalized moments (GM) estimation for the spatial error model (Kelejian and Prucha, 1998a), and generalized methods of moments (GMM) estimation of the autoregressive parameter in the spatial error model in the presence of heteroskedasticity as well as when endogenous variables are included in the specification (Kelejian and Prucha, 2010). GM and GMM estimation of combination models including both a spatial lag term and a spatial autoregressive term are included (Kelejian and Prucha, 1998b; Arraiz et al., 2010; Drukker et al., 2013). Finally, a family of spatial regime specifications is supported which incorporate spatial heterogeneity for all included estimation methods with Chow tests for spatial coefficient heterogeneity (Anselin, 1990).

In addition to state of the art estimation methods, spreg includes an array of non-spatial diagnostics, including a multicollinearity condition number, Jarque–Bera test for normality and tests for heteroskedasticity including Breusch–Pagen, Koenker–Basset and White's test. The spreg module supports the use of a rich set of spatial weights including various contiguity criteria, distance bands, *k* nearest neighbors and inverse distance as well as kernel-based weights required by the heteroskedasticity and autocorrelation consistent estimators, and the module handles non-spatial endogenous variables.

Other PySAL modules

There are several additional PySAL modules not portrayed in Figure 14.1. The *inequality* module includes the classic Gini and Theil inequality indices as well as spatially explicit versions of these that can be used to analyze interregional inequality. These include decomposition-based statistics for Theil along with approaches to inference (Rey, 2004) as well as a spatial Gini index (Rey and Smith, 2013).

PySAL also includes a number of so-called 'contributed' modules. These are not part of the core library itself, but rely on optional dependencies that a user may have installed for particular types of analyses. Contributed modules currently exist for *Shapely* (Gillies et al., 2008) and a visualization module which supports choropleth mapping through matplotlib (Hunter, 2007). The latter will be demonstrated in the empirical illustration below.

PySAL use cases

By design PySAL as a library is intended to support a variety of delivery mechanisms and use cases. This is a recognition of the diversity of end users and of computing platforms that spatial analytical services are consumed on. Below I outline the different use cases supported by PySAL.

Interactive computing

In many areas of scientific investigation, often one does not have a clear hypothesis in mind and instead adopts an exploratory, or data-driven, approach to the analysis. Here the use of an interactive prompt is invaluable as the ultimate scientific workflow is not readily apparent, and instead the next computational task that the research will apply is only known after the results of the previous step are generated. PySAL supports interactive computing using either the built-in Python interpreter or the more powerful IPython shell (Pérez and Granger, 2007).

Graphical user interface clients

A second use case that PySAL supports is the wrapping of components of the library in rich desktop clients which provide access to the underlying functionality through a user-friendly GUI. One such example is the package *Crime Analytics for Space–Time* (CAST). Figure 14.2 shows one selected CAST window

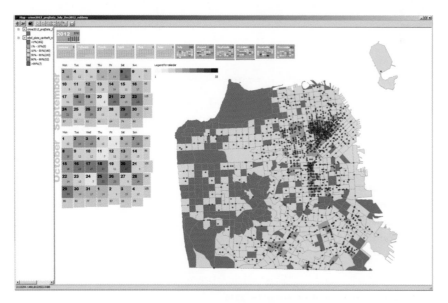

FIGURE 14.2   Crime Analytics for Space–Time (CAST)

which is illustrative of the kind of functionality it supports. CAST enables the joint consideration of multiple types of spatial supports (polygon, point, network) in a powerful and flexible set of fully interactive dynamic graphics. Also shown are calendar maps that provide insights into the temporal distribution of crime events.

The specialized nature of CAST is emblematic of a development philosophy at the Center where end user applications are tightly focused on the spatial analytical

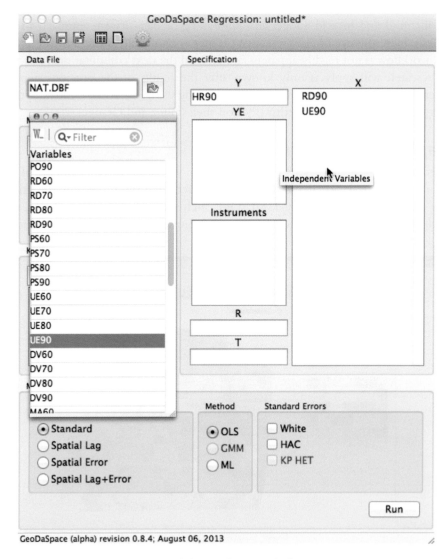

FIGURE 14.3 GeoDaSpace model specification dialog

functionality most appropriate to a substantive problem domain, in this case the spatial dynamics module. Rather than attempting to develop a one-size-fits-all GUI-based application, the wide scope of methods in the PySAL engine can be selected from to develop tailored applications in a time-efficient manner.

Another prominent example of a standalone application built around a PySAL core is the *GeoDaSpace* package, which is based on the *wxPython* module for its graphics and provides the end user with easy access to the advanced functionality in the spreg module. GeoDaSpace shields the user from many of the low-level operational and technical implementations while focusing on the most common operations and options. An example of one of the model specification dialogs for GeoDaSpace is shown in Figure 14.3.

GIS toolkits

In addition to interactive shells and GUI clients, users can interface with PySAL through toolkit architectures of geographic information systems (GIS) such as ArcGIS and QGIS (see O'Brien, Chapter 17). Figure 14.4 displays an example of an early version of a toolbox for ArcGIS 10.1.[2]

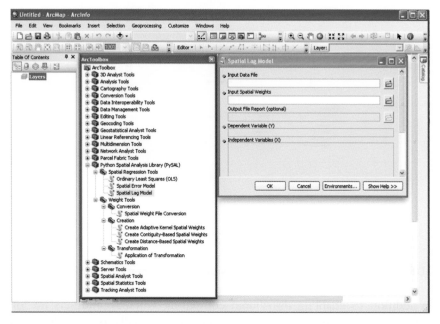

FIGURE 14.4   ArcGIS PySAL toolbox

---

[2]At the time of writing the ArcGIS toolbox is in alpha, with a stable release planned for spring 2015.

Web services

A final delivery mechanism for PySAL is through web services. These make a selection of the spatial analytic functionality available to the end user via a browser. An example is *CGPySAL*, which can be seen in Figure 14.5. This provides an interface to spreg on the CyberGIS Gateway (Wang, 2010). Here the user can upload their own data and then, using a flexible drag-and-drop interface, select variables to specify a model.[3]

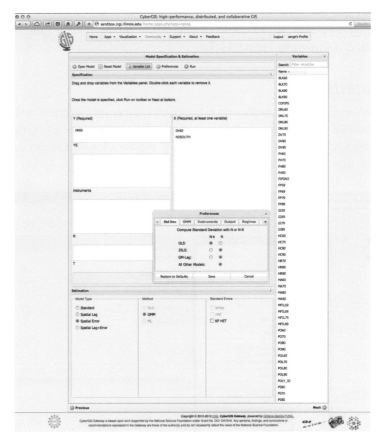

FIGURE 14.5   CGPySAL spatial regression in the CyberGIS Gateway

In addition to enabling distributed computing whereby endusers no longer require local installation of software, the centralized installation of PySAL as part of the CyberGIS Gateway allows us to implement a highly optimized version of PySAL that fully exploits the characteristic of the underlying hardware. This results

[3]The main site for CGPySAL is https://sandbox.cigi.illinois.edu/home/.

in computational gains that are generally not available in the version of PySAL that is released for users to install locally since we focus on general portability in that version rather than targeting specific hardware.

Illustration

Given the scope of the modules in PySAL, space limitations prevent an exhaustive set of illustrations. Instead I focus on one particular module, *spatial dynamics*, and a case study exploring the dynamics of homicide patterns in 1412 southern US counties using data developed as part of an earlier broader project (Baller et al., 2001; Messner et al., 1999). The focus here is on illustrating the use of PySAL with an empirical dataset. A detailed substantive investigation is beyond the scope of this chapter.

Spatial distribution of homicide rates

Figure 14.6 shows choropleth maps for the homicide rate (HR = homicides per 100,000) using a quintile classification for the decades 1960–1990. The classification method is one of the options in the PySAL map classification module and it is used here with the contributed visualization module mentioned previously.

Examination of the class boundaries indicates that the lower four quintiles are fairly stable over the four decades, while the fifth quintile is highest in the first

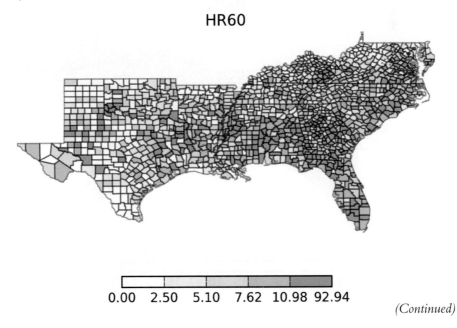

HR60

0.00   2.50   5.10   7.62   10.98 92.94

(Continued)

(Continued)

HR70

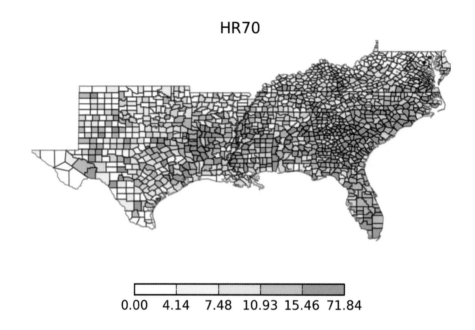

0.00   4.14   7.48   10.93 15.46 71.84

HR80

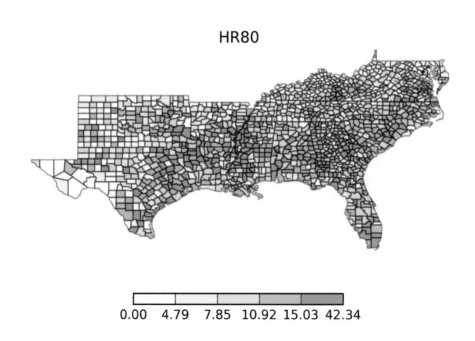

0.00   4.79   7.85   10.92 15.03 42.34

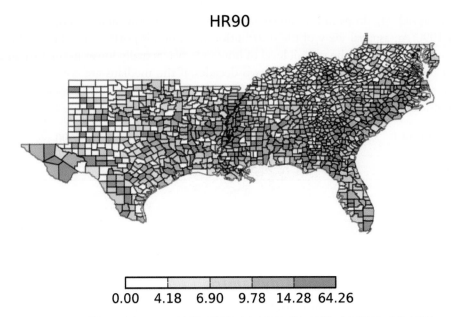

FIGURE 14.6   Homicide rate, 1960–1990: (a) 1960; (b) 1970; (c) 2980; (d) 1990

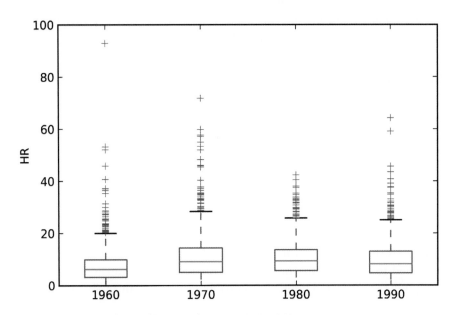

FIGURE 14.7   Boxplots of homicide rate, 1960–1990

decade (92.94), drops in the intermediate two decades, and then rises up to 64.26 in 1990. An aspatial view of the distribution dynamics is portrayed in Figure 14.7 which suggests that the overall level of homicide was actually lower in the first and last decade, relative to the middle two decades, if the medians are considered. At the same time, applying Moran's I to the homicide rate for each decade reveals significant positive spatial autocorrelation in all periods (Table 14.1).[4]

TABLE 14.1 Moran's I (queen contiguity) for homicide rate, 1960–1990

Rate	I	$p - value$
HR60	0.216	0.001
HR70	0.262	0.001
HR80	0.252	0.001
HR90	0.257	0.001

Spatial distributional dynamics

The global measures of autocorrelation are similar over this period, suggesting that the pattern of homicide activity may be relatively stable. Similarly, the stability of the majority of the quintiles may also be interpreted as evidence of distributional stability. However, both sets of measures are global, or whole map, measures that may mask more complex dynamics at work within the distribution. PySAL's spatial dynamic module has a number of space–time analytics that consider the role of space in the evolution of distributions over time.

Discrete Markov chains

The first of these is based on a classic discrete Markov chain which uses the quintiles to define the states of the chain. More specifically, the homicide rate in each county is viewed as a sample chain that can take one of five discrete values corresponding to its position in the quintile distribution in a given year. By pooling all the sample chains, the probability transition matrix can be estimated via maximum likelihood as

$$\hat{P}_{i,j} = \frac{\Sigma_t \, n_{i,j,t}}{\Sigma_t \Sigma_j \, n_{i,j,t}} \tag{14.1}$$

where $n_{i,j,t}$ is the number of times a sample chain started in state i in period t and transitioned to state j in the next period. Applying this estimator to our sample chains gives the estimated transition probability matrix P in Table 14.2.

---

[4]The global results are robust to the choice of the spatial weights matrix (rook versus queen).

TABLE 14.2 Homicide rate transition probabilities

t	Q1	Q2	t+10 Q3	Q4	Q5
Q1	0.431	0.253	0.128	0.099	0.088
Q2	0.240	0.291	0.241	0.145	0.083
Q3	0.154	0.214	0.277	0.221	0.135
Q4	0.106	0.150	0.215	0.271	0.258
Q5	0.071	0.091	0.138	0.263	0.438
π	0.200	0.200	0.200	0.200	0.200

There is strong evidence of mobility in the homicide rate distribution as the probability of remaining in the same quintile over sequential decades is less than 0.50 for all quintiles. The mobility is higher for the intermediate quintiles than is the case for the first and last quintiles. Lower mobility for the fifth class is to be expected given the skewed nature of the homicide rate distribution in each decade which results in a wider interval for that state in the Markov chain. Interval width alone, however, does not account for the lower mobility in the first quintile as its width is similar to that of the fourth quintile.

The last row in Table 14.2 is the estimated long-run steady-state distribution π, which suggests a uniform distribution holds when the chain reaches equilibrium. Note that although the states of the chain are defined using the quintiles of the distribution, this does not imply that the ergodic distribution will necessarily be uniform. In other words, there is no evidence of a convergence of homicide rates in particular parts of the distribution in the long run.

Spatial Markov chains

The classic discrete Markov chain provides a first view of the distributional dynamics; however, it does not consider the spatial location of the sample chains and how the local context of a county might affect the chain's movement in the distribution and transitions across states. One approach to this is the spatial Markov chain which conditions the transition dynamics of a county's homicide rate on the spatial lag of homicide rates. In other words, rather than a single transition probability matrix P, counties may face a different transition matrix depending on their spatial context.

Use of the spatial Markov class in PySAL estimates the conditional transition probability matrices reported in Table 14.3. The matrices are ordered according to the value of a chain's spatial lag at the beginning of the transition period, so that the first conditional matrix P (1) is for chains that had neighbors with homicide rates in the lowest quintile, and the last matrix P (5) is for chains with spatially lagged homicide rates falling in the upper quintile. These are estimated using

$$\hat{p}(l)_{i,j} = \frac{\Sigma_t\, n(l)_{i,j,t}}{\Sigma_t\, \Sigma_j\, n(l)_{i,j,t}} \tag{14.2}$$

where $n(l)i,j,t$ is the number of times a sample chain with a spatial lag in quintile l started in state i in period t and transitioned to state j in the next period.

TABLE 14.3   Spatially conditioned transition probability matrices

t		Q1	Q2	t+10 Q3	Q4	Q5
P(1)	Q1	0.528	0.218	0.106	0.095	0.053
	Q2	0.382	0.259	0.195	0.109	0.055
	Q3	0.271	0.243	0.250	0.194	0.042
	Q4	0.288	0.192	0.219	0.137	0.164
	Q5	0.352	0.074	0.259	0.111	0.204
π(1)		0.408	0.217	0.176	0.122	0.076
	Q1	0.473	0.268	0.123	0.077	0.059
	Q2	0.219	0.321	0.232	0.161	0.067
P(2)	Q3	0.235	0.246	0.278	0.182	0.059
	Q4	0.218	0.155	0.211	0.261	0.155
	Q5	0.096	0.233	0.192	0.178	0.301
π(2)		0.281	0.256	0.203	0.159	0.101
	Q1	0.299	0.328	0.164	0.104	0.104
	Q2	0.218	0.326	0.249	0.130	0.078
P(3)	Q3	0.076	0.240	0.292	0.263	0.129
	Q4	0.091	0.217	0.202	0.293	0.197
	Q5	0.053	0.127	0.167	0.273	0.380
π(3)		0.142	0.250	0.222	0.215	0.170
	Q1	0.210	0.309	0.173	0.123	0.185
	Q2	0.143	0.286	0.316	0.150	0.105
P(4)	Q3	0.119	0.206	0.299	0.216	0.160
	Q4	0.045	0.104	0.270	0.284	0.297
	Q5	0.042	0.083	0.125	0.287	0.463
π(4)		0.094	0.173	0.237	0.230	0.265
	Q1	0.286	0.161	0.143	0.161	0.250
	Q2	0.118	0.211	0.250	0.237	0.184
P(5)	Q3	0.073	0.127	0.253	0.253	0.293
	Q4	0.047	0.118	0.171	0.289	0.374
	Q5	0.048	0.053	0.104	0.284	0.511
π(5)		0.078	0.108	0.165	0.265	0.383

Examination of the table reveals that the spatial context of a chain can influence its transition dynamics over a decade. Counties that have homicide rates in the fifth quintile face different probabilities of remaining in that quintile depending on whether their surrounding counties also have rates in the upper quintile ($p(5)5,5 = 0.511$), or if the neighbors fall in the fourth quintile ($p(4)5,5 = 0.463$). At the other end of the spectrum, counties with the lowest homicide rates face a higher probability of remaining in the first quintile when their neighbors are also in the first quintile ($p(1)1,1 = 0.528$) relative to when the neighbors' rate falls in the second quintile ($p(2)1,1 = 0.473$). A formal test for the heterogeneity of the transition probabilities across lag quintiles rejects the null ($H_0: P = P(l)$ for all $l = \{1,2,\ldots,k\}$) of a single homogeneous transition probability matrix $\chi^2_{(80)} = 454.27, p < 0.001$.

The spatial heterogeneity in the transition probabilities also has implications for the estimated long-run distribution of homicide rates. Recall that under the homogeneity assumption, the long-run distribution is uniform across the five classes. By contrast, there are five separate estimated conditional ergodic distributions in Table 14.3, none of which are uniform. The mass of the distribution moves towards the tail of the distribution reflected in the value of the conditional lag – for example, the distribution conditioned on the first lag quintile $\pi(1)$ is right skewed, while the distribution conditioned on the fifth lag quintile $\pi(5)$ is left skewed. Moreover, the long-run distribution conditioned on the central lag quintile $\pi(3)$ departs from uniformity as the mass moves out of both tails and into the central three classes.

LISA Markov chains

The spatial Markov chain provides insight into the role of regional context at the beginning of a transition period in influencing the movement of a county's homicide rate within the distribution over time. In other words, different spatial contexts display different distributional dynamics. Those conditional dynamics, however, generate new realizations of the spatial lag for each county in the next period, and the question of how the county's homicide rate may co-evolve with its spatial lag naturally arises.

The spatial dynamics module includes several analytics designed to address this question. One way to visualize the co-movement of a county's homicide rate with that of its spatial lag is to consider the origin standardized movement vector (Rey et al., 2011) obtained from comparing Moran scatterplots (Anselin, 1996) from sequential decades. Figure 14.8 pools all the movement vectors over the three transitions (1960–1970, 1970–1980, 1980–1990), which provides an impression of the directional tendencies in the distributional dynamics. One striking feature is the relative asymmetry of the lengths of the movement vectors when considering the focal unit dimension (x-axis) versus the spatial lag dimension (y-axis). This is due to the spatial lag operator being a weighted average of homicide rates, whereas for the focal unit only the raw rate is considered. Thus the latter will tend to have variances that are greater than or equal to those of the lag.

It is important to note that the movement vectors reflect *relative* movement of a LISA within two scatterplots, and not *absolute* movements. In other words, movements to the southwest in Figure 14.8 indicate a county's homicide rate was declining in concert with a decline in value for its spatial lag. Similarly, movements to the northeast represent increases in the focal county's homicide rate and that found in its neighboring counties between two periods. These do not necessarily represents movements between quadrants in the Moran scatterplot.

Here we slightly abuse the notion of an 'absolute move' to define it as a movement across one of the four quadrants of the Moran scatterplot. From this, we can define a LISA Markov chain where the states of the chain are taken as the four quadrants of a Moran scatterplot in a given period (HH = 1, LH = 2, LL = 3, HL = 4), where HL indicates that the crime rate in the county was above the average for that period while its spatial was below average. Between any two decades a county's position in the Moran scatterplot may change to transition between the quadrants. Collecting all these transitions allows for the estimation of LISA Markov transition probabilities reported in Table 14.4.

Examination of these probabilities reveals several interesting characteristics about the spatial dynamics of homicide rates. First, the staying probabilities (i.e, probability of remaining in one state of the space) are highest for quadrants 1

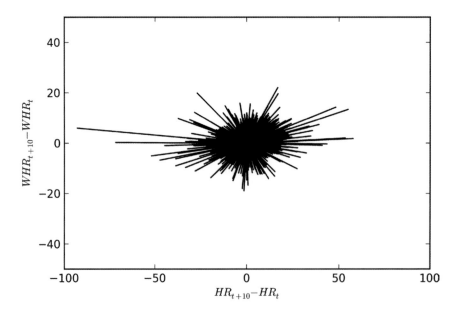

FIGURE 14.8 Origin standardized LISA movement vectors for homicide rates

TABLE 14.4 LISA Markov transition probabilities

t	HH	t+10 LH	LL	HL
HH	0.615	0.210	0.083	0.093
LH	0.327	0.336	0.246	0.091
LL	0.063	0.107	0.675	0.156
HL	0.190	0.124	0.411	0.275
π	0.287	0.180	0.390	0.143

(HH) and 3 (LL) of the scatterplot. This is indicative of relative stability in positive spatial autocorrelation that we encountered in the application of the global Moran's *I* statistic earlier. Second, the estimated equilibrium distribution for the LISA chain is to have the mass of the distribution more concentrated in these two classes (amounting for almost two-thirds of the distribution), again reflecting the dominant pattern of spatial clustering.

A third pattern seen in Table 14.4 is that when movements out of a quadrant do occur they are more likely to reflect the contribution of the movement of the homicide rate in the focal county than they are the movement of the spatial lag. For example, considering the chains in the initial state HH, the movement to LH, which involves an absolute change for the focal unit rate but not the spatial lag, occurs more frequently than movement to HL, which involves a change in the absolute position of the lag but not the focal rate.

Similarly, for an initial state of LL, moves to HL are more frequent than moves to LH. These patterns reflect the asymmetry in the magnitudes of the movement vectors in Figure 14.8 alluded to previously.[5]

A formal test for co-movement dependence can be derived by decomposing the LISA chain into two marginal chains, one for the focal unit (i.e. the own chain O) and one for the lag chain (i.e. for the neighbors N). Each of these marginal chains has two states, H for high and L for low, relative to the mean value in a given period. Under the null that these two chains are independent the expected transition probability matrix for the joint chain (i.e., the LISA chain) is given by

$$P(LISA) = P(ON) = P(O) \otimes P(N) \tag{14.3}$$

where $\otimes$ is the Kronecker product operator, $P(O)$ is the transition probability matrix for homicide rates, and $P(N)$ is the transition probability matrix for the spatial lag of homicide rates. Table 14.5 reports the observed and expected joint

---

[5]The LISA Markov chain also has options for conditioning on the significance of the static LISA in each period of the transition. These are not reported here due to space limitations.

spatial chain transitions. A formal test of the difference between these two estimated transition probability matrices resulted in $\chi^2_{(9)} = 2619, p < 0.001$, indicating the movement of a county's homicide rate in the distribution is not independent of the movement of homicide rates in neighboring counties.

TABLE 14.5   Observed versus expected joint spatial chain transitions

t		t+10 HH	LH	LL	HL
Observed	HH	756	258	102	114
	LH	244	251	184	68
	LL	103	175	1105	255
	HL	118	77	255	171
Expected	HH	338	310	278	303
	LH	173	221	198	155
	LL	357	457	463	361
	HL	161	148	150	163

Summary

Application of PySAL's spatial dynamics module to the case of homicide data has revealed clear evidence of spatial contextual effects in shaping the evolution of homicide activity distributions over space and time. At the same time it is important to keep in mind that the patterns identified are indirect evidence of spatial diffusion processes and do not imply a particular causal structure. The latter requires confirmatory modeling that reflects specific structural mechanisms giving rise to the patterns identified here. Nevertheless, detection of the patterns via the application of exploratory space–time methods suggests that such investigation is warranted.

Conclusion

This chapter has presented an update, overview, and illustration of the Python Spatial Analysis Library. Given the scope of the modules contained in PySAL, the illustration focused by necessity on only a subset of the analytical functionality provided by the library, but this should serve to provide the reader with a flavor of how PySAL can be used in practice.

One of the strengths of open source spatial analysis projects is that through the efforts of development teams, new state-of-the-art methods appearing in the scientific literature are often implemented in libraries that serve an important dissemination function. In this regard, PySAL shares in this effort, yet a unique feature of the

PySAL project is that several of the core developers have also been the creators of new spatial analytical methods that then become components of the library. By building on the contributions of others, this has allowed us to focus on the parts of the spatial analytical research stack where our efforts can have the most impact.

PySAL is well situated to contribute to an evolving CyberGIS infrastructure, and several ongoing projects are focused on embedding PySAL in high-performance environments. The notion of spatial analytical workflows requires frameworks for tracking the provenance of a sequence of analytical operations that are chained together to produce a result. We are currently exploring a prototype of such a provenance framework that captures the workflow in a distributed context and would support full replication of results. Closely related to this work is the development of a spatial econometrics workbench (Anselin and Rey, 2012) that would target spatial econometrics in particular but bring the methods from the spreg module of PySAL into a distributed application available to researchers to access through web browsers.

Finally, it is important to mention that our targeting of the Python language and scientific community is sometimes questioned by spatial scientists who are more familiar with R and the ecosystem of associated spatial packages, as the implication is that we are diverting energies that should otherwise be directed towards R. While we understand the sentiment behind such questions, we think they are somewhat misplaced for several reasons. Firstly, at the time of PySAL's conception, Python was beginning to make major inroads into scientific computing, yet spatial analysis was largely absent. We felt that focusing on Python would serve an important dissemination mechanism whereby leading developments in spatial analysis could be brought to new communities. The inclusion of PySAL in Anaconda is evidence that we have achieved some success in this regard. Secondly, we do not see our efforts as duplicative of the excellent work in R-related projects. Moreover, with new tools like the IPython notebook, it is now possible to use R and PySAL together in the same workflow. From this perspective, developments in either project R or PySAL can serve to benefit the other.

FURTHER READING

PySAL has extensive documentation available for both end users as well as developers. See `http://pysal.org` for full details.

ACKNOWLEDGEMENTS

Recent development of PySAL has been supported by Award No. 2009-SQ-B9-K101 from the National Institute of Justice, Office of Justice Programs and National Science Foundation Grant OCI–1047916.

15

REPRODUCIBLE RESEARCH: CONCEPTS, TECHNIQUES AND ISSUES

Chris Brunsdon and Alex Singleton

Reproducibility in research

The term *reproducible research* (Clærbout, 1992) has appeared in the scientific literature for nearly two decades (at the time of this writing) and has gained attention in a wide range of fields such as statistics (Buckheit and Donoho, 1995; Gentleman and Temple Lang, 2004), econometrics (Koenker, 1996) and signal processing (Barni et al., 2007). The aim of reproducible research is that full details of any results reported and the methods used to obtain these results should be made available, so that others following the same methods can obtain identical results. Clearly, this proposition is more practical in some areas of study than others — it would not be a trivial task to reproduce the chain of events leading to samples of lunar rock being obtained, for example! However, in the area of geocomputation, and particularly spatial data analysis, it is a reasonable goal.

To some, the justification of reproducible research may be self-evident. It may even be seen as a necessary condition for well-founded scientific research. However, if a more concrete argument is required, perhaps the following scenarios could be considered:

1. You have a data set that you would like to analyse using the same technique as described in a paper recently published by another researcher in your area. In that paper the technique is outlined in prose form, but no explicit algorithm is given. Although you have access to the data used in the paper,

and have attempted to re-create the technique, you are unable to reproduce the results reported there.

2. You published a paper five years ago in which an analytical technique was applied to a dataset. You now discover an alternative method of analysis, and wish to compare the results.

3. A particular form of analysis was reported in a paper; subsequently it was discovered that one software package offered an implementation of this method that contained errors. You wish to check whether this affects the findings in the paper.

4. A dataset used in a reported analysis was subsequently found to contain rogue data, and has now been corrected. You wish to update the analysis with the newer version of the data.

Each of the above scenarios (and several others) describe situations that cannot be resolved unless explicit details of data and computational methods used when the initial work was carried out are available. A number of situations may arise in which this is not the case. Again, some possibilities are listed:

1. You do not have access to the dataset used in the original analysis, as it is confidential.

2. The dataset used in the original study is not confidential, but is available for a fee, and you do not already own it.

3. The dataset used in the original study is freely available, but the original study does not state the source precisely, or provide a copy.

4. The steps used in the computation are not explicitly stated.

5. The steps used in the computation are explicitly stated, but require software that is not free, and that you do not already own.

6. The steps used in the computation are explicitly stated, but the software required is not open source, so that certain details of procedures carried out are not available.

Addressing the problems

All of the situations above stand in the way of reproducible research. In situation 1 this state of affairs is inevitable unless the researcher interested in reproducing the results obtains consent to access the data. Situations 2 and 5 can be resolved by financial outlay if sufficient funds are available, but situations 3, 4 and 6 cannot be resolved in this way. For this last set of situations, it is argued that resolution is achieved if the author(s) adopt certain practices at the time the research is executed and reported. Situation 6 is in some ways a variant of situation 4 where non–disclosure of computational details is due to a third party rather than the author of the research, and can be resolved if, whenever possible, open source software is used.[1]

However, attention in this discussion is focused on situations 3 and 4. Both of these situations arise if exact details are not made widely available. In most cases, this is not done with malice aforethought on the part of researchers – few journals insist that such precise details are provided. Although, in general, researchers must cite the sources of secondary data, such citations often consist of acknowledge-ment of the agency that supplied the data, possibly with a link to a general website, rather than an explicit link (or links) to a file (or files) that contained the actual data used in the research. Similarly, the situation described earlier in which com-putational processes are described verbally rather than in a more precise algorith-mic form is often considered acceptable for publication.

Software barriers to reproducibilty

Another source of uncertainty in identifying exact data sources or code used is that this information is not necessarily organised by the researchers in a way that enables complete recall. Typically this occurs when software used to carry out the analysis is interactive – to carry out an analysis a number of menu items were chosen, buttons clicked and so on, before producing a table or a graph that was cut and pasted into a word-processing document. Unfortunately, although interactive software is easier to use, its output is less reproducible. Some months after the original analysis it may be difficult to recall exactly which options were chosen when the analysis took place. Cutting and pasting the results into another document essentially broke the link between the analysis itself and the reporting of that analysis – the final document shows the output but says nothing about how it was obtained.

In general, unless interactive software has a recording facility, where commands associated with mouse clicks are saved in some format, and can be replayed in order, then graphical user interfaces and reproducible research do not go well together. However, even when analysis is carried out using scripts, reproducibility cannot be guaranteed. For example, on returning to a long-completed project, one may find a number of script files with similar content – but no information about which one was actually run to reproduce the reported results – or indeed, whether a combination of 'chunks' of code from several different files were pasted into a command-line interface to obtain reported results.

---

[1]The distinction is made here between *open source* software such as that discussed in O'Brien, Chapter 17, where the source code is distributed regardless of whether there is a distribution fee; and *zero cost* software, which is obtained without fee, but may not have openly available source code.

Literate programming

To address these problems, one approach proposed is that of literate programming (Knuth, 1984). Originally, this was intended as a means of improving the documentation of programs – a single file (originally called a WEB file) containing the program documentation and the actual code is used to generate both a human-readable document (via a program called Weave) and computer readable content (via a program called Tangle) to be fed to a compiler or interpreter. The original intention was that the human-readable output provided a description of the design of the program (and also neatly printed listings of the code), offering a richer explanation of the program's function than conventional comment statements. However, WEB files can also be used in a slightly different way, where rather than describing the code, the human-readable output reports the results of data analysis performed by the incorporated code. In this way, information about both the reporting and the processing can be contained in a single document. In this case, rather than a more traditional programming language (e.g. Pascal or C++), the code could be scripts in a language designed specifically for data processing and analysis, such as R or SAS. Also, the Weave part of the process could also incorporate output from the data processing into the human-readable part of the document.

Examples of this approach are the NOWEB system (Ramsey, 1994) and the Sweave package (Leisch, 2002). The former incorporates code into LaTeX documents using two very simple extensions to the markup language. The latter is an extended implementation of this system using R as the language for the embedded code.

An example of Sweave

This chapter was created using Sweave, and the simple example of analysis of variance in Table 15.1 was created with some incorporated code.

In this case, the code is echoed in the document, and the result is incorporated in the final document. A number of other options are possible - for example, the code need not be echoed, and the output can be interpreted directly as LaTeX code, giving the output in Table 15.2.

Note that from the echoed code, this document contains information about the data (via assignments to the variables ctl, trt, Group and Weight) as well as the steps used to analyse the data. If this document is passed to a third party, they will be able to reproduce the analysis by applying the Sweave program. Also, by using a further program, Stangle, they will be able to extract the R code.

TABLE 15.1   R ANOVA: example of incorporated code

```
> ##   From:   Annette Dobson (1990)
> ##   "An Introduction to Generalized Linear Models".
> ##   Page 9: Plant Weight Data.
> ctl <- c(4.17,5.58,5.18,6.11,4.50,4.61,5.17,4.53,5.33,5.14)
> trt <- c(4.81,4.17,4.41,3.59,5.87,3.83,6.03,4.89,4.32,4.69)
> Group <- gl(2,10,20, labels=c("Ctl","Trt"))
> Weight <- c(ctl, trt)
> lm.Dobson.p9 <- lm(Weight ~ Group)
> anova(lm.Dobson.p9)

Analysis of Variance Table

Response: Weight
            Df Sum Sq Mean Sq F value Pr(>F)
Group        1 0.6882 0.68820          1.4191   0.249
Residuals   18 8.7292 0.48496
```

TABLE 15.2   Formatted ANOVA output

	Df	Sum Sq	Mean Sq	F value	Pr(>F)
Group	1	0.69	0.69	1.42	0.2490
Residuals	18	8.73	0.48		

A geocomputation example

The previous example demonstrates the general principle of Sweave but does not involve geographical data. The choropleth map in Figure 15.1 is produced using the GISTools package. This demonstrates not only the use of spatial data, but also that the system is capable of including graphical output from the incorporated code. As before, the R commands are echoed to the document to illustrate the method, although in more usual situations this would not be the case.

The previous example showed a reproducible exercise in standard map creation in R. Figure 15.2 shows the result of applying pycnophylatic interpolation (Tobler, 1979) to another variable from the 1990 US Census. Here the variable considered is the number of children aged one year and under, on a per-county basis. This variable is obtained from the same source as the previous example. Pycnophylactic interpolation is a technique used to estimate population density as a continuous surface, given a list of population counts for a set of supplied corresponding geographic zones. Here, the population counts are numbers of children aged under one year old, and the zone boundaries are obtained from the shapefile used in the

```
> library(GISTools)
> data(georgia)
> par(mar=c(0.5,0.5,1.5,0.5))
> shd <- auto.shading(georgia$MedInc/1000,n=7,
+             cols=brewer.pal(7,'YlOrBr'))
> choropleth(georgia,georgia$MedInc/1000,shading=shd)
> choro.legend(-81.4,34.5,shd)
> text(-80.9,34.6,'Income $1000\'s')
> text(-83.6,30.3,'Source: 1990 US Census')
> title('Median Household Income (Georgia Counties)')
> box("outer",lwd=2)
```

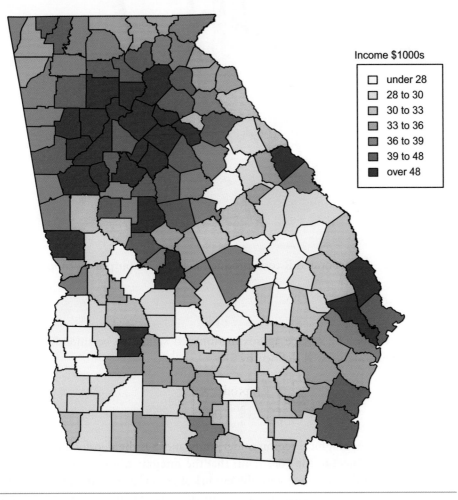

Median Household Income (Georgia Counties)

FIGURE 15.1 Incorporated R code producing graphical output

Source: 1990 US Census

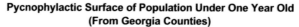

FIGURE 15.2 Incorporated R code producing graphical output: pycnophylactic surface

Source: 1990 US Census

previous example. The surface approximation is designed to be smooth in terms of minimising roughness R defined by

$$R = \int_A \left(\frac{\partial^2 S}{\partial x^2} + \frac{\partial^2 S}{\partial y^2} \right)^2 dA \tag{15.1}$$

where A is the geographical region under study and S is the density, a function of location (x, y), subject to the constraints that the integrated population totals over the supplied set of zones agree with the empirical counts provided. In practice this is achieved using a finite element approximation, so that population counts for small rectangular pixels are computed. These calculations are carried out using the R package pycno (Brunsdon, 2011).

In the example, graphical output from the code is used once again, although this time a three-dimensional surface is shown rather than a conventional map.

Implications for geocomputation

The above example demonstrates the incorporation of documentation, analysis and data in a single document. However, the example was not geographical. The key question here is how practical is this kind of approach in a geocomputation context. A number of issues arise:

1. Geocomputational data sets can sometimes be large, or in other ways impractical to reproduce.
2. Geocomputational computation times can be large.
3. Can combinations of text and code be expressed in other formats than R and LaTeX?

The first two of these are perhaps harder to address. In the example here the data consists of around 20 numbers, but in many situations the data is much larger than this, and it becomes impractical to incorporate the data in a WEB type file. One approach may be to incorporate code that reads from a specific file – although ultimately this still implies that code and data must be distributed together to allow reproducible research. An alternative might be to provide details (such as URLs) of where the data were downloaded – for instance, by incorporating code used to obtain the data if R was used to access data directly from a URL. However, in this case reproducibility depends on the remote data not being modified.

The second problem is not so much one of reproducibility, but one of practicality. Some simulation-based approaches (and other methods using large datasets) may require code taking several hours to run, and therefore major resources are

```
> library(pycno)
> Under1Pyc <- pycno(georgia,Under1,0.05)
> pycmat <- matrix(Under1Pyc$dens,
+     data.frame(getGridTopology(Under1Pyc))[,3])
> par(mar=c(0.3,0.3,1.5,0.3))
> persp(pycmat,axes=FALSE,box=FALSE,phi=40,theta=150,
+        col='wheat',shade=0.3,border=rgb(0,0.2,0,0.1),
+        )
> main.title <- 'Pycnophylactic Surface of Population
  Under 1 Years Old'
> sub.title <- '(From Georgia Counties)'
> title(sprintf('\n%s\n%s',main.title,sub.title))
> text(0.2,-0.38,'Source: 1990 US Census')
> box('outer',lwd=2)
```

required for reproduction; this is a scaled-down version of the 'lunar rock' example in the introduction. Reproduction may therefore be difficult, but not impossible; this is simply in the nature of such research. Another issue with simulation-based studies is the use of random numbers – unless the software being used gives explicit control of the generation method and specification of a seed, distinct runs of the same code will give different results. Fortunately, in R such control is possible (see Table 15.3); however, this is not possible in Microsoft Excel, for example.

TABLE 15.3 Demonstration of seed control in R

```
> set.seed(123,kind="Mersenne-Twister")
> # Generate some uniform random numbers with seed as
  above
> runif(4)

[1] 0.2875775 0.7883051 0.4089769 0.8830174

> runif(4)

[1] 0.9404673 0.0455565 0.5281055 0.8924190

> set.seed(123,kind="Mersenne-Twister")
> # Resetting seed should reproduce the first set of
  numbers
> runif(4)
[1] 0.2875775 0.7883051 0.4089769 0.8830174
```

The final issue listed is the use of alternative formats for either the text or code in literate documents. This is of key practical importance for the GIS community, many of whom do not typically use either LATEX or R. If reproducibility is as fundamental a concept as is being argued here, then ideally an inclusive approach should be adopted – simply insisting that all geographical information practitioners learn new software is unlikely to encourage uptake. A further literate programming tool, StatWeave (Lenth, 2012), offers more flexibility, allowing a number of different programming languages and statistics packages to be embedded into LATEX, including SAS, Maple, STATA and flavours of Unix shells. Unfortunately at this stage, there are is no GIS software embedded in this way, although StatWeave does offer the facility to incorporate new 'engines' into its portfolio, so that other command-line-based software tools can be used. A number of possibilities exist here – for example, incorporating Python would allow much of the functionality of either ArcGIS or QGIS to be incorporated. Also, returning to Sweave, there are a number of R packages for handling geographical information.

For others, a bigger barrier to adopting either Sweave or StatWeave is the use of LaTeX as a typesetting tool, when they are more accustomed to a word processor. However, it is possible to use `StatWeave` to process .odf files (the XML-based format for OpenOffice files) with embedded code, and there is an R package, odfweave, offering the same functionality (provided the embedded language is R). In both cases, the embedded code is typed directly into a document, which is then saved and post-processed to replace the embedded code with the output that it generates in a new .odf file. OpenOffice is then capable of saving the files into .doc or .docx formats, although obviously it is important to distribute the original .odf files with embedded code if the documentation is to be reproducible.

The difficulty for some is perhaps a move away from a GUI-based cut-and-paste approach to producing documents. Unfortunately, in its current from this approach is the hardest to reproduce, as was stated earlier. There are some possible future developments that could address this, however. If, when images or tables were cut and pasted, the object contained information about the process that created it (perhaps by journalling a set of Python or other commands recording a user's interaction with the GUI), then these could be embedded in a document (e.g. an .odf file), although this would require a more fundamental redesign of existing GIS software.

However, the adoption of such practices offers major advantages:

- It is helpful for authors who can reproduce figures and calculations in the revisions of a paper or report.
- It is helpful to others who want to carry out research in the field can work from the current state of the art, without having to 'reverse engineer' algorithms and techniques.
- It provides a framework for verification if the methods used in a given paper or report are called into doubt.

Although the adoption of these practices may require a little initial effort, the above factors suggest that the long-term payback easily justifies this.

FURTHER READING

As well as the items in the References, there are a number of very useful websites that discuss reproducible research. A few of these are listed below:

- `http://adv-r.had.co.nz/Reproducibility.html` provides some useful tips for carrying out reproducible research in R.
- `http://blog.stodden.net/category/reproducible-research/` is Victoria Stodden's blog about reproducible research.
- `http://reproducibleresearch.net/blog/` is a very comprehensive link site for work on reproducibility.

PART V

ENABLING INTERACTIONS

16

USING CROWD-SOURCED INFORMATION TO ANALYSE CHANGES IN THE ONSET OF THE NORTH AMERICAN SPRING

Chris Brunsdon and Lex Comber

Introduction

Although the term *volunteered geographical information* (VGI) was coined by Goodchild (2007), the activity which it is used to describe has been in evidence for a long time. There is a long tradition of volunteers collecting and reporting different types information about the environment we live in (Brown, 2011). Robert Marsham started to formally note the arrival of the first swallow in 1736. A recent description notes that 'This popular science really took off when he reported his records to the Royal Society in 1789 and many other country gentlemen took up the pastime' (Penn, 2011).

More recently, the rise of the internet – and in particular the idea of web 2.0 where 'ordinary' users are encouraged to upload information – has led to the much wider availability of geographical information collated from a broad group of private citizens, typically without formal training, and on a voluntary basis. A related idea is that of citizen science (CS) – see, for example, Cooper et al. (2007) or McCaffrey (2005). Here, information is collated from a large group of citizens – and in several cases this information is also geographically referenced. A third, related concept is that of public participation GIS – see Chapter 18. These concepts are not identical – but in general, both activities are seen as activities involving the collection of data by the public at large, rather than by officially sanctioned agencies. A notable distinction is that the degree of prior understanding required of volunteers in a CS-based project can vary greatly. In some cases the situation is quite similar to that of VGI with virtually no skills required of the volunteers, whereas in others some degree of volunteer instruction – and possibly selection – is necessary, so that the input of information has some degree of formal control. This is generally the case when the collection of information requires some degree of understanding, rather than basic observation or the use of easy-to-use sensors, such as the GPS facility in a smartphone.

There is an increasing amount of such data that could be, and in some cases is being, incorporated into formal scientific analyses. This includes spatially referenced and geolocated data such as the environmental data referred to above, as well as purely locational data (e.g. the location of streets in OpenStreetMap). Also, much historical volunteered information is held by public organisations and agencies which have some kind of obligation or pledge to make their data holdings publicly available (Lister and the Climate Change Research Group, 2011).

However, in many of the above situations the data collection process differs from that of a prescribed scientific experiment – and this should be taken into consideration when analysing the data, calibrating models or testing hypotheses. In science, a key mode of operation is the designed experiment (Myers et al., 2010) – and this is often regarded as a desirable situation for reliable – and more easily interpretable – statistical modelling. Furthermore, in a fortunate situation in which one has a great deal of control over the process of data collection, it is often possible to deduce strategies for data collection, giving optimal calibrations of statistical models, and put these into practice. However, VGI and CS both provide a very different situation from this, as researchers rely on individuals volunteering observations, and lack the control and coordination over data-collection strategies that are needed for such optimally designed experiments.

Despite this disadvantage, there are other benefits to using the public participation approach. The greatest of these is arguably that data are collected by a potentially very large unpaid workforce – for example, although noting the importance of the correct training of staff, Cohn (2008) observe that the use of CS has resulted in the saving of $30,000 per year on one particular project, and at the time of writing, a large amount of data for a diverse range of applications is collected in this way (Hand, 2010). This may seem a small sum compared with the scale of funding of some projects, but for others it could imply the difference between a project being viable and not. The aim of this chapter is not simply to report that this phenomenon is occurring – by now discussion of this is quite widespread, and several examples appear elsewhere in this book – but to consider the issues arising when analysing data of this kind. In particular, attention will be given to a project in the area of phenology – see, for example, Schwartz (1994). Phenology is the study of periodical plant and animal life-cycle events and their relation to climate.

In recent years, phenology has been used as an instrument to assess evidence of climate change. In particular, Cayan et al. (2001) and Schwartz and Reiter (2000) use observations of the first bloom and first leaf dates of particular plants recorded over a number of years to assess gradual advances in the onset of spring. The data analysed in these studies contains the recorded first bloom and first leaf dates of lilacs (*Syringa*) from a series of observation locations, with the data provided by a mixture of trained scientists and others.

Here we provide a demonstration of how data of this kind may be analysed, and in addition to reporting preliminary findings, we discuss and outline some of the issues that were encountered when carrying out the analysis. In the next section, the dataset used is described in detail. The following section demonstrates a problem of misinterpretation that may occur if the data-collection process is not considered. An approach allowing for this is then proposed, and implications are considered.

The lilac data

The data here was downloaded from the website provided by Schwartz and Caprio (2003),[1] and is derived from two studies, both spanning a number of decades. The first takes place in the western states of the USA. The second is based mainly in the eastern states, with a small number of locations in Canada also included. The former is described in detail in Cayan et al. (2001), the latter in Schwartz (1994, 1997). In each case, the dataset records the first leaf and first bloom dates of the common lilac, expressed as a number of days since the start of the year. The Schwartz and Reiter (2000) paper combines data from both of these studies to obtain a dataset for the whole of the USA. The locations are shown in Figure 16.1 on a backdrop of national boundaries.[2] Clearly, the east–west divide is not an exact one. Also note that the density of observations also changes in the two studies, and that particularly in the Midwest states density is relatively low.

Both studies were implemented via networks of observers. The most detailed description of this is given by Cayan et al. (2001) – who state that the western survey was initiated by Caprio (1957) who describes how the Montana Agricultural Experimental Station set up a network of observers, with contributors from the US Weather Bureau and local garden clubs, to monitor various stages in the annual cycle of the lilac – a very early example of data collected using the CS paradigm. A history of this and the eastern network is provided by the USA National Phenology Network (2011). Over time this activity extended geographically, and

---

[1] ftp://ftp.ncdc.noaa.gov/pub/data/paleo/phenology/north_america_lilac.txt.
[2] http://www.naturalearthdata.com/downloads/50m-cultural-vectors/50m-admin-0-countries-2/.

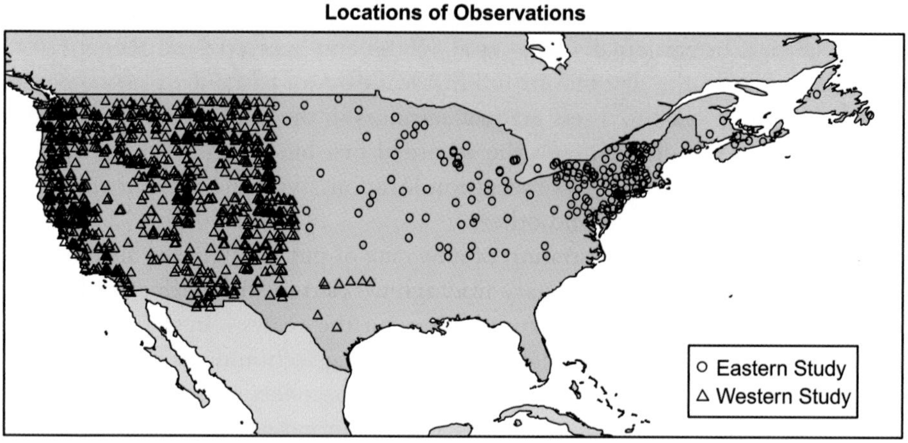

Locations of Observations

FIGURE 16.1   Locations of lilac observation points

continued until 1994. There was a subsequent revival of interest in 1999, until the last observations of the data recorded in this dataset in 2003. The eastern network was initiated after the western one – in 1961 - and saw most activity during the 1960s and 1970s. The numbers of observations from each network in five-year periods are tabulated in Table 16.1. In total there are 15,072 observations (although only 14,265 have recorded the first bloom date) from 1126 distinct observation locations.

From the table it can be seen that the dominance between eastern and western observations in the dataset changes over time – in the period from 1995 onwards, the majority of the data is composed of eastern observations – while in the first

TABLE 16.1   Numbers of observations by five-year period for eastern and western lilac phenology networks

	Eastern	Western
1955–1959	0	1997
1960–1964	97	2548
1965–1969	449	2420
1970–1974	640	1965
1975–1979	643	1049
1980–1984	578	778
1985–1989	248	664
1990–1994	231	411
1995–1999	183	8
2000–2004	120	43

five-year interval, the entirety of the data is provided by the western network. A more balanced pattern in data collection would be desirable. Going back to the notion of experimental design, the ideal situation would be a uniform coverage of all observation points across the USA over the entire time period. In contrast, this is an example of 'real-world' data, where events such as the cessation of public funding, the loss of a key organising individual, or the emergence of a new key player can bring about unexpected changes in the pattern of data collection. Whilst not achieving the ideals of the designed experiment, the collection and distribution of this kind of data is at least achievable in terms of resources.

Some problems with an oversimplified analysis

Caprio (1957) suggests that the initial motivation for the data collection was to map the geographical variation in the date of onset of spring, rather than in changes in this date over time. However, currently temporal change in climate is an important research topic, and the motivation here is that these data can be used to investigate climate change, if analysed appropriately.

In Schwartz and Reiter (2000) these data were analysed on a site-by-site basis, with a regression model applied at each location. However, since no site can have more than 46 observations (on for each year between 1957 and 2003) – and many have fewer – the aim here is to attempt an analysis of the data pooled for all of the sites. Another issue with this approach is that of multiple hypothesis testing, if regression coefficients are tested for significance – see Brunsdon and Charlton (2011) for an outline of this issue. An initial approach to analysis is to fit a simple linear model of the form $B_i = a + bY_i + \varepsilon_i$ to the pooled data, where B_i is the day of first bloom for observation i, Y_i is the year in which the ith observation was made[3] and ε_i is a normally distributed random error term for each observation i, with variance σ^2. a is the intercept term of the model, and b is the slope, which may be interpreted as the rate of advance (if $b < 0$) or retreat (if $b > 0$) of the date of first bloom, in days of advance per year. Given that a relatively slow rate of change is likely, one would expect b to be a fairly small quantity – quite likely fractional. The results of the analysis are given in Table 16.2.

TABLE 16.2 Regression analysis for the model $Bi = a + bYi + \varepsilon i$

	Estimate	Standard Error	t-statistic	Two-tailed p-value
a	126.39	0.25	514.39	0.00
b	0.22	0.02	12.11	0.00

[3] Y_i is centred on 1980, the mid-point of the time interval, as this reduces rounding error when calibrating the model.

This result suggests that b differs significantly from zero – suggesting that there is evidence that the day of first bloom varies over the study period. However, and rather surprisingly, the estimate for b is positive, indicating that the first bloom date is retreating. That is, it is getting later in the year. A plot of Y_i and B_i is given in Figure 16.2. The plot format is a binned hexagon plot, in which scatterplot points are allocated to small hexagonal regions, and the size of hexagon drawn in each region is proportional to the count of points contained there – this method is preferable to a standard scatterplot when there are a large number of points (in this case more than 10,000), and can be created relatively easily using the R statistical programming language (R Core Team, 2013). The regression line is superimposed on the plot.

Both the plot and the regression analysis show the first bloom day to be getting later – suggesting that the accumulation of thermal time, as the driver of plant development, is getting *slower* over time. This runs counter to expectation but, more importantly from the viewpoint of data analysis, also contradicts the findings of the Schwartz and Reiter analysis of the same data mentioned earlier. Their conclusions from the analyses of each of the individual observation locations suggest a general trend in which the first bloom day gets earlier.

A reason for this discrepancy may be deduced by further consideration of the data-collection process, and in particular the change in geographical concentration seen over the data-collection period. In Figure 16.3 boxplots are given for the first bloom dates of both the eastern and western data. One notable pattern is that the western data has earlier first, second and third quartiles for the first bloom date than the eastern data. Recall, however, that the eastern data are more prominent towards the end of the study. The extreme values are more expanded for the western data, but this could be attributable to the fact that, in general, there were more samples provided by this network. This suggests that one possible explanation for the surprising estimate for b is the fact that the later blooming eastern data dominate the pooled set towards the end of the study period. Put geographically, the 'centre of gravity' of the volunteered data collection process has drifted eastwards over time, towards areas where spring tends to arrive later. Therefore, if change in location is not considered as a conditioning factor, the average first bloom date does indeed increase with year of observation for this particular pooled dataset.

This is an example of Simpson's paradox (Simpson, 1951). The paradox is very succinctly stated by Appleton et al. (1996), who refer to 'the dangers of ignoring a covariate that is correlated to an outcome variable and an explanatory one'.

In this case, the covariate being ignored is the location of the observation; the explanatory variable is Y_i and the outcome is B_i. The situation here is unusual, in that examples of the paradox more usually involve probabilities or rates estimated using categorical data rather than regression analysis applied to continuous data (see, for example, Wagner, 1982), but nevertheless it clearly fits the situation described by Appleton et al.

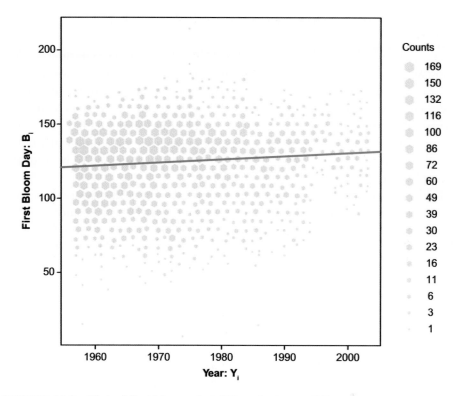

FIGURE 16.2   Plot of first bloom day (B_i) against year (Y_i)

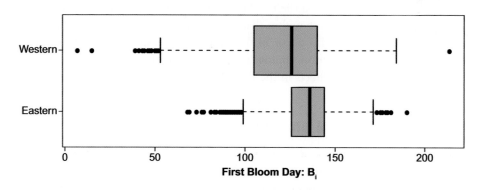

FIGURE 16.3   Boxplots comparing first bloom dates for eastern and western data

One way to address this problem may be to include an indicator variable in the regression model, giving the updated model

$$B_i = a + b_Y Y_i + b_N N_i + \varepsilon_i \tag{16.1}$$

where, in addition to the previously defined variables, b_Y is the regression coefficient for the year, N_i is the network indicator variable for observation i (0 for eastern, 1 for western), and b_N is the regression coefficient for this variable. Thus, this coefficient is a measure of the difference in first bloom date (on aggregate) between the eastern and western networks. Implicit in this model is a uniformity of rate of change of first bloom date across both networks, and a uniformity of the intercept term within each network, if the intercept is considered to be a for the eastern network, and $a + b_N$ for the western one.

The results of fitting this model are given in Table 16.3. In this case, the slope for Y_i is still positive, but no longer significantly different from zero since $p > 0.05$. This suggests the possibility that Simpson's paradox is contributing to the results of the basic model calibration here – as allowing for geography even in a fairly crude way does change the results relating to the slope term.

TABLE 16.3 Regression analysis for the model $B_i = a + b_Y Y_i + b_N N_i + \varepsilon_i$

	Estimate	Standard Error	t-value	Two-tailed p-value
a	134.01	0.41	331.20	0.00
b_Y	0.036	0.02	1.918	0.06
b_N	−11.59	0.50	−23.44	0.00

However, it could be argued that some geographical effects are still ignored. In the paper by Caprio (1957) maps of the first bloom date for the western area show notable geographical patterns exist within that area – and it is a reasonable expectation that similar variation may also occur in the eastern area – surely this at least merits investigation. In the next section, approaches to modelling the data allowing for more comprehensive variational effects in the model coefficients will be considered.

Proposed alternative analyses

The counter-intuitive result due to Simpson's paradox in the previous section arises essentially from the failure to incorporate sufficient information about geographical variation in the model. As a starting point to address this, we propose a model where the slope (rate of advance/retreat in the onset of spring) is the same everywhere, but that each observation station has a different intercept. That is, we assume there is a 'green wave' (Schwartz, 1998) across the USA so that some regions see the first bloom of lilac before others, but the rate of change of onset of this wave is uniform. We can model this by

$$B_{ij} = a_j + b_Y Y_{ij} + \varepsilon_{ij} \tag{16.2}$$

where the extra subscript j denotes a quantity relating to observation i at station j – thus B_{ij} is the ith first bloom date at station j, which was observed in the year Y_{ij}, and so on. Note that only a_j has just the j subscript, suggesting that there is a different first bloom date associated with each station, but not a unique one for each year at each station in the model proposed by equation (16.2).

Such a model could be calibrated using ordinary regression, treating the a_j as series coefficients for dummy variables indicating which station each observation occurred at. However, recalling that there are l distinct locations in the data, this would require a large number of coefficients to be calibrated, with a resulting increase of degrees of freedom in the model, and a resultant increase in the standard error of the estimate for b_Y. An alternative approach is to use a random coefficient model (Longford, 1993) where the a_j are themselves assumed to be random variables: for example,

$$a_j \sim N\left(a, \sigma_a^2\right) \tag{16.3}$$

where a is the distribution mean of the a_j and σ_a^2 is the variance. Thus the likelihood of the observed data can be written in terms of just four parameters: a, σ_a^2, b_Y and σ^2, rather than over 1100 parameters as in the model specified in equation (16.2). Another justification of this approach is that since the focus is on the estimation of b_Y, rather than attempting to calibrate every a_j exactly, the aim here is more simply to take into account the fact that a_j *does* vary, and to characterise this variability using a small number of parameters, namely a and σ_a^2.

Note that, by writing a_j as the sum of a and a zero-centred random variable υ_j (with variance σ_a^2), equation (16.2) can be rewritten as

$$B_{ij} = a + b_Y Y_{ij} + \upsilon_j + \varepsilon_{ij} \tag{16.4}$$

This is very similar to initial pooled linear model, except that now the random part of the model consists of two terms, representing variation at the observation level and at the location level. For this reason, the model can also be described as a multilevel model (Goldstein, 1986, 1987a, 1987b). In this chapter the convention is adopted that random terms in models predicting the first bloom date will be denoted by Greek letters, and fixed terms will be denoted with Latin letters.

In the basic pooled model the random part of the model is independent for each observation, but for model (16.4) it can be checked that for two observations ij and kl (i.e. the first observation is the ith at station j, and the second is the kth observation at station l), the correlation between the random terms, $\rho_{ij:kl}$, is given by

$$p_{ij:kl} = \begin{cases} \dfrac{\sigma_a^2}{\sigma_a^2 + \sigma^2} & \text{if } j = l \\ 0 & \text{otherwise} \end{cases}$$

$$(16.5)$$

Without focusing too much on this equation, it is sufficient to note that if two observations are made at the same location (but in different years) there will be some positive correlation. If they are made at different locations there will be no correlation. Arguably, this is a crude representation of Tobler's first law (Tobler, 1970): 'Everything is related to everything else, but near things are more related than distant things.' Here, 'near things' are considered to be observations taken from the same location, regardless of time.

The results of fitting this model are shown in Table 16.4. From this it may be seen that the estimate for b_Y is now negative, and is significantly different from zero. With an estimated value of around −0.18, this suggests that the onset of spring advances by around one day every six years.

Although the focus of this study has been the estimation of b_Y, it is still possible to estimate the individual aj via the multi-level model. Effectively, the estimate for each aj is achieved by computing an estimate of the expected value of aj given the respective values of all Bij and Yij observations, and the estimates of a, b_Y, σ^2 and σ_a^2 obtained when calibrating model (16.4). These are shown in map form in

TABLE 16.4 Regression analysis for the model $B_{ij} = a + b_Y\, Y_{ij} + v_j + \varepsilon_{ij}$

	Estimate	Standard Error	t-value	p-value
		Two-tailed		
a	121.89	0.65	188.56	0.00
b_Y	−0.18	0.01	−17.18	0.00

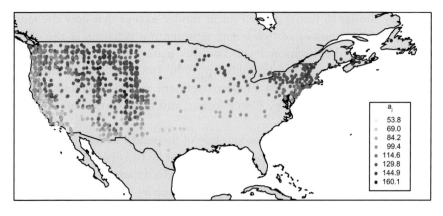

FIGURE 16.4 Map showing estimated a_j.

Figure 16.4. From this it can be seen that earlier first bloom dates tend to occur along the west coast of the USA, and also that superimposed on this there is a north–south trend, with spring arriving later in the north.

By obtaining a map of the 'green wave' as experienced through the first bloom dates of lilacs in this way, we suggest that even the approach of the model of equation (16.1) fails to reflect the full geographical variability in first bloom dates – Figure 16.4 suggests that geographical variation also occurs within both networks, whereas the model in equation (16.1) allows only for between-network variation.

Thus, in this study an estimate of b_Y (with standard error) taking into account the variability of an intercept term is created. However, this assumes that the relationship between the first bloom date and the year of observation is linear, so that the change in B per year is fixed over the entire study period. A more flexible approach is to estimate a general time effect, so that rather than modelling the temporal change in first bloom date with the regression term $b_Y Y_{ij}$, an alternative model replaces this by an effect for each year, say c_i for each year indexed by i. As with locations, these yearwise effects can be modelled as random effects, so that

$$c_i = c + \tau_i, \quad \text{where } \tau_i \sim N(0, \sigma_c^2) \tag{16.6}$$

and therefore

$$Bij = A + \tau_i + \upsilon_j + \varepsilon_{ij} \tag{16.7}$$

where the single term A replaces $a + c$ to avoid redundancy in the model (any pair of a and c adding up to A would give the same model likelihood for a given data-set). This model is still a random coefficient model, but it can no longer be described as a multi-level model, as the effects are no longer nested – effects for time do not nest within locations. Models of this kind are referred to as *crossed-effects models* (Baayen et al., 2008). The result of fitting a model of this kind is shown

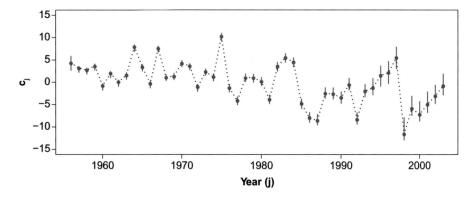

FIGURE 16.5 Graph showing estimated τ_j against time, with upper and lower pointwise confidence intervals for the individual τ_j values

in Figure 16.5. This plot shows the variation in τ_j over the time period 1956–2003, also giving bands showing standard errors for the estimates, based on bootstrapping techniques (Davison and Hinkley, 1997) – see the next section for detail.

The pattern seen in Figure 16.5 shows a general advance in the first bloom date over the study period, although it does suggest that the trend is more complex than the linear model used until now. In particular, in the second half of the study period, it appears that there is some degree of oscillation around a general downward trend, with the oscillation period being just over a decade or so. Also of note is the fact that the standard error bands are notably larger from around 1986 – this was the time when funding of these networks begin to reduce, and in turn, the numbers of observations were also reduced. In terms of the model calibration, this is reflected in greater uncertainty of parameter estimation.

One final parameter of interest may be of use here – this is used to measure the trend of τ_i over time. Considering this trend strictly, it is not the case that τ_i is decreasing *every* year – the oscillatory effect suggests that there will be some pairs of consecutive years when τ_i increases. However, it is helpful to consider whether there is an *overall trend* towards lower τ_i values. A straightforward way to measure this is to compute the difference in mean values of τ_i for the first and second halves of the study period. That is, the respective periods 1956–1979 and 1980–2003. If this statistic is called Δ, then in mathematical terms

$$\Delta = \frac{1}{24} \left[\sum_{i=1,24} \tau_i - \sum_{i=25,48} \tau_i \right] \tag{16.8}$$

Thus, positive values of Δ suggest a trend of spring getting earlier, and negative values suggest it is getting later. The estimate of Δ is 4.65, with 95% bootstrap confidence limits 3.71 and 4.76. From this there is fairly strong evidence of a trend of spring getting earlier. That is, despite some fluctuation around the trend, the average first bloom date in the second half of the study period is earlier than that of the first half.

Computational considerations

In this section, some more detail is supplied about the software tools and techniques that were used to carry out this analysis. All of the statistical modelling was carried out using the R statistical programming language (R Core Team, 2013). In particular, the random coefficient models were calibrated using the lme4 package.

The functions supplied in the R base library and lme4 were sufficient for all of the computations, except for the standard errors associated with the τ_i values, and Δ. For these, a regression bootstrap approach as set out in Davison and Hinkley (1997) is used. Briefly, this estimates the sampling variation of parameters of interest by simulating datasets drawn from the model that is being fitted to the data (in this case the model given by equation (16.7)). The sampling variation simulated is just that due to the variability in ε_{ij} – so that rather than randomly assigning new values for the τ_j and υ_i for each simulated sample, it is assumed they are fixed at the estimated values.

By simulating a large number of datasets in this way (say, 1000), and applying the random coefficient estimation function supplied by lme4 to each simulated dataset, an estimate of the sampling variability of the τ_j is obtained.

Conclusions and discussion

In this chapter, a number of models have been used to estimate the degree of change in the onset of spring over the period 1956–2003, calibrated with data on first bloom dates of lilacs collated from a number of networks of volunteers. These dates are strongly linked to the accumulated thermal time of the plants, and hence act as a proxy for patterns in seasonal temperature. From an initial analysis that was distorted due to Simpson's paradox, the final analysis takes into account the changes in geographical distribution of the networks of people collecting data, and identifies both general trends and fluctuations in the onset of spring.

The uncertainties and biases associated with analysing such data due to the voluntary nature of its collection have long been recognised. For example, hoopoe are birds that are occasionally seen in the UK with a preference for long grass habitats with tree cover. Much historical information in the UK relating to the sighting of these birds records them in the gardens of vicars – a combination of habitat and a nineteenth-century predilection among the clergy for recording nature. However, as yet little work has explored the sensitivity and reliability of crowd-sourced phenology data of the kind considered here. Some ecological research has explored the variation and uncertainties associated with the use of phenological data. Robbirt et al. (2010) compared plant specimens (herbarium data) with field observations and found the response of flowering time to variation in mean spring temperature to be identical and much of the variation in the results to be due to the geographic location of the collection sites – a factor which we have also found to be important in the analysis above. Also, there are other factors which may need to be considered: Miller-Rushing et al. (2006) compared herbarium data with phenological events as recorded in dated photographs. They suggested that first flowering dates may not be ideal measures of plant responses to climate change due to the extremes of flowering distributions being more susceptible to confounding effects than central values – such as the date at which exactly half of the lilacs in a particular site had bloomed. This is perhaps another situation where there is a trade-off between the ideal situation and what may be achieved in practice. Central values, such as means or medians, would need all bloom dates at a given location to be recorded, which may require more observational effort than can be realistically provided by a volunteer network. A compromise may be to obtain a central measure such as the mid-point between the first and last blooms (although this may still suffer from the problem of being sensitive to extremes as it depends on these two extremes). Of course, any such recommendations can only apply to future data, as recording the first bloom date is already a well-established convention – and a great deal of data using this convention already exists.

A further issue relates to the linkage between phenological event timing and temperature: van Oort et al. (2011) explored the sensitivity of phenological events and the possible correlation between temperature and phenology prediction error of rice, and found that phenological models were not as sensitive as thought at the higher end of the temperature range. As this study concentrates more on the timing of the phenological events, this finding perhaps has less direct bearing on the analysis; however, it does perhaps have implications when interpreting the observed patterns. In general it is hoped that one of the main messages in this chapter is that the analytical techniques used when working with VGI and CS data need to reflect issues arising from this unique kind of data acquisition. Any data of this kind are a reflection of both the underlying natural process and the process of data collection and organisation, and reliable analysis of such data needs to reflect this and account for it in any conclusions.

FURTHER READING

The web is a very useful resource for discovering how crowd-sourced phenology information may be collected. The following sites provide some insight into organisations collecting such data:

- `https://www.usanpn.org` gives some useful discussions relating to the USA National Phenology Network, and the organisation of the data collection process.
- `http://www.naturescalendar.org.uk/research/phenology.htm` provides a similar set of discussions and examples of how similar data are collected in the UK.

17

OPEN SOURCE GIS SOFTWARE

Oliver O'Brien

Introduction

This chapter reviews some of the most popular software packages in the well-developed and wide-ranging open source GIS software field. This is not intended to be a comprehensive review, as this is already well covered (Steiniger and Bocher, 2009; Donnelly, 2010; Tsou and Smith, 2011; Steiniger and Hunter, 2013). It should be borne in mind that, as with all actively developing software fields, and in particular those where open source has a strong contribution, such as GIS, the literature can quickly be superseded by developments.

Open source GIS software has a long history, but is also currently rapidly evolving. For example, Geographic Resources Analysis Support System (GRASS: http://grass.osgeo.org/), one of the earliest open source GIS software packages, was initially released in 1984; by contrast, QGIS (http://qgis.org/), another GIS software package, is much more recent, and its development is becoming increasingly active. Furthermore, in some sense, modern GIS software is increasingly difficult to define, and in addition to traditional desktop-based software packages such as GRASS or QGIS, there are web map servers, such as MapServer (http://mapserver.org/) and GeoServer (http://geoserver.org/), along with geospatial web (geoweb) client libraries to display spatial data online. Furthermore, a plethora of GIS functions for the storage, manipulation or analysis of spatial data are incorporated in other classes of open source software such as spatial databases or statistical programming languages such as R (http://r-project.org/). There are also various library packages, often coupled to many GIS and other applications, and specialist tools aimed at solving specific problems. The website http://opensource gis.org now lists over 300 applications and libraries.

What is GIS software?

Geographic information systems (GIS) software allows the storage, simultaneous display, analysis and visualisation of multiple pieces of geographic information.

Figure 17.1 shows a typical three-step software workflow between GIS data sources, GIS software and the final presentation of the content. Core desktop GIS software capabilities are, however, increasingly divergent, for example, with the addition of new functionality related to the end-presentation of geographic information. For example, by including functionality enabling enhanced cartographic styling and adornments of maps, these encapsulate a role that was traditionally carried out by conventional (non-georeferenced) graphics programs.

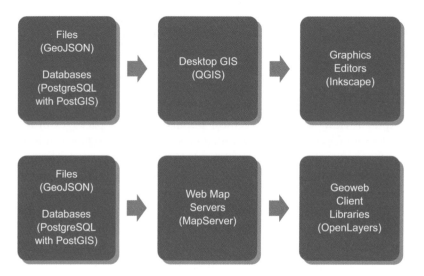

FIGURE 17.1   Typical GIS software flows. Top: Desktop-based GIS workflow for map production. Bottom: Server/client GIS workflow for web display of geospatial information. Examples in brackets

Furthermore, some databases now implement spatial analysis functionality directly, which can be useful when volumes of data are very large, and subsumes the analytical functions that may traditionally have been completed within a desktop GIS package. For example, the PostGIS (http://postgis.net/) extensions to the PostgreSQL (http://www.postgresql.org/) open source database can carry out a wide range of spatial operations.

What is open source software?

Software can be considered to be open source if it is distributed with its source code files, and with a licence that allows the free use, adaptation and redistribution of the source code. Users can examine the code to understand it, and potentially enhance it. The licence is often carried forward, so that if enhancements are made and distributed, the licence applies to them as well. Open source software is also typically (but not exclusively) made available for free. Copyright often remains with the creator, and attribution to the original and subsequent authors of the code may have to be retained. There are numerous variants of open source software licences. Popular examples include the Berkeley Software Distribution (BSD) licence, which simply asserts copyright and allows redistribution and use of the software as long as the creator is attributed (but not used to promote the software), and the General Public License (GPL), which ensures that modified versions of the software will always remain open for further modification and free use.

A key benefit of having access to source code means that if a limitation or problem is encountered, a user with the appropriate skills can examine the underlying source code, either to understand why it is behaving the way it is, or even to fix an issue or adapt it. The results of such corrections can then be shared with the user community. A wide pool of developers, with many different backgrounds, abilities and resources, can help improve a project by approaching problems in different ways, improving the overall quality by collaboration, consensus and visibility – the principle of 'many eyes'. An important body within the open source software field is the Open Source Initiative (http://opensource.org/) which considers open source software to be software distributed without the traditional copyright restrictions, where source code is included and can be viewed and adapted.

The open source GIS community

Many open source GIS packages have come from academic institutions and public bodies seeking to address a specific need. As their software has continued to evolve, a worldwide community has been built up to take on, manage and promote such projects. The Open Source Geospatial Foundation (OSGeo) is the principal organisation that promotes and organises the major open source GIS projects around the world. It has a number of established projects, and also an 'incubator' where smaller but promising projects are built up. Most of the software packages outlined in this chapter come under the auspices of OSGeo. OSGeo produces a DVD, OSGeo Live, that contains all its software projects together in one place, suitable for trying out. The DVD can be created from disk images at http://live.osgeo.org/.

OSGeo also organises an annual conference, Free and Open Source Software for GIS (FOSS4G). The conference takes place at locations around the world and typically has up to 1000 participants. It is often used to launch new versions of open source GIS software projects. For example, in the 2013 conference in Nottingham, UK, major software launches and announcements included QGIS 2, GeoServer 2.4, the first beta of OpenLayers 3 and OSGeo Live 7.

A further organisation, Open Geospatial Consortium (OGC), is concerned primarily with defining standards for spatial data formats. Such standards are generally widely supported within open source GIS software, either natively or often through the GDAL library, an OSGeo project which acts as a translation layer between many geospatial formats, including those defined by the OGC. Some of the well-known standards that have been defined by the OGC include KML (originally from Google Earth) which describes layers and viewpoints and annotations when looking at a map, Web Map Service (WMS) which is a standard way of requesting and presenting map images, and Styled Layer Descriptor (SLD) which is a format for defining how to represent features cartographically.

Desktop GIS

Desktop GIS software is a key interface for moderate and expert users to expose geospatial data to investigation, analysis and visualisation. The two most prominent open source desktop GIS software are GRASS and QGIS.

GRASS was originally developed by the US Army and is over 30 years old; it is now a core project administered by OSGeo. GRASS is a command-line application, but has a graphical user interface (GUI) called wxGUI. In addition, GRASS functionality has been linked into QGIS. GRASS contains a huge number of modules, which have a particular naming convention – for instance, modules that perform vector operations start with a 'v.'. For example, v.hull takes a vector map and produces a corresponding convex hull polygon. By contrast, r.slope.aspect creates slope and aspect raster maps from an elevation map. As GRASS is natively operated from the command line, it has a relatively steep learning curve for those unfamiliar with operating a GIS. More information on GRASS can be found at the project's website (http://grass.osgeo.org/).

QGIS, formerly also known as Quantum GIS, is an open source desktop GIS, designed from the start to be easy to use and cross-platform. It has a large and active community of developers, whose activities have recently accelerated, as it builds out its core features and becomes more widely adopted by the desktop GIS-using community. It is tightly integrated with the Python scripting language and PostGIS database servers; and versions are available for various operating systems. QGIS includes many standard spatial analysis functions (contained with a package formerly called SEXTANTE but recently renamed Processing) and can also integrate with a GRASS installation, thus allowing access to the many GRASS modules from within QGIS. QGIS can read from and write to a wide variety of spatial file and database formats. To access many of the formats, QGIS uses ogr2ogr, part of the GDAL library of tools.

The QGIS user interface design consists of a main window showing the mapping data, with a reorderable and togglable layer list. Operations are carried out using menus or one of numerous toolbars. A separate Print Composer window can be used to place the layered and styled map on a virtual sheet of paper, and adornments can be added. The resulting map can then be saved as a vector PDF, SVG or a number of other formats. The Print Composer is intended to replace the traditional final step of producing a map from GIS data where a graphics editor may be used.

The extensibility of QGIS has led to a number of complementary projects, including QGIS Cloud (a geodata infrastructure service), QGIS Server (allowing views of QGIS maps and layers to be accessed by multiple users) and a QGIS for Android project, bringing the application to a major mobile device ecosystem. The application itself can be driven through an API, making custom versions of QGIS with specific features and workflows branded and streamlined. The availability of the API also has enabled an array of plugin extensions to be created. For example, the Time Manager plugin, by Anita Graser, allows temporal geodata to be easily viewed in the application, using a time slider (Figure 17.2). Further information on Time Manager can be found at http://anitagraser.com/projects/time-manager/.

FIGURE 17.2   Time slider in the Time Manager plugin in QGIS

There are numerous other open source desktop GIS applications, which are similar to GRASS or QGIS, but typically present alternative user interfaces or different feature emphasis, such as concentrating on raster/field operations. For example, uDig (http://udig.refractions.net/) is a Java-based desktop GIS that shares QGIS's goals of being user-friendly and multiformat. gvSIG (http://www.gvsig.org/) is another Java-based desktop GIS that, like QGIS, uses the SEXTANTE package for spatial analysis. GeoDa (https://geodacenter.asu.edu/ogeoda) specialises in sophisticated exploratory spatial data analysis, autocorrelation and modelling of point and polygon data, looking at the links between spatial and non-spatial statistics of the data. Opticks (http://opticks.org/) focuses more on processing and analysing remote sensing and imagery data. Both gvSIG and Opticks have the status of being incubator projects within OSGeo.

Spatial databases

Storing geospatial data effectively and efficiently has always been a key aspect of open source GIS, particularly as the two- or three-dimensionally varying nature of geographical information means that data volume can become extremely large.

While flat file formats, such as Shapefile or GeoJSON, are good for storing some geospatial data, a dedicated database server can often be necessary. Typically, the ability to manage or analyse spatial data is provided by extensions to traditional relational database servers (RDBMS). For example, MySQL (http://www.mysql.com/) is an open source RDBMS and is part of the 'LAMP' (Linux Apache, MySQL, PHP) software stack that is widely used for simple dynamic webpages. However, core MySQL has been extended with spatial functions, enabling the storage of spatial data classes, alongside analytical functions.

Perhaps the most established and actively developed open source spatial database is PostgreSQL, with the PostGIS extension. PostgreSQL is a general purpose open source relational database server that was first released in 1986. PostGIS is a more recent package of stored procedures, definitions and extensions that enhances PostgreSQL by allowing it to deal with spatial structures, defined through special geometry and geography data types. The server was known for its complex installation and configuration process, particularly relating to memory requirements, and while care is still needed to install and specify a server that will perform optimally for the type and use of data stored within it, it is now more functional 'out of the box'.

Mapnik, an open source graphical renderer for maps, can use PostGIS-enhanced PostgreSQL databases as data sources. An appropriate query for certain features is defined by the programmer, and Mapnik modifies the query by adding spatial parameters. In the following example SQL query, the section in italics has been specified by the programmer, and Mapnik wraps additional SQL around it to capture the data for only a particular area. 900913 is a spatial reference ID (SRID) for the coordinate reference system (CRS) that the bounding box is specified for. PostGIS stored procedures generally have an 'ST_' prefix.

SELECT ST_AsBinary("geom") AS geom, "metric" FROM *(select geom, gid, geographycode as name, QS606EW0048/(QS606EW0001*1.0) as metric from c11_ew_oa_bfe_augm_google as b, c11_ew_oa as d where d.geographycode = b.oa11cd and QS606EW0001 > 0)* as foo WHERE "geom" && ST_SetSRID ('BOX3D(-46473.71319738716 6684876.745708373,-22013.86414613076 6709336.594759629)'::box3d, 900913)

PostGIS/PostgreSQL databases can also be added as data sources to QGIS, and a further extension to PostgreSQL, called pgRouting (http://pgrouting.org/), can be accessed through QGIS by using the PgRouting Layer plugin. pgRouting allows PostgreSQL/PostGIS to be used for complex routing queries over any topologically connected network of any kind, such as the UK's mainline railway network and other metro networks available in OpenStreetMap (Singleton, 2014a).

Web map servers

Web map servers access layers of geospatial content from databases and files, and serve them to web clients, either as simple HTML for display without further processing by standard web browsers, or in a format that can be handled by JavaScript (or similar) 'geoweb' client libraries such as OpenLayers, viewed within the web browser. They generally use one of two standards specified by the OGC – WMS (raster images) or Web Feature Service (WFS: geospatial objects as vectors). WMTS is a variant of WMS that serves WMS data in standardised square tiles, that can, if necessary, be pre-rendered for speed. TMS and XYZ are alternative specifications for naming and structuring sets of tiled map images, they share somes similarity to WTMS but are simpler to use. Web map servers are not always required in a typical map-based website toolchain, as web clients can do limited file processing, database access (via a simple proxy) and layering. Similarly, web map servers do not necessarily require a geoweb client library, as they can pre-render, annotate and deliver maps as images that can be displayed in a web browser using simple HTML.

There are two major open source web map servers, MapServer (http://mapserver.org/) and GeoServer (http://geoserver.org/), both of which are OSGeo projects. MapServer is a long-standing project, written in C, that allows users to browse GIS data and web application developers to create geographic image maps that direct users to content. It supports many standards for geographic data, and various vector and raster data formats. It is frequently used to provide WMS mapping images and was originally developed in conjunction with NASA, to make its satellite imagery more easily accessible. GeoServer is a Java-based server that, like MapServer, can process and display mapping data to a wide variety of standards and in many formats. It can work with various Java Servlet containers.

Geoweb client libraries

Geoweb client libraries are typically JavaScript-based pieces of software that are loaded into a user's web browser when they visit a website that includes an interactive map, allowing that map to be displayed with layers, augmented and navigated by the user. Web map servers can deliver simple images to a web browser with no geoweb client libraries needed; however, use of a geoweb client library allows the user much more flexibility to pan/zoom around a map, request additional function and view vector data, feature pop-ups and so on. The first popular 'modern' geoweb client library was the Google Maps API. Although not open source, and with its access limited to API calls rather than deep integration with client code, it gave access for web map developers to Google's extensive, and free, mapping and aerial imagery, while allowing the maps to be easily augmented with vector features – the famous 'red pins' being the simplest example. The service initially did not have any usage or load restrictions. Allowing anyone to augment a map in such a simple way, without requiring a dedicated service with a web map server on it, worrying about the cost, or needing much in the way of programming or installation knowledge, meant the service became hugely popular and inspired a number of open source peers.

OpenLayers (http://openlayers.org/) is perhaps the oldest and best established open source geoweb client library. It was for many years used to power the default rendering of the OpenStreetMap project, as shown at http://www.openstreetmap.org/, although that recently switched to an alterative library, Leaflet (http://leafletjs.com/). OpenLayers 3 was recently released, although the widely deployed version 2 (http://openlayers.org/two/) remains available for download. There are significant differences in functionality between the two versions and both are likely to remain key geoweb client libraries for a while. This new version of the library is a major rewrite, which aims to simplify the object structure, and ease adoption of the library, which can have a steep learning curve. OpenLayers can read in numerous formats, including WMS, GeoJSON and Shapefile, as well as TMS and XYZ map tiles used on Google Maps, Bing Maps, and other sources. The library, being pure JavaScript, is easily installed on a web server and hooked into webpages via a simple call to load the JavaScript.

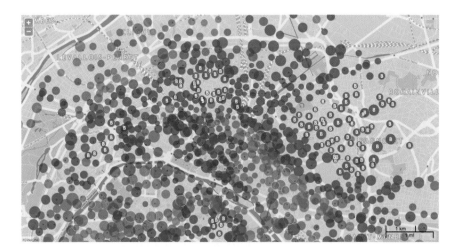

FIGURE 17.3 Example of an OpenLayers 2 map, showing styled vector data as circles of various sizes, colours, border highlights and labels. Two OpenLayers adornments, a zoom control and an automatic scale bar, are included. The background map is copyright OpenStreetMap contributors

The OpenLayers website includes annotated examples, wiki-based general documentation and JavaScript API generated documentation. One of the strongest features of OpenLayers is its advanced vector styling capabilities (Figure 17.3). This includes rule-based styling, allowing vector symbols to have varying colours, opacities, label display and orientation. OpenLayers partially implements the SLD specification from the OGC.

Leaflet, shown in Figure 17.4, is a more recent JavaScript geoweb client library. It shares many of the same features as OpenLayers and has similar default functionality, displaying maps in a draggable or 'slippy' pan/zoom manner with a minimal user interface (by default) of a pair of zoom buttons. However, it was written from scratch to present a straightforward implementation of 'slippy maps' for developers. It also, from the beginning, focused on mobile web applications, dealing well with touch-based (as well as mouse-based) interaction, although OpenLayers has also more recently included this as core functionality. It is very lightweight, requiring a download by the viewer of just 33 kb of compressed and minified JavaScript code. The small code size means it does less 'out of the box' but has a plugin architecture to incorporate additional file formats and other layer

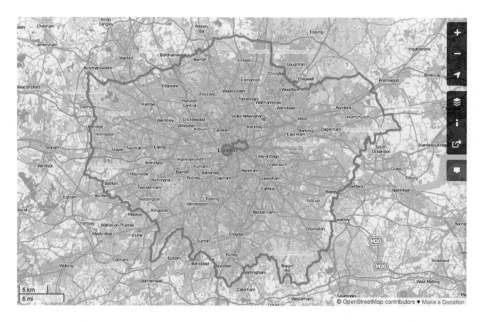

FIGURE 17.4   Example of the Leaflet geoweb client library, here presenting a rendering of OpenStreetMap data that appears on the project's main website. Highlighted as vectors are the Greater London and City of London boundaries, and a nominal location for the 'London' label

types. For example, TileLayer.GeoJSON implements support for the increasingly popular GeoJSON spatial file format, while Leaflet.GeoSearch allows a Leaflet-based map to easily integrate with a number of third-party geocoding services such as Google's.

Other open source GIS software

This section has only touched upon of the huge variety of open source GIS software, many fulfilling particular aspects of a geospatial toolchain. One key software technology not covered so far is R (Cheshire and Lovelace, Chapter 1). This scripting language has its roots in the academic/scientific community, as a powerful and comprehensive mechanism for performing statistical analyses programmatically, but its open source status and design, together with its ease of manipulating data in sophisticated structures (such as matrices), has allowed it to find many uses. A large and active community have developed geospatial-specific functions for R, many of which are contained in the sp package maintained by Roger Bivand.

One final package of note is Routino (http://www.routino.org/). This is a routeing software package that is specifically geared to using OpenStreetMap data for routeing. The software can be run as a website, or from the command line, and can be used to easily and quickly generate routes across the road and path network for a parcel of OpenStreetMap data that is converted to a format suitable for fast processing. Different traffic profiles can be specified, such as obeying one-way directions on particular road segments, using steps, maximum speeds and road type desirabilities, and detailed route descriptions can be created.

Applying open source GIS

The following section illustrates how to create an informational map of rail transport accessibility in London with the QGIS open source GIS application, integrating freely available (and open) spatial data sources. The intention is to identify those built-up areas in the city that are not well served by railway or metro stations.

Because the area is within Great Britain, we take advantage of the Ordnance Survey Open Data datasets (http://www.ordnancesurvey.co.uk/opendatadown load/). The Ordnance Survey is the institutional mapping agency for the country. Its datasets can be considered to be generally complete and authoritative, and certainly sufficient for this task. The following data are needed for this task: the administrative boundary of London; the extent of built-up areas within the city; and the location of railway and metro (e.g. London Underground) stations.

The following datasets need to be downloaded from the Ordnance Survey, they are generally large files (40–400 MB):

- Boundary-Line – from which the London boundary can be derived. At the time of writing, the name of the file when downloading is bdline_gb.zip.
- Ordnance Survey Strategi – which includes data on urban extent. At the time of writing, the name of the file when downloading is strtgi_essh_gb.zip.
- Ordnance Survey VectorMap District (vector version) – for myriads TQ and TL. From this data, railway/metro stations can be obtained. At the time of writing, the names of the two files when downloading are vmdvec_tq.zip and vmdvec_tl.zip.

These datasets can be obtained directly from the Ordnance Survey webpage referenced above, but may also be downloaded from the MySociety mirror at http://parlvid.mysociety.org:81/os/ which does not require registering and waiting for an email link to the data. Ordnance Survey data in Great Britain is split up into a grid of 100 km × 100 km squares, known as myriads, which are assigned a sequential two-letter code. The TQ myriad covers almost all of London, but TL is also needed for completeness, covering a small section in the far north of the city, which has some building development and a single station, Crews Hill.

You can install QGIS by downloading it from http://qgis.org/ and following the installation instructions. Depending on your version of QGIS, the following instructions and screenshots may vary slightly.

Loading spatial data layers into QGIS

Before loading in the required data, change the project projection to British National Grid – choose Project Properties from the Project menu, select Enable 'on the fly' CRS transformation, type 'British National Grid' into the Filter box, select the resulting item and press OK. This sets up the map projection for the map, as shown in Figure 17.5.

Choose Add Vector Layer from the Layer menu. Navigate to the uncompressed bdline_essh_gb folder, then under the Data subfolder, choose greater_london_const_region.shp.

Choose Add Vector Layer from the Layer menu. Navigate to the uncompressed strtgi_essh_gb folder, then under the data subfolder, choose urban_region.shp.

Choose Add Vector Layer from the Layer menu. Navigate to the uncompressed OS VectorMap District (ESRI Shape File) TQ folder, then under the data subfolder, choose TQ_RailwayStation.shp. Repeat this step for the TL folder.

QGIS should now look like Figure 17.6 – note that the colours used to display each layer are randomly picked so will be different if the steps above are followed. Because the boundary layer was added first, the QGIS window will have zoomed to its extent. You may need to alter the colour of the railway station dots so that they stand out from the building outlines sufficiently – this can be done by right-clicking on the layer in the layer list on the left, choosing Properties from the

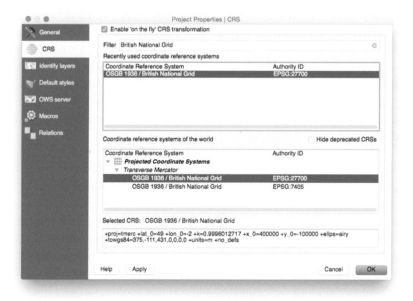

FIGURE 17.5 Setting the project's coordinate reference system

pop-up menu, then Style and then Color.

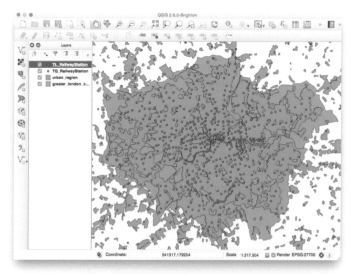

FIGURE 17.6 The layers loaded into QGIS

Preparing the data for analysis

We are focused on the area of London, so we need to prepare layers to show only this extent. First we need to create a polygon containing the simple administrative boundary of London. We carry this out by performing a Dissolve operation. Under the Vector menu, choose Geoprocessing Tools, then Dissolve. Make sure the correct layer is selected, choose FILE_NAME as the attribute to dissolve by (as it is the same for all the polygons in the layer) and choose an appropriate name for the new file. Tick the 'add the result to canvas' checkbox, so that the new layer is added straight in to QGIS. Finally, you can delete the old layer by right-clicking on it in the list and choosing Remove. See Figure 17.7.

FIGURE 17.7   Preparing the data: the Dissolve operation

We then need to clip the other three layers to this new layer just containing a London polygon. This is done using the Clip operation which is under the Geoprocessing Tools. Be sure to select the new London boundary layer as the layer to clip to. You need to repeat this step for all three of the other layers. Each time, you can remove the original layer after you add the new layer. This is shown in Figure 17.8.

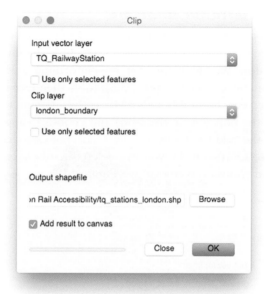

FIGURE 17.8   Preparing the data: the Clip operation

FIGURE 17.9   Preparing the data: merging shapefiles

Finally, we want to merge the TL and TQ layers together. This will reduce the four layers to three. This is done in a similar way to the above – but using the Data

Management Tools submenu and then the Merge shapefiles to one menu item (Figure 17.9).

On completion of these spatial operations, the window should look something like Figure 17.10.

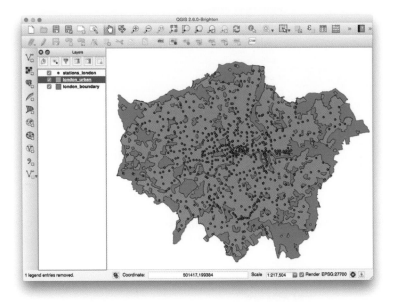

FIGURE 17.10   The merged map

Performing spatial analysis on the data

We want to show built-up areas (i.e. those with significant concentrations of build-ings) that are far from railway stations. We define such areas as those that are further than 1 mile (approximately 1610 m), in a straight line, from any station.

We do this in two steps. First, we add a one-mile buffer around the stations. This is done by the Buffer option, in the Geoprocessing Tools submenu. The Buffer distance should be set to 1610 as our projection (British National Grid) has units in (real) metres. Add the new layer to the map as normal and tick Dissolve buffer results. The result is shown in Figure 17.11. Finally, we subtract the buffer from the urban area layer. To do this, choose Difference from the Geoprocessing Tools sub-menu. The input layer is the urban area layer, clipped for London. The difference layer is the new buffered stations layer. The resulting layer is once again added to the list of layers, and all others, apart from the London boundary, can be removed. See Figure 17.12.

It is necessary to name the areas identified. To do this, add in another layer from the Strategi data – settlement_seed.shp. Clip this to the areas layer that was created, then edit the properties for the resulting clipped labels layer, by right-clicking and

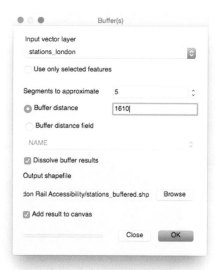

FIGURE 17.11   Performing the spatial analysis: adding a buffer

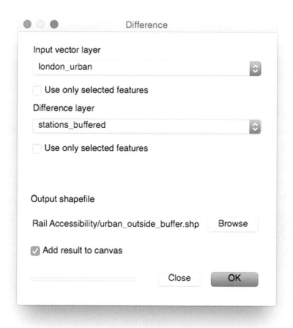

FIGURE 17.12   Performing the spatial analysis: subtracting the buffer from the urban area layer

choosing Properties. Choose Labels, tick Label this layer with and choose NAME. Add some appropriate styling – for example, a 1 mm buffer is recommended as this will make the text stand out more from the background. Figures 17.13–17.15 show this sequence.

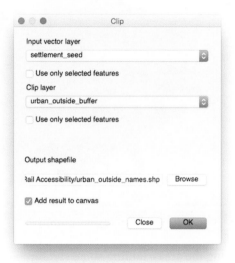

FIGURE 17.13   Naming the areas: Clip operation

FIGURE 17.14   Naming the areas: editing layer properties

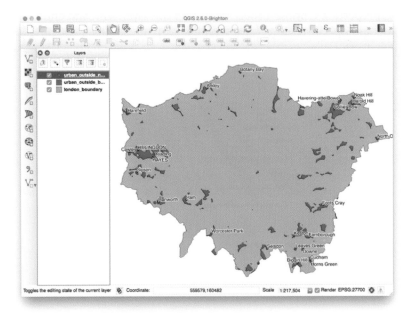

FIGURE 17.15   Naming the areas: adding some styling

Creating a map

Our analysis has been performed, and labels have been added. Now we just need to produce the final map. Choose New Print Composer from the Project menu and specify an appropriate name. Choose Add Map from the Layout menu and drag out an area for the main map, leaving some space around the edges of the virtual sheet of paper for adornments. In order to ensure that the full London area is shown, you may need to go to the Item properties tab, then click on the Set to map canvas extent button. See Figure 17.16.

Finally, you can add in adornments, such as a scale bar or a title. These are again accessed under the Layout menu, then by dragging each item onto the map. Options are displayed in the Item properties tab on the right of the Print Composer. When you are happy, you can generate a PDF, SVG or other formats, with the Export menu items on the Composer menu. Both PDFs (Figure 17.17) and SVGs generated from this map will be saved as vectors, so will print to a high quality.

Conclusion

This chapter has illustrated some of the main categories of open source GIS software, and community infrastructure supporting its development, promotion and use.

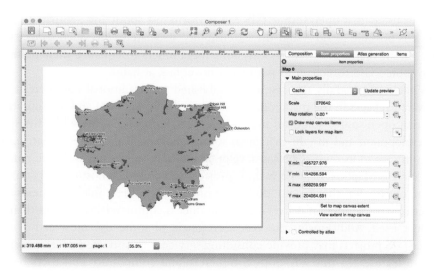

FIGURE 17.16 The Print Composer

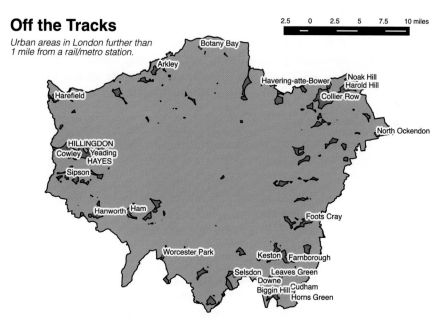

FIGURE 17.17 The final map saved in PDF format

An example of spatial analysis of transport accessibility is given using QGIS to demonstrate how open source GIS can be utilised in an applied context. It is just one task possible with using just one of many open source GIS software packages.

It is fair to say that open source GIS can perform just about any geospatial operation it is possible to do – and the extensibility associated with open source software means that it is extremely easy for programmers to enhance their preferred open source GIS toolchain to perform a novel task at hand.

While it is generally the case that established and traditional commercial GIS software packages in the market have a support, documentation and training structure that open source solutions may not yet match, this is countered by the speed of development and insight that open source allows and promotes. Open source GIS software in its own right has developed business opportunities. Some companies, such as Mapbox, manage various projects and provide a hosting and configuration option for those who do not wish to learn the toolchain themselves. Effectively, the service of delivering the product is commercialised rather than the product itself. Additionally, the core developers of some the most active open source GIS projects may provide consulting services or contracted development of custom features. In this way, both commercial and community end-users benefit from the accelerated improvement in the space.

FURTHER READING

Books and online documentation for open source GIS software are invaluable as a reference for understanding the software and becoming familiar with its structure and usage. There are a range of good books on many of the applications mentioned in this chapter, published by Packt Publishing, and available as eBooks or as physical printed books. For example:

- Corti, P., Mather, S.V., Kraft, T.J. and Park, B. (2014) *PostGIS Cookbook*. Birmingham: Packt Publishing.
- Graser, A. (2013) *Learning QGIS 2.0*. Birmingham: Packt Publishing.
- Gratier, T., Hazzard, E. and Spencer, P. (2015) *OpenLayers 3 Beginner's Guide*. Birmingham: Packt Publishing.
- Iacovella, S. and Youngblood, B. (2013) *GeoServer Beginner's Guide*. Birmingham: Packt Publishing.

As with all well-used and modern open source software, there is also a wealth of information available on the web.

18

PUBLIC PARTICIPATION IN GEOCOMPUTATION TO SUPPORT SPATIAL DECISION-MAKING

Richard Kingston

Introduction

The extent to which computational modelling has influenced and been a part of decision-making in geography and urban planning has a long and distinguished history, as outlined by Geertman and Stillwell (2009) and Batty and Brail (2008) amongst others. With more recent advances in visualisation techniques and the ability of computers to manipulate, process and analyse increasingly complex and growing amounts of spatially referenced data, the application of geocomputation to support decision-making within an online environment continues to grow. GIS as the primary visualisation tool over past decade has matured into a tool that can be increasingly utilised by a wider range of users. Ever since the criticisms of GIS in the mid-1990s (cf. Pickles, 1995), GIS has moved from specialised to lay use, alongside a growth in the ability of a broader range of users to use and understand computerised mapping. While 20 years ago it may not have been envisaged that there could be public participation in geocomputation, today we can build, provide models, and allow access via easy to understand user interfaces that to some extent are starting to replicate systems that were only available to advanced users in the past. With the ever-increasing availability of spatial data there has been a growing demand for translation into usable information that supports many aspects of spatial decision-making. This chapter provides an overview of some of those methods used, as well as two illustrative examples of online geocomputational mapping tools that support public participation in environmental planning. The chapter concludes with a look towards the future and the relationship between geocomputation, public participation and the ways in which modelling within online environments allows for improved access to geocomputational models.

Participation and decision support

While there are growing numbers of examples illustrating the use of geocomputation to support public participation in urban planning, many still remain one-off examples, as illustrated in the use of planning support systems (PSSs). Vonk et al. (2005) recognised this problem some time ago when they addressed some of the bottlenecks for implementation, and still today there are very few known cases where computational modelling is used to support routine decision-making in urban planning on a day-to-day basis. Arguably, a major reason for this shortfall is the lack of skills within the planning profession to utilise the appropriate techniques, and the simple fact that much of the technology tends to be developed as one-off bespoke software. As Singleton (2014b) has recently argued, the need for computer programming to be an integral component of geography degrees begs the question whether this should be the case for planning. Well, perhaps as a specialist option so students at least have some understanding of the methods and techniques that planners could draw upon in practice. As Klosterman (2008: 174) argues, 'we are witnessing a second revolution of computer use in planning' where planners will need to know less about specific programs to support their decision-making and instead 'future planners will not have to develop GIS-based tools and models from scratch, but will obtain them from the web and link them together to perform particular tasks' (p. 175). It is now very easy to set up map-based public participation tools to support public engagement in decision-making. With a Google account, combined with its mapping application programming interface (API) and some knowledge of its Fusion tables, pretty much anyone can quickly set up an interactive map with absolutely no programming experience. Furthermore, customisable tools such as CartoDB (http://cartodb.com) make such a task even easier.

The notion of a PSS was coined by Britton Harris in the late 1980s when he highlighted concerns about the need for strategic support of routine planning tasks. Harris (1989) gave planners new insights into the uses and design of (geo) computation models to support strategic plan making and development of management associated with specific planning tasks. The term 'planning support system' in this context refers to the combined use of GIS, spatial decision support systems, spreadsheets, databases and other ICT-based tools. Over the past 15 years these have increasingly been made available over the internet as PSSs have essentially grown into online computer-based systems aimed at integrating a vast array of tasks to support the work of planners, associated decision-makers and increasingly the public (Harris and Batty, 1993; Kingston et al., 2000; Kingston, 2007). At the heart of many of the tools to support the decision-making process are geocomputational models and methods. These can be for a broad range of application areas, for example optimising site location for large-scale retailers, or modelling the flood risk potential of a new residential housing development using techniques such as location allocation and spatial interaction models (Chapters 11 and 13). The focus in this chapter, though, is to highlight how we can provide public access to geocomputational methods that support decision-making, particularly when making decisions about the environment around us.

The involvement of the public in decision-making has a long tradition in planning going back to the 1960s, coincidentally around the same time as when computation was introduced into the profession. Spatial planning systems across the world influence the places and spaces where people work and live. Planning is deemed to operate in the public interest in order to ensure that development and the use of land result in more sustainable communities, but this is a complex task given the diverse nature of public, private and community interests affected by and involved in the planning system and process. Whether it be for ensuring development occurs where communities need it, or for the protection and enhancement of the natural and historic environment and the countryside, planning affects us all. During the 1990s the notion of collaborative planning (Healey, 2005) began to emerge together with the suggestion that to understand participation in public policy it is necessary to examine the broader governance strategies that institutions may utilise to achieve different levels of engagement. These can be categorised in to three such strategies. First, *top-down strategies* start and finish with the government agencies that plan such schemes, and the local authorities that implement them. Such initiatives occupy the therapeutic and manipulative levels of Arnstein's (1969) ladder, and at best there can be an element of negotiation. Second, *limited dialogue strategies* involve a stronger commitment to a two-way communication process. The method and policy framework are set by national and local public bodies, but with dialogue that can lead to substantial change in decisions and thinking. Finally, *bottom-up strategies* are distinguished by their commitment to open discussion and the active encouragement of excluded or marginalised groups to become involved. In more recent times the use of ICTs to empower local communities has begun to take hold through the use of online participatory GIS to complement well-established techniques. It is this approach that is the focus of the remainder of this chapter. To what extent can online geocomputational tools support public engagement in spatial planning?

First, though, it is worth noting that the use of GIS and computational modelling to support spatial planning has witnessed a patchwork of success globally for a number of reasons. Often the success or failure of the implementation of geocomputational tools in urban planning is predicated not just on the institutional factors of GIS adoption (Budic and Godshalk, 1994), but on the underlying 'system' within which planning operates. Globally, planning systems tend to adhere to one of two types of system referred to as 'regulatory' (e.g. USA, Canada, France, Germany, Spain) or 'discretionary' (e.g. England, Scotland, Wales, Northern Ireland), with some countries using a hybrid of the two. It is suggested that the differing natures of these two types of system mean that the regulatory approach lends itself to being more amenable to utilising geocomputational modelling than the discretionary system, and hence geocomputational support for planning is more widespread in those countries utilising the former approach. Nevertheless it still requires planners in practice not necessarily to learn new programming skills, but to think about how they organise themselves and how they communicate their plans and proposals to their stakeholders. One of the key findings of a recent research project, outlined in the next section, was that GIS began to help in the breaking down of 'silos' between departments in order to gather evidence of the need to act on climate change adaptation. It brought together different people in different agencies who had 'ownership' of different types of data that were needed to support the planning policy-making process. Without this collaboration the problems they are facing would not be solved.

In the next section two examples are given showing how different stakeholders make use of web-based spatial data analysis to support decision-making. The first is from an EU-funded project to develop climate change adaptation action plans across Europe, and the second is a prototype system to support the connection of decentralised energy production to the UK's National Grid network.

Participation in understanding local climate change impacts

Climate change is a problem globally and within a European context, and being able to understand that Europe's climate is changing is important not just for decision-makers, but also the public at large. Recent extreme weather events in North America, Australia and parts of Europe illustrate how ill prepared our built and natural environment is. In 2012 we witnessed a record warm winter in the US Midwest, followed by a record cold winter in 2013; record heat in Australia, when normally ocean currents would have cooled the land; and drought in the UK, followed by floods in 2014. The two Spot 6 satellite images in Figure 18.1 show the River Parrett near Bridgewater, Somerset in south-west England on 8 June 2013 and again on 11 February 2014, illustrating the extent of the flooding.

Building on research originally carried out at the University of Manchester, the GRaBS project (http://www.ppgis.manchester.ac.uk/grabs/) developed an innovative, cost-effective and user-friendly risk and vulnerability assessment tool to aid the strategic planning of climate change adaptation responses in towns, cities and

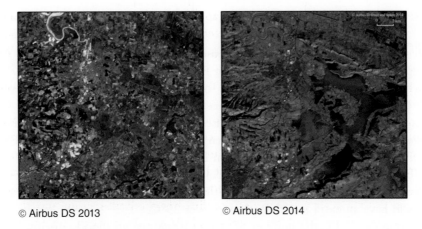

© Airbus DS 2013   © Airbus DS 2014

FIGURE 18.1   River Parrett before and during flooding (http://
airbusdefenceandspace.com/satellite-images-reveal-uk-flood-impact/)

regions across Europe. The main aim of the GRaBS Assessment Tool was to assess
current vulnerability of urban areas to climate change impacts, with an additional
assessment of relative patterns of spatial risk where suitable data were available. The
tool is based on a GIS and has been built on top of the Google Maps interface using
a range of spatial data for the whole of Europe, together with specific spatial data
from project partners – in total over 325 different map layers. From the start of the
project there was a desire to develop a tool that would be easily accessible to a wide
range of stakeholders involved in climate change adaptation, including the general
public. With 11 partners responsible for developing adaptation action plans, creating
a tool that would meet the needs of all project partners has been quite a challenge.

The GRaBS project looked in detail at the potential impacts of climate change
across the European continent. This understanding helped to provide a context for
the assessment tool from the perspective of vulnerability to climate change in the
partner cities and regions. Two key issues are as follows:

- *Temperatures* are projected to increase over Europe at levels higher than the
  global average. Increases of 2–6°C are expected, depending on the
  greenhouse gas emissions scenario considered. The north is projected to
  warm at a slightly faster pace than the south.
- *Precipitation* patterns are expected to change significantly. Recent seasonal and
  geographic changes in rainfall are projected to intensify. Northern Europe has
  become wetter over the last century, a trend that is likely to continue, and
  particularly over the winter. However, southern Europe has become drier, with
  further reductions in precipitation expected. Increases in precipitation extremes
  are also likely.

These changes in temperature and precipitation will generate impacts including:

- *Flooding* events are expected to increase in frequency and magnitude, especially over northern and western Europe during the winter months where river discharges are projected to increase. Extreme rainfall events will increase the threat of flash floods across most of Europe.
- Lower rainfall and associated river discharge/aquifer recharge will increase *drought* risk, particularly over southern and eastern Europe. The threat of longer dry spells is lower in northern Europe.
- Central and eastern Europe in particular are likely to experience summer *heatwaves* of increasing intensity and duration. By 2050, the Mediterranean could have an extra month with days over 25°C.
- During the 1990s, the number of *extreme weather events* (floods, storms, droughts) doubled. This increase in frequency has persisted into the twenty-first century and is projected to continue to do so. It is extreme events, not gradual changes in climate, that cause the most significant impacts on human and natural systems.
- *Sea-level rise* is a key threat to coastal communities and ecosystems. Globally, increases of 18–59 cm are projected, with regional variations also likely. The prospect of more significant increases cannot be ruled out, for example relating to the melting of the polar ice caps.
- Climate change is expected to bring with it changes in *wind speeds*, with significant regional variation. Increases of 10–15% in average and extreme wind speeds are possible over northern Europe, with a 10–15% decline in speeds in southern Europe. An increase in storminess over western Europe during the winter months is also possible.

(Adapted from European Environment
Agency (EEA), 2008)

At the time of the project, the most up-to-date information on recent and pro-jected future change to Europe's climate could be found in the European Environment Agency publication *Impacts of Europe's Changing Climate – 2008 Indicator-Based Assessment* (EEA, 2008). With this overview setting the context for the project, and feeding into the requirements of the tool, individual partners set about assessing their own climate impact priorities and deciding what their own priorities were within their own municipalities.

The tool is based on, but further develops, the findings of the University of Manchester's Adaptation Strategies for Climate Change in the Urban Environment (ASCCUE) project (Lindley et al., 2007); and development followed the principles of an online public participation GIS (Kingston et al., 2000). Here the main aim was to develop web-based GIS to enhance public involvement and participation in environmental planning and decision-making. These systems are referred to as PPGIS and are a form of PSS. The main objective is based on the belief that by providing all stakeholders, including citizens, with access to information and data in the form of maps and visualisations they can make better-informed decisions about the natural and built environment around them. The nature of the GRaBS project with a pan-European user base meant that the tool had to be built using a common map base for all partners. Using Google Map's API, which is 'free', generic, highly customisable, and, probably most importantly from a user perspective, very simple to use and navigate, meant that we were able to build a single generic tool for all partners.

At the start of the project it was difficult to anticipate the amount and complexity of the data layers that would be used in the tool. Indeed we involved all partners in assessing what were the key climate impact factors and climate-related issues in their areas so that the tool was fit for purpose in each of the municipalities and regions across Europe. Thus the project team embarked on a collaborative process of tool development with all the partners in order to shape the exact form and function of the final tool (see Figure 18.2). Through the use of a user needs and requirements analysis (UNRA), partners were asked to identify the types of functionality and their anticipated requirements of the tool; what their climate change impact priorities were; a series of questions on the

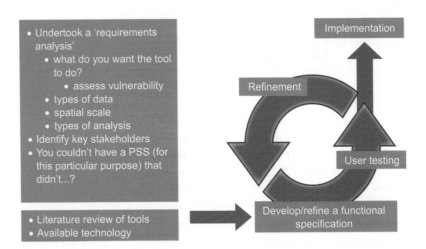

FIGURE 18.2   Assessment tool development

availability of data to input into the assessment tool; and the origin and source of the spatial data alongside any access rights issues on the data provided. There are currently 325 different spatial layers in the tool, with partner level data ranging from 7 to 50 layers (see Table 18.1). This reflects the extent to which different municipalities have been collecting relevant data and their ability to work across departmental boundaries to obtain necessary data. In the UK, the Cabinet Office's Policy Action Team 17 and 18 reports (Department of the Environment, Transport and the Regions, 2000; Office for National Statistics, 2000) have

TABLE 18.1 Data types

Partner Level	No. of layers
Social infrastructure (schools, hospitals, emergency services, care homes, prisons, etc.)	52
Civil infrastructure (utility services, railway stations, roads at risk of flooding, waste disposal facilities, etc.)	49
Population structure (population density, areas of deprivation)	14
Green and blue space (parks, trees, playgrounds, sports fields, water bodies, cemeteries, etc.)	54
Vulnerability (population ages over 75 and under 4, people with poor health, people on low income, people vulnerable to high temperatures and to flooding, properties vulnerable to flooding, population living on ground floor or basement and living on third floor or higher)	60
Hazards (flood zones, historic flooding events, surface water flooding, groundwater flooding, sewer flood risk)	39
Urban development (current and proposed residential and industrial development, historic and cultural buildings)	7
European Level	
Climate (including the 2080 IPCC A2 emissions scenarios and PRUDENCE 1961–1990 mean temperature and precipitation)	15
Climate hazards (drought, extreme temperature, river flooding, forest fires, storm surge, storm events)	5
Civil infrastructure (airports, mayor roads and railway lines)	3
Population structure (density, growth, 0–14 and 60+ years)	4
Urban development (degree of urbanisation, urban areas)	2
Green and blue space (urban green space, blue space, protected areas)	3

helped UK local authorities to put in place robust systems for the collection and storing of spatial data. At the EU level there are 32 spatial layers available.

To complement the UNRA, a review of climate change vulnerability, risk and adaptation assessment tools was undertaken (Lindley, 2009). This review concluded that a wide range of online tools are available for risk assessment and risk management, but that there was debate surrounding what comprises such a tool in relation to climate adaptation. The review also identified that there was no existing tool specifically for vulnerability assessment, and there was a scarcity of tools specifically focused on cities and urban areas. This was further enhanced by a review of the available technology, and at the time of development, only the Google Maps platform provided an API, allowing software developers to customise and build other tools on top of a common map base (see sample code below for an example of how to query a map feature). Given the popularity and familiarity of the Google Maps interface, this was deemed a suitable platform to build the tool on, although it could be supplemented for a range of alternative solutions. The combination of the above aspects of the tools development revealed that the tool needed to:

- map risk and vulnerability to climate change hazards and show spatial distributions;
- allow map overlay to enable identification of priority areas for adaptation actions;
- identify relative patterns of risk in urban environments;
- show the severity of the impacts;
- generate a list of climate hazards and receptors;
- inform preparation of planning documents to enable accounting of climate change impacts on people, property, assets and infrastructure;
- be fully accessible to all stakeholders online, including residents, developers, local business, utilities and councils.

The tool was built in a single technical environment that could be used seamlessly across Europe – both during the project and after its lifetime. The project budget and timetable did not allow for 12 individual partner approaches to the underlying technical development of the tool, but was devised to enable the tool to provide a common framework for use by each of the partners and in its application across Europe. This also provided flexibility for other municipalities to implement the tool with their own data. A pilot version of the tool was tested in March 2010 by three of the partners. Most of the initial user testing focused on the design of the interface in terms of the 'look and feel' and this fed into an initial redesign of the tools interface.

Sample extract of JavaScript for the Google Maps API version 2

This function allows the user to click on a map feature and display the attribute information in a pop-up bubble.

```javascript
// Query a map feature function

this.toggleQueryLayer = function(MapFileName, LayerName, legendTitle) {
        // remove previous listeners
        GEvent.clearListeners(map, "click");
        // Query feature when click on
        queryMapfileName = MapFileName;
        queryLayerName = LayerName;

        GEvent.addListener(map, "click", function(overlay, currentlatlng,
        overlaylatlng) {


                        //transfer latlon into a pixel
                        var bounds = map.getBounds();
                var currentSouthWest = bounds.getSouthWest();
                var currentNorthEast = bounds.getNorthEast();

                var currentWidth = map.getSize().width;
                var currentHeight = map.getSize().height;

                // return the LatLon when clicked
                if (currentlatlng) {
                        var currentpixel = map.fromLatLngToContainerPixel(c
                        urrentlatlng);

                        //query the map feature
                        var currenturl = "http://130.88.68.64/cgi-
                        bin/mapserv?map=/home/common/grabs_mapfile/" +
                        queryMapfileName + "&BBOX=" + currentSouthWest.
                        lng() + "," + currentSouthWest.lat() + "," +
                        currentNorthEast.lng() + "," + currentNorthEast.
                        lat() + "&SERVICE=WMS&VERSION=1.1.1&REQUEST=G
                        etFeatureInfo" + "&LAYERS=" + queryLayerName
                        + "&query_layers=" + queryLayerName +
                        "&STYLES=&SRS=EPSG:4326&WIDTH=" + currentWidth +
                        "&HEIGHT=" + currentHeight + "&info_format=text/
                        plain" + "&x=" + currentpixel.x + "&y=" +
                        currentpixel.y;

                        //get the value of query item
                        $.get(currenturl, function(data) {
                                var string = data.split("=");
                                var queryvalue = legendTitle + " = " +
                                string[1];

                                //popup window to display the attribute
                                info
```

```
                                     map.openInfoWindow(currentlatlng, document.
                                     createTextNode("You have queried: " +
                                     queryvalue));
                          });
              }
          });
}
```

The tool was then introduced to all other partners at a workshop in Graz, Austria. This part of the testing made use of the concept of a 'storyline' where all partners followed a scenario, presenting a realistic climate change impact situation demanding action from decision-makers. This included: the risk of flooding in the London Borough of Sutton; protection of critical infrastructure from climate impacts in Styria, Austria; and risk of high temperatures in Catania, Sicily. These storylines contained step-by-step instructions on how to use the tool, its functionality, data and information within the tool, and how this could be used to support their applications. Partners then filled in a questionnaire to provide further feedback on the tool's menu system, its functionality and data display capabilities. Being present at the workshop to observe the partners using the tool also provided valuable feedback which fed into further improvements.

The feedback from the Graz workshop provided over 60 specific comments on ways the tool could be improved, and this approach formed an important part of the methodology of engaging the project partners in the development and refine-ment of the tool. The majority of comments were related to the tool's interface rather than fundamental problems with the underlying approach. This further sup-ported the design process adopted in the early stages of the tools development (Cavan and Kingston, 2012).

At the start of the GRaBS project there was some scepticism amongst many of the partners over the usefulness of a web-based tool to support climate change adaptation. Following a defined methodology of collaborative development with all the partners and extensive user testing, the application of the Assessment Tool by partners was implemented (see Figure 18.2). Users of the tool reported:

> It has been used internally by officers, already familiar with climate related risks, to help inform strategy work that is currently underway. In particular, it has been used to test the validity of assumptions about areas of the city that were believed to experience problems.

This supports the case for using the tool to aid strategic decision-making, and in supporting the development of adaptation action plans by all partners. From the outset of the project it was envisaged that the tool would be very useful in supporting stakeholder engagement:

We have included the tool as an activity to be used with community groups in the training material we have developed. At present, this training material includes a reference to the tool which directs users to the above web address; we intend to add further guidance to this web address to help community practitioners in using the tool with their community groups.

This aspect of the tool interacts closely with a further aim of the GRaBS project: to improve stakeholder and community understanding and involvement in planning, delivering and managing green infrastructure in new and existing urban mixed use development, and based upon positive community involvement techniques. For the next partner, a lack of access to relevant data in relation to climate change vulnerability has been a pertinent issue, and part of the tools job has been to identify the gaps in necessary data as much as it is to support adaptation responses.

Ease of use is the main positive [aspect] of the system and awareness raising may be the most important use. The general lack of data on our behalf grabbed the attention of the attendees.

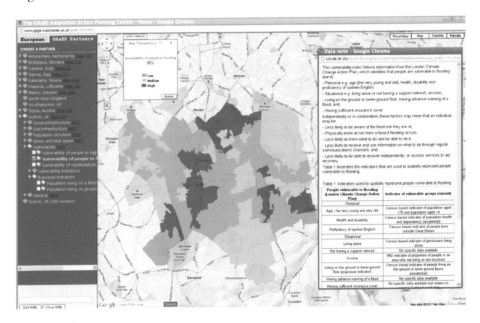

FIGURE 18.3   Final interface

While we did manage to build the tool,[1] the final stages of the project opened up more questions. Now that some of the partners were demonstrating the tool to

---

[1]It was recently rewritten using OpenStreetMap for mainland Europe and OS OpenSpace for the UK due to Google turning off version 2 of its mapping API.

their colleagues and wider stakeholder groups, further issues were raised relating to how the tool would be sustained in the future. One partner highlighted that 'it would be useful to have a tool that individual authorities could be fully in control of'. The tool is hosted as an open source project at the University of Manchester and is available for use in other local authorities by changing the spatial data and associated metadata – although this required knowledge of the Google Maps API and JavaScript programming. Further work in the future will, however, investigate how the tool could be easily updated and edited by non-specialists so as to keep pace with changing knowledge, updates of the underlying data and applications within alternative contexts. Others raised questions in relation to apparent missing data that were not identified at the time of the UNRA, showing that not knowing who to 'talk to' within your own municipality and partner organisations can leave gaps in the knowledge base. Other issues raised related to the lack of climate mod-elling capabilities within the tool – an extremely tricky nut to crack in its own right, and even more so with a web-based tool. Since the tool was developed in 2011, improvements in web-mapping services mean that it is now possible to access an increasing amount of spatial data directly from source. For example, the Environment Agency in England provides its flooding data via web services, and the European Environment Agency has a data clearing house (http://www.eea. europa.eu/data-and-maps/).

Finally, the process of developing the tool has been valuable in terms of partners building bridges between departments during data collection and bringing together different datasets. A key finding of the project has been the need to break down silos between departments in order to gather evidence of the need to act on adaptation. While the tool in its current form could only deal with current vulnerability to climate change, it helps to improve aware-ness amongst decision-makers of what the issues are within their locality, and stimulates conversations around resources required to act. It also highlights the need for a common approach throughout Europe on identifying and collect-ing datasets that enable successful spatial planning for adaptation to climate change. This also has cost benefits – a key consideration within the current financial climate.

Decentralised energy production – siting and connection

Traditionally electrical power networks have been designed as passive systems where there is generally no intervention required under normal operating circum-stances. Electrical demand (load) comes from population and business centres, while electrical supply (generation) has always come from centralised generating power stations, with the flow of electricity largely one-way. Over recent years there has been an acceleration of the number of low-carbon technologies that can generate and connect onto the electrical network. However, one problem that this creates relates to the fact that the electricity networks in many countries were not originally designed for such scenarios, and electrical distribution network operators increasingly need to identify where there may be capacity problems and manage them accordingly.

Electrical networks typically have different parts of the network with different capacities both in terms of the substations and the cables between them. The cables can be under or above ground. When new decentralised energy producers wish to connect to the network they need to establish if the current network can cope with the extra energy being added to the network. If not, the next step is to assess what needs to be done to 'balance' supply and demand without necessarily having to 'reinforce' the network (i.e. increase capacity within the substations and cables) as this is very expensive.

As we attempt to move towards a less carbon-intensive environment there is a continued appetite to connect renewable generation. However, some connections cannot be progressed without significant reinforcement on account of limitations in the existing network. Network reinforcement is both a costly and time-consuming solution. Predictions suggest that there could be over 1.5 million fuel cell electric vehicles (FCEVs) powered by hydrogen in the UK by 2030, by which date 51% of the required hydrogen will be produced by electrolysers (UK H2 Mobility, 2014). Assuming the above, and that FCEV take-up is consistent with existing cars, this would add 355 MW of demand to the electrical network in Scotland alone. Energy companies in Scotland are therefore considering the implications of the installation of electrolysers on the electrical network in support of the hydrogen economy initiatives from the government. Electrolysers could therefore offer significant benefits by providing grid-balancing services to enable more renewable generation where the existing network is constrained. This could allow energy companies to provide cheaper and quicker connections of distributed generation, rather than providing connections requiring more traditional reinforcement as outlined above. In addition, there are the implications for energy providers of large demand, such as electrolysers, and generation being connected in constrained network environments, and particularly in urban environments where demand is high.

One viable solution to this problem is to apply geocomputational techniques to these problems with the potential to identify situations where the supply and demand conundrum can be balanced effectively. Such an approach also requires a range of stakeholders to assess the problem, not just the energy companies. The local authority, land owner and the individual owner of the decentralised generation facility also need to know if they will have a problem with connecting to the network. Our solution here has been to provide a publicly accessible

online mapping tool that undertakes geocomputational modelling of new generation and demand scenarios on the fly. Thus, users can test different scenarios
and combine these with a range of spatial environmental constraint variables that
may inhibit the location of such supply and demand infrastructure. Figure 18.4
shows the user interface with the electrical network showing balanced supply
(green traffic lights), lack of capacity (purple traffic light) and close to critical
(orange traffic light). The user can then turn on/off a range of environmental
constraints; in this case the orange circles are 500 m buffers around schools.

FIGURE 18.4   Optimising supply and demand with environmental factors

Source: http://www.tellus-toolkit.com/video.html

By selecting a substation or power line a user can increase or decrease supply
or demand on the network until they have balanced the network (i.e. changed
the traffic light colour from purple or orange to green). This approach allows
energy providers to move from a static electrical model based approach for connecting low-carbon technologies, to a dynamic geospatial approach by identifying where there is an opportunity to deploy what is called an active network
management scheme on the network. In addition, the geospatial approach, as
opposed to a traditional schematic electrical network approach, allows for a set of
spatial constraint maps such as flooding, telecommunications availability, environmental or archaeological factors. This enables improved engagement with stakeholders by providing indicative information on network availability and
connection costs. By offering stakeholders a service allowing them to register a
proposed scheme, they can geographically begin to look at best-value combined
solutions (i.e. where we can control new generation and flexible demand for a
lower cost and quicker connection).

Sample extract of the Leaflet open source JavaScript Library

This function sets up OpenStreetMap as the base map and creates a legend with a series of spatial constrain maps.

```
//#region Initializes the map.
this.InitializeMap = function () {
    // Move to the initial area.
                        SSE.Spatial.Map.setView(
            new L.LatLng(SSE.Spatial.MapControlModel.InitialLatitude,
                SSE.Spatial.MapControlModel.InitialLongitude),
                SSE.Spatial.MapControlModel.InitialZoomLevel
                                );

    // Create the OpenStreetMap Layer.
    SSE.Spatial.Layers.OpenStreetMap = new L.TileLayer(
        SSE.Spatial.MapControlModel.OsmUrl,
        {
                minZoom: SSE.Spatial.MapControlModel.MinZoom,
                maxZoom: SSE.Spatial.MapControlModel.MaxZoom,
                attribution:        SSE.Spatial.MapControlModel.
                OsmAttribution

        }
    );

    // Add the OpenStreetMap Layer in the map.
        SSE.Spatial.Map.addLayer(SSE.Spatial.Layers.OpenStreetMap);

    // Render the spatial constraints on the map.
    SSE.Spatial.RenderSpatialConstraints();

    // Set the base maps group layer.
    SSE.Spatial.Layers.BaseMapsGroupLayer = {
        "OpenStreetMap": SSE.Spatial.Layers.OpenStreetMap
                                };

    // Set the constraints group layer.
    SSE.Spatial.Layers.ConstraintsGroupLayer = {
        "<img src='assets/img/sse/legend-awi.png'/><span>Ancient Woodland
        Inventory</span>": SSE.Spatial.Layers.AwiConstraints,
        "<img src='assets/img/sse/legend-lnr.png'/><span>Local Nature
        Reserves</span>": SSE.Spatial.Layers.LnrConstraints,
        "<img src='assets/img/sse/legend-sac.png'/><span>Special
        areas
        of conservation</span>": SSE.Spatial.Layers.SacConstraints,
            "<img src='assets/img/sse/legend-sssi.png'/><span>Sites of
                Special Scientific Interest</span>":
                SSE.Spatial.Layers.SssiConstraints,
                "<img src='assets/img/sse/legend-
                school.png'/><span>Schools</span>":
                SSE.Spatial.Layers.SchoolConstraints,
```

```
            "<img src='assets/img/sse/legend-school.png' /><span>200m
            School
                              Buffers</span>":
                SSE.Spatial.Layers.SchoolBuffer200mConstraints,
            "<img src='assets/img/sse/legend-school.png' /><span>500m
            School
            Buffers</span>": SSE.Spatial.Layers.SchoolBuffer500mConstraints
        };
};
//#endregion
```

This approach can inform the user of the best sites for locating facilities that add generation or demand onto the network. Such an approach allows the user to apply a method and create a model to identify limitations on the electrical distribution network, and enables the uploading and analysis of the impact of planned additions to the network in an interactive manner. This geocomputational approach allows the user to apply potential solutions to the limitations on the electrical distribution network by using an online GIS environment; and to apply different factors and constraints that can be manipulated in an interactive manner using the online mapping interface. It allows the user to add (remove) of a range of site characteristics and contextual data to (from) the map – for example, flood maps, proximity to schools as appropriate. Such an approach gives more certainty to potential generator,s as they get a better indication of what capacity exists on the current network. Without this certainty they risk the possibility of building their generating scheme without knowing if they can connect to the network, and, in many cases, not bringing forward more sustainable modes of energy production.

Conclusions

This chapter has discussed two examples of the application of online geocomputational mapping tools that support public participation in spatial decision-making. The first example showed how stakeholders can participate in adapting urban areas to deal with a changing climate. The second example showed how, as we move towards a less carbon-intensive environment, there is a continued appetite to connect renewable generation. The problem here, though, lies in the fact that electricity generation companies have not planned for this decentralisation of energy supply and there are problems associated with connecting such supply to the current electricity network.

Recent developments in web-based technologies mean that it is now possible not only to provide public stakeholder access to interactive mapping but also to interrogate the underlying computational models. Users can interact with and undertake their own scenario analysis in ways that were previously not possible in an online environment. Using various scripting languages (e.g. PHP, Javascript, MySQL), online spatial databases using web mapping services, AJAX (e.g. jQuery) and online mapping libraries (e.g. Leaflet, D3), developers can implement high-end interactive geocomputational methods to support complex spatial decision-making problems.

As these technologies become more mature over the coming years it should become easier for decision-makers to adopt geocomputational approaches to support decision-making more easily than ever before. In the past, these approaches have been too complex to be adopted in main stream decision-making, but, as Klosterman (2008: 174) observed, 'we are witnessing a second revolution of computer use in planning'. Such approaches will become even more widespread as the technology develops further in increasingly flexible and well-designed packages that are easily customisable with limited programming knowledge. The key to their success will be for decision-makers and other stakeholders to understand that good decision-making is both an art and a science, and that geocomputational tools are there to support and aid them to help improve the outcomes.

FURTHER READING

Pickles, J. (1995). *Ground Truth: The Social Implications of Geographical Information Systems*. New York: Guilford Press.
Although this is now quite an old book, many of the criticisms in it still hold true today. Pickles discusses the social implications of GIS use which led to the PPGIS response from the GIS community. The book provides what was the first in-depth engagement of GIS with critical social theory.

Ramasubramanian, L. (2008) *Geographic Information Science and Public Participation*. Berlin: Springer-Verlag.
This book argues that many positive planning and policy outcomes can result when GIS experts and community organisers work together to address problems affecting our neighbourhoods and cities. The case studies are drawn from multiple scales and represent stakeholders, although the focus is on North America. It highlights the pathways to facilitate social change through consensus building and provides practical guidelines for policy-makers seeking to implement consensus-building approaches that are supported by GIS and other digital tools.

Svennerberg, G. (2010) *Beginning Google Maps API 3*. New York: Apress.
This is the techie book for those interested in trying to build their own web-based mapping tools using Google Maps API 3. It provides some of the skills and knowledge necessary to incorporate Google Maps version 3 on web pages in both desktop and mobile browsers. It also describes how to deal with common problems that most map developers encounter at some point, such as performance and usability issues with having too many markers and possible solutions to that. The

book is aimed at web designers/developers with a basic knowledge of HTML, CSS and JavaScript. If you have API 2 applications there is help here to easily transfer to the new API.

There are a number of other techie books out there, but as things change all the time it is best to go direct to their websites to access the APIs and help files. Useful links include:

Google Maps https://developers.google.com/maps/documentation/javascript/

Leaflet http://leafletjs.com/

OpenStreetMap http://wiki.openstreetmap.org/wiki/API

OpenLayers http://openlayers.org/

OS OpenSpace http://www.ordnancesurvey.co.uk/business-and-government/products/os-openspace/

ACKNOWLEDGEMENTS

I would like to thank many of my PhD students who over the years have provided the space in which to discuss and elaborate on these ideas in my research. Particular thanks go to those of you who have focused on the many aspects of web-based PPGIS and PSS, including Khai Zhou, Jessica Pineda-Zumaran, Félix Aponte-González and current students Seyedehsomayeh Taheri Moosav, Moozhan Shakeri and Vasilis Vlastaras.

CONCLUSION

THE FUTURE OF APPLIED GEOCOMPUTATION

Chris Brunsdon and Alex Singleton

This book has presented 18 chapters, arranged into the thematic groupings describing how the world looks through visualisation and exploratory data analysis; how movements in space can be represented across a variety of spatial and temporal scales; how geographical decisions can be made through the application of spatial algorithms; how geocomputational techniques can be used to explain spatial processes; and finally, how new methods of enabling interactions are impacting upon the field of geocomputation. Inevitably there will be overlap, and chapters could potentially be arranged in multiple ways. As such, in this final chapter of the book, we draw a series of themes together from the chapters and place these within the context of our speculations about those emerging and future directions in applied geocomputational research.

Open data, tools and reproducibility through open practices in science

An increasingly important theme for the future of geocomputation will be reflected in a more open paradigm of conducting research (Brunsdon and Singleton, Chapter 15), comprising a tighter coupling of data, software, analytical methods and final presentations. Sui (2014) usefully places such developments within a broader anatomy of GIS (Longley et al., 2005), and a number of themes related to open practices of science are echoed within this book.

Contemporary GIS adopt either an open source or closed source model (Steiniger and Bocher, 2009), with open source applications available in a source code format within the public domain, and under a number of licences that permit different permutations of reuse, adaptation and redistribution. Such software is prevalently available at no cost, and development work is often completed by a community of programmers, although some have debated the extent to which open source software development is a group effort (Krishnamurthy, 2002). An introduction to open source GIS software was presented by O'Brien (Chapter 17), and some specific applications were illustrated with the statistical programming language R (Cheshire and Lovelace, Chapter 1; Alexiou and Singleton, Chapter 8;

Spielman and Folch, Chapter 9; and Brunsdon and Comber, Chapter 16) and Python (Lewis, Chapter 10; Rey, Chapter 14). However, it was interesting to note that the use of open source tools was not universal in this book; for example, others adopted either traditional GIS software such as ArcGIS from ESRI (Tomintz, Clarke and Alfadhli, Chapter 11); and other software such as MATLAB (Rohde and Corcoran, Chapter 7) or Excel (Morrissey, Chapter 13). Some authors developed bespoke software applications for their work (Batty, Chapter 3; Harland and Birkin, Chapter 5), with still others developing open source tools that aim at supporting a broad community of users, techniques and application areas. For example, Crooks (Chapter 4) introduces the MASON multiagent simulation library, and Rey (Chapter 14) describes PySAL, which provides a set of Python tools that implement a variety of spatial analytical methods. Contemporary geo-computation is reflected by a hybridisation of software tools, both in terms of where functionality is situated (e.g. spatial analysis functions within databases), and in terms of licensing frameworks (e.g. commercial GIS providers making some content open source). We would argue that such trends and cross-fertilisation will continue. However, as open source software becomes more mature, these will also likely have an impact on the traditional markets of commercial GIS and their implementation of geocomputational functionality.

Open data represents a general movement towards public data being released with less restrictive licences, enabling a new paradigm in data sharing and reuse. Within a UK context, the term 'open data' has an explicit definition, and relates to those data that have been released under the Open Government Licence for Public Sector Information (http://www.nationalarchives.gov.uk/doc/open-government-licence/). However, this term is also used more flexibly to refer to those other data that are free from reuse restrictions or financial cost of acquisition. Alexiou and Singleton (Chapter 8) gave an example of using open data in the development of a geodemographic classification. We postulate that open data will grow in prevalence over the next decade, with support building as a result of enhanced economic benefits derived from the exploitation of new business models, alongside both perceived and measured societal gains through applications linked to education, policing, the environment, health and well-being. Many of the platforms currently used to disseminate open data can be considered in their infancy, and the structure of such data released has been variable in terms of scale (e.g. anonymised individual vs. aggregated data) and format (e.g. XML, CSV, Excel). We argue that geocomputation will play a role in enhancing such cyberinfrastructure, enabling more flexible dissemination through the openness of formats, data linkage, disclosure control, sharing, visualisation and web-based services coupling to third-party applications.

As outlined by Brunsdon and Singleton (Chapter 15), reproducible research is enabled when work is published, and full details about the methods and data used to obtain the reported results are transparent. In much current social science research this is ensured through the peer review process; however, we argue that such mechanisms should be tightened in the future. As Brunsdon and Singleton (Chapter 15) outlined, and was illustrated in the assembly of a number of other chapters for this book (e.g. Cheshire and Lovelace, Chapter 1; Brunsdon and Comber, Chapter 16; and Spielman and Folch, Chapter 9), data, methods and geocomputational analysis can be embedded in analytical workflows linking to written interpretation. Examples within this book included the use of Sweave, a tool for embedding R code into LaTeX, which is a document processing language. We believe that such tighter coupling will begin to appear as criteria for submission to some social science journals, as is becoming common practice in a number of science publications. There are of course geocomputational challenges, for example, related to data with restrictive licences, how to manage stochastic models and, more generally, how analytical work involving high-performance computing could be reproduced.

Turning Big Data into big information

In addition to open data mentioned in the previous section, the other most common contemporary prefix to the word data is the term big. "Big Data" relate to those new data sources that are very large in size, making manipulation and analysis complex with traditional tools and methods. Such data are generated through a variety of mechanisms including, but not limited to earth observation satellites, transportation networks, consumer data recording and social media activity. For those engaged in geocomputation, such new data present potentially interesting new areas of research. However, it is worth noting that contemporary smaller data are in fact 'big' within a historical context (Graham and Shelton, 2013); and we assert that caution is needed with the term. Furthermore, accompanying the growth in use of the term 'Big Data', there have been some grand claims about the impact that such resources will have on a new data-intensive form of science. Hey et al. (2009) argue that science has been characterised by a series of paradigms including the empirical/descriptive, theoretical (testing models) and computational (simulation). They argue that we are entering a 'fourth paradigm' of science, characterised by statistical exploration and data mining. Others, and rather dangerously in our view, have speculated an 'end of theory', where correlation in "Big Data" is enough, relative to more traditional modelling approaches involving hypothesis testing (Anderson, 2008). While we argue that such boosterish claims should be treated with caution (see, for example, Brunsdon, 2014), we would make the pragmatic case that both traditional and new approaches offer strength, and that for problems relating to a spatial dimension, a framework of geocomputation provides much potential for integration.

However, it is clear to us that as the prevalence of Big Data grows, this will create new research challenges for geocomputation. For example, how could those models presented by Nakaya (Chapter 12) be implemented for data-rich, rather than data-poor, applications (Kitchin, 2013)? Or to what extent could the spatial and temporal data warehousing technology outlined by Miller (Chapter 6) be made efficient, robust and resilient under the constraint of such increased volumes of new data? One area that is likely to see increased activity is the area of cloud computing, referring to both distributed storage and processing provided by organizations such as Google (https://cloud.google.com/products/compute-engine/) and Amazon (http://aws.amazon.com/ec2/). To some extent, such developments are not new, and the use of high-performance computing has been one of the hallmarks of geocomputation since the 1990s (see Turton and Openshaw, 1998). However, in recent years, we have seen an evolution from traditional 'grid' computing infrastructure (Harris et al., 2010), perhaps housed by a single or group of institutions, to those in the 'cloud', a typically more distributed assemblage of computers often run by commercial organisations. Services comprise a mixture of storage and processing, including new technologies such as graphics processing units, and billing is flexible to meet a range of potential usage scenarios. Furthermore, for services such as ec2 from Amazon, operating system images are available for running both commercial (e.g. https://aws.amazon.com/solutions/global-solution-providers/esri/) and open source GIS tools within the cloud (http://boundlessgeo.com/2010/09/opengeo-suite-community-edition-on-amazon-web-services/); or more generic operating system images enabling simple deployment of bespoke software. Such advances are making high-performance computing more accessible through ease of access and cost, and as such we would anticipate the utility of such services will likely be exploited more explicitly by the geocomputation community in the future. However, there are constraints that require further research, such as how sensitive data might be managed within such settings where the place of storage and computation is not necessarily known.

Fast versus slow temporal dynamics and the measurement and conceptualizations of place

Activities on the earth's surface are situated in both space and time, and in this book a variety of the authors tackle such dynamics across a range of spatial and temporal scales from the aggregate (e.g. Batty, Chapter 3; Crooks, Chapter 4; Morrissey, Chapter 13) to the personal and/or micro scale (Torrens, Chapter 2; Rohde and Corcoran, Chapter 7; Harland and Birkin, Chapter 5). In some sense, these examples begin to address those concerns raised by Goodchild (2013) who discusses how the use of maps as a conceptual framework for most

GIS software has resulted in time being considered in only very course resolution: the distribution of a population from a census, or the location and extent of physical features such as a mountain range. The examples presented within this book consider dynamics at varying geographic and temporal scales. However, this should perhaps not be surprising, and indeed Longley (1998: 3) notes: '[t]he hallmarks of geocomputation are those of research-led applications which emphasise process over form, dynamics over statics, and interaction over passive response.'

Critical within this early definition of geocomputation was an emphasis on dynamics and interactivity, and although some have offered paths for integration into desktop GIS (Yu and Shaw, 2008), or extension of geographical indicators (Cheng et al., 2012), such concepts typically remain an adjunct rather than an integral component. Some have gone further; for example, Batty (2012a: 193) says that such '[n]ew data begets new theory'; and there is a challenge to explore how such short-term Big Data may be linked with more traditional longer-term data such as censuses or surveys. Some progress in this area was reported by Singleton and Spielman (2014) in their discussion of a requirement to shift emphasis from a variables to a contextual paradigm in the social sciences, and is illustrated in the chapters by Alexiou and Singleton (Chapter 8) and Spielman and Folch (Chapter 9) in this book. As such, we would argue that future analysis will continue to exploit the dynamics of time in addition to location, thus remaining a key facet of geocomputation research.

A related challenge for geocomputation research will be to derive new methods that address what Sui and Goodchild (2011: 1744) refer to as the 'world of place (social media)' rather than the 'world of space (traditional GIS)'. This moves the ontology of place from one which focuses on Cartesian space (e.g. x and y coordinates) to something more flexible; for example, 'London' might be defined as a place, rather than an (x, y) pair for the centroid of the city, or the boundary of an administrative unit reflecting official extent. This differentiation in part relates to burgeoning volumes of new data created through processes of volunteered geographic information, defined as 'the widespread engagement of large numbers of private citizens, often with little in the way of formal qualifications, in the creation of geographic information' (Goodchild, 2007: 212). A useful disambiguation of VGI, is provided by Wilson and Graham (2013). Furthermore, there are increasing volumes of network-based data generated through social media, but also other forms of large data generation in cities such as mobility tracking through transit networks, or prospectively through RFID, or low-power Bluetooth devices; and these are also eliciting a growing relevance of network-based ontologies (Sui and Goodchild, 2011) for applied geocomputation. As such, we would argue that a further key future direction will be in the area of geographic network analysis (see O'Sullivan, 2014).

Cities and the scale of geocomputation

Modelling and simulation have a rich history of application within the context of cities, and maintain contemporary relevance (Clarke, 2013). Batty (2009, 2010, 2013) describes a new science of cities, building on developments that have emerged from the wider-complexity sciences that represent a shift in thinking about how cities function, from a concept of top-down and central ordering, through to more organic and bottom-up processes where cities structure and function are a result of the dynamic actions of millions of individual and group decisions, typically occurring with limited central control (Batty, 2012b). In such applications, it is not uncommon to create analytical frameworks that mix a range of the geocomputational methods presented within this book, for example micro-simulation (Harland and Birkin, Chapter 5) to create synthetic populations, through spatial interaction models to calibrate flows (Morrissey, Chapter 13) and agent-based simulation (Crooks, Chapter 4; Torrens, Chapter 2) to manage inter-actions and emergence across scales and contexts.

Much of this new science is data-rich, embedding those resources discussed earlier in the context of 'Big Data' into modelling frameworks for either calibra-tion or validation. Within this context, visualisation utilising tools such as those presented by Cheshire and Lovelace (Chapter 1) becomes important; as do those methods of conflating disparate data sources into meaningful indicators (Spielman and Folch, Chapter 9; Alexiou and Singleton, Chaper 8) alongside the storing of such resources within spatio-temporal databases (Miller, Chapter 6). Within this context, geocomputation in the future will not only have a role in facilitating a deeper embedding of the production of data within the fabric of cities (Dodge and Kitchin, 2005), but also play a significant role in enabling the consumption of data, and information derived through modelling, and particularly so for those services designed to engender citizen participation (Kingston, Chapter 18). Within this setting we will likely see the scale of geocomputation become more detailed (see Torrens, Chapter 2), enabled by enhanced computational power and also an increasing prevalence of data available about how individuals move within cities. Furthermore, we would argue that geocomputation will impact the scale of data consumption within cities, and will increasingly be embedded within those tech-nologies enabling the ambient supply of spatial data and informational resources, for example, through mobile devices such as phones or augmented reality eyewear.

Concluding comments

It is always difficult to pull together a text designed to represent all activities within a field, and particularly so for one such as geocomputation where there has been such a great deal of activity since the 1990s. Perhaps the most enduring feature of geocomputation, and testament to its relevance today, is how embedded it has become in many of those activities we now consider as routine to our daily lives: for example, navigation between places utilising distributed routing facilities enabled on mobile devices. We hope that through the chapters presented, these highlight the essence of geocomputational methods, and those applied contexts in which they are implemented. In this final chapter, we have aimed to provide some of our own speculations about the areas in which we see geocomputation having future impact. It is clear to us that there are a great deal of opportunities for researchers to engage with geocomputation, and the relevance of these methods will increase, as they have done since the definition of the field in the 1990s.

REFERENCES

Abdul-Rahman, A. and Pilouk, M. (2008) *Spatial Data Modelling for 3D GIS*. Berlin: Springer.

Agarwal, P. and Skupin, A. (2008) *Self-organizing Maps: Applications in Geographic Information Science*. Chichester: Wiley.

Agostinelli, C. and Lund, U. (2011) R package 'circular': Circular Statistics (version 0.4-3). https://r-forge.r-project.org/projects/circular.

Allen, J.F. (1984) Towards a general theory of action and time. *Artificial Intelligence*, 23: 123–154.

Alonso, W. (1964) *Location and Land Use: Toward a General Theory of Land Rent*. Cambridge, MA: Harvard University Press.

Alonso, W. (1967) A reformulation of classical location theory and its relation to rent theory. *Papers in Regional Science*, 19(1): 22–44.

Anagnostopoulos, S.A. (1988) Pounding of buildings in series during earthquakes. *Earthquake Engineering & Structural Dynamics*, 16(3): 443–456.

Anderson, C. (2008) The end of theory: The data deluge makes the scientific method obsolete. *Wired*, 23 June. http://www.wired.com/science/discoveries/magazine/16-07/pb_theory.

Anderson, J.R. and Lebiere, C. (1998) *The Atomic Components of Thought*. Mahwah, NJ: Lawrence Erlbaum Associates.

Andrienko, G., Malerba, D., May, M. and Teisseire, M. (2006) Mining spatio-temporal data. *Journal of Intelligent Information Systems*, 27: 187–190.

Andrienko, G., Andrienko, N., Dykes, J., Fabrikant, S. and Wachowicz, M. (2008a) Geovisualization of dynamics, movement and change: Key issues and developing approaches in visualization research. *Information Visualization*, 7: 173–180.

Andrienko, N., Andrienko, G., Pelekis, N. and Spaccapietra, S. (2008b) Basic concepts of movement data. In F. Giannotti and D. Pedreschi (eds), *Mobility, Data Mining and Privacy* (pp. 15–38). Berlin: Springer-Verlag.

Andrienko, G., Andrienko, N., Bak, P., Keim, D. and Wrobel, S. (2013) *Visual Analytics of Movement*. Heidelberg: Springer.

Anselin, L. (1980) Estimation methods for spatial autoregressive structures: A study in spatial econometrics. PhD thesis, Cornell University.

Anselin, L. (1988) *Spatial Econometrics: Methods and Models*. Dordrecht: Kluwer.

Anselin, L. (1990) Spatial dependence and spatial structural instability in applied regression analysis. *Journal of Regional Science*, 30(2): 185–207.

Anselin, L. (1995) Local indicators of spatial association – LISA. *Geographical Analysis*, 27(2): 93–115.

Anselin, L. (1996) The Moran scatterplot as an ESDA tool to assess local instability in spatial association. In M. Fischer, H.J. Scholten and D. Unwin (eds), *Spatial Analytical Perspectives on GIS*. London: Taylor & Francis.

Anselin, L. and Kelejian, H. (1997) Testing for spatial error autocorrelation in the presence of endogenous regressors. *International Regional Science Review*, 20: 153–182.

Anselin, L. and Rey, S.J. (1991) Properties of tests for spatial dependence in linear regression models. *Geographical Analysis*, 23: 112–131.

Anselin, L. and Rey, S.J. (2012) Spatial econometrics in an age of CyberGIScience. *International Journal of Geographic Information Science*, 26: 2211–2226.

Anselin, L., Bera, A., Florax, R.J.G.M. and Yoon, M. (1996) Simple diagnostic tests for spatial dependence. *Regional Science and Urban Economics*, 26: 77–104.

Applebaum W. (1965) Can store location research be a science? *Economic Geography*, 41(3): 234–237.

Appleton, D.R., French, J.M. and Vanderpump, M.P.J. (1996) Ignoring a covariate: An example of simpson's paradox. *American Statistician*, 50(4): 340–341.

Arnstein, S. (1969) A ladder of citizen participation. *Journal of the American Institute of Planners*, 35(4): 216–224.

Arraiz, I., Drukker, D.M., Kelejian, H.H. and Prucha, I.R. (2010) A spatial Cliff-Ord-type model with heteroskedastic innovations: Small and large sample results. *Journal of Regional Science*, 50(2): 592–614.

Arthur, W.B. (1994) Inductive reasoning and bounded rationality. *American Economic Review*, 84(2): 406–411.

Ashby, D.I. and Longley, P.A. (2005) Geocomputation, geodemographics and resource allocation for local policing. *Trans GIS*, 9(1): 53–72.

Atkinson, P.M., German, S.E., Sear, D.A. and Clark, M.J. (2003) Exploring the relations between riverbank erosion and geomorphological controls using geographically weighted logistic regression. *Geographical Analysis*, 35: 58–82.

Augustijn-Beckers, E., Flacke, J. and Retsios, B. (2011) Simulating informal settlement growth in Dar es Salaam, Tanzania: An agent-based housing model. *Computers, Environment and Urban Systems*, 35(2): 93–103.

Aveyard, P., Manaseki S. and Chambers, J. (2002) The relationship between mean birth weight and poverty using the Townsend deprivation score and the super profile classification system. *Public Health*, 116(6): 308–314.

Avolio, M., Di Gregorio, S., Mantovani, F., Pasuto, A., Rongo, R., Silvano, S. and Spataro, W. (2000) Simulation of the 1992 Tessina landslide by a cellular automata model and future hazard scenarios. *International Journal of Applied Earth Observation and Geoinformation*, 2(1): 41–50.

Ayeni, B., Rushton, G. and McNulty, M.L. (1987) Improving the geographical accessibility of health care in rural areas: A Nigerian case study. *Social Science & Medicine*, 25(10): 1083–1094.

Baayen, R.H., Davidson, D.J. and Bates, D.M. (2008) Mixed-effects modeling with crossed random effects for subjects and items. *Journal of Memory and Language*, 59: 390–412.

Bainbridge, W.S. (2007) The scientific research potential of virtual worlds. *Science*, 317 (5837): 472–476.

Ball, P. (2003) *Critical Mass: How One Thing Leads to Another.* London: Arrow Books.

Ballas, D. and Clarke, G.P. (2001) The local implications of major job transformations in the city: A spatial microsimulation approach. *Geographical Analysis*, 33: 291–311.

Ballas, D., Clarke, G., Dorling, D., Rigby, J. and Wheeler, B. (2006) Using geographical information systems and spatial microsimulation for the analysis of health inequalities. *Health Informatics Journal*, 12(1): 65–79.

Baller, R.D., Anselin, L., Messner, S.F., Deane, G., and Hawkins, D.F. (2001) Structural covariates of U.S. county homicide rates: Incorporating spatial effects. *Criminology*, 39(3): 561–588.

Banerjee, A., Dhillon, I.S., Ghosh, J. and Sra, S. (2005) Clustering on the unit hypersphere using von Mises–Fisher distributions. *Journal of Machine Learning Research*, 6: 1345–1382.

Banos, A., Godara, A. and Lassarre, S. (2005) Simulating pedestrians and cars behaviours in a virtual city: An agent-based approach. In *Proceedings of the European Conference on Complex Systems*, Paris.

Barabási, A.-L. and Albert, R. (1999) Emergence of scaling in random networks. *Science*, 286(5439): 509–512.

Barnes, T., Encarnação, L.M. and Shaw, C.D. (2009) Serious games. *Computer Graphics and Applications*, 29(2): 18–19.

Barni, M., Perez-Gonzalez, F., Comesaña, P. and Bartoli, G. (2007) Putting reproducible signal processing into practice: A case study in watermarking. *Proceedings of the IEEE International Conference on Acoustics, Speech and Signal Processing*, 4, 1261–1264.

Barros, J. (2012) Exploring urban dynamics in Latin American cities using an agent-based simulation approach. In A.J. Heppenstall, A.T. Crooks, L.M. See and M. Batty (eds), *Agent-Based Models of Geographical Systems* (pp. 571–590). Dordrecht: Springer.

Barthelemy, M. (2011) Spatial networks. *Physics Reports*, 499: 1–101.

Batey, P. and Brown, P. (1995) From human ecology to customer targeting: The evolution of geodemographics. In P.A. Longley and G. Clarke (eds), *GIS for Business and Service Planning*. Cambridge: GeoInformation International.

Batey, P. and Brown, P. (2007) The spatial targeting of urban policy initiatives: A geodemographic assessment tool. *Environment and Planning A*, 39: 2774–2793.

Batty, M. (1976) *Urban Modelling: Algorithms, Calibrations, Predictions.* Cambridge: Cambridge University Press.

Batty, M. (1997a) The computable city. *International Planning Studies*, 2(2): 155–173.

Batty, M. (1997b) Predicting where we walk. *Nature*, 388(6637): 19–20.

Batty, M. (2005) *Cities and Complexity: Understanding Cities with Cellular Automata, Agent-Based Models, and Fractals*. Cambridge, MA: The MIT Press.

Batty, M. (2006) Rank clocks. *Nature*, 444: 592–596.

Batty, M. (2008) The size, scale, and shape of cities. *Science*, 319(5864): 769–771.

Batty, M. (2009) Cities as complex systems: Scaling, interactions, networks, dynamics and urban morphologies. In R. Meyers (ed.), *Encyclopedia of Complexity and Systems Science* (Vol. 1, pp. 1041–1071). Berlin: Springer.

Batty, M., (2010) Visualising space–time dynamics in scaling systems. *Complexity*, 16(2): 51–63.

Batty, M. (2012a) Smart cities, big data. *Environment and Planning B*, 39: 191–193.

Batty, M. (2012b) Building a science of cities. *Cities*, 29(Suppl 1): S9–S12.

Batty, M. (2013) *The New Science of Cities*. Cambridge, MA: MIT Press.

Batty, M. and Brail, R.K. (2008) Planning support systems: progress, predictions, and speculations on the shape of things to come. In R.K. Brail, *Planning Support Systems for Cities and Regions* (pp. 3–30). Cambridge, MA: Lincoln Institute of Land Policy.

Batty, M. and Longley, P. (1994) *Fractal Cities: A Geometry of Form and Function*. London: Academic Press.

Batty, M., Desyllas, J. and Duxbury, E. (2003a) The discrete dynamics of small-scale spatial events: Agent-based models of mobility in carnivals and street parades. *International Journal of Geographical Information Science*, 17(7): 673-697.

Batty, M., Desyllas, J. and Duxbury, E. (2003b) Safety in numbers? Modelling Crowds and Designing Control for the Notting Hill Carnival. *Urban Studies*, 40(8): 1573-1590.

Baudrillard, J. (1994) *Simulacra and Simulation*. Ann Arbor, MI: University of Michigan Press.

Baxter, G.W. and Behringer, R. (1990) Cellular automata models of granular flow. *Physical Review A*, 42(2): 1017.

Beaven, R., Birkin, M., Crawford Brown, D., Kelly, S., Thoung, C., Tyler, P. and Zuo, C. (2014) Future demand for infrastructure services. In J. Hall and M. Tran (eds), *Planning Infrastructure for the 21st Century: System-of-Systems Methodology for Analysing Society's Lifelines in an Uncertain Future*. Farnham: Ashgate.

Bédard, Y. and Han, J. (2009) Fundamentals of spatial data warehousing for geographic knowledge discovery. In H.J. Miller and J. Han (eds), *Geographic Data Mining and Knowledge Discovery*, 2nd edn (pp. 45–68). Boca Raton, FL: CRC Press.

Benenson, I. and Torrens, P.M.. (2004) *Geosimulation: Automata-Based Modeling of Urban Phenomena*. Hoboken, NJ: Wiley.

Benenson, I., Martens, K. and Birfir, S. (2008) PARKAGENT: An agent-based model of parking in the city. *Computers, Environment and Urban Systems*, 32(6): 431–439.

Benguigui, L., Blumenfeld-Lieberthal, E. and Batty, M. (2008) Macro and micro dynamics of city size distributions: The case of Israel. CASA Working Paper 139, University College London.

Berens, P. (2009) CircStat: A MATLAB toolbox for circular statistics. *Journal of Statistical Software*, 31: 1–21.

Bernoulli, D. (1808) *Recherches physiques et astronomiques sur le problème proposé pour la seconde fois par l'Académie royale des sciences de Paris*. Paris: Chez Bachelier.

Berry, B.J.L. (1972) The goals of city classification. In B.J.L. Berry and K.B. Smith (eds), *City Classification Handbook: Methods and Applications* (pp. 1–26). New York: Wiley-Interscience.

Berry, B.J.L. and Kasarda, J.D. (1977) *Contemporary Urban Ecology*. New York: Macmillan.

Birkin, M. (1995) Customer targeting, geodemographics and lifestyle approaches. In P.A. Longley and G. Clarke (eds), *GIS for Business and Service Planning*. Cambridge: GeoInformation International.

Birkin, M. (2013) Big data challenges for geoinformatics. *Geoinformatics and Geostatistics: An Overview*, 1(1): 1–2.

Birkin, M. and Clarke, M. (1988) SYNTHESIS – a synthetic spatial information system for urban and regional analysis: methods and examples. *Environment and Planning A*, 20: 1645–1671.

Birkin, M. and Clarke, M. (1989) The generation of individual and household incomes at the small area level using synthesis. *Regional Studies*, 23: 535–548.

Birkin, M. and Clarke, G.P. (1991) Spatial interaction in geography. *Geography Review*, 4: 16–24.

Birkin, M. and Clarke, G.P. (2012) The enhancement of spatial microsimulation models using geodemographics. *Annals of Regional Science*, 49(2): 515–532.

Birkin, M., Turner, A., Wu, B., Townend, P., Arshad, J. and Xu, J. (2009) Moses: A grid-enabled spatial decision support system. *Social Science Computer Review*, 27: 493–508.

Birkin, M., Harland, K. and Malleson, N. (2013) The classification of space–time behaviour patterns in a British city from crowd-sourced data. In B. Murgante et al., *Computational Science and Its Applications – ICCSA 2013*, Lecture Notes in Computer Science 7974 (pp. 179-192). Berlin: Springer.

Bissonnette, L., Wilson, K., Bell, S. and Shah, T.I. (2012) Neighbourhoods and potential access to health care: The role of spatial and aspatial factors. *Health and Place*, 18: 841–853.

Bivand, R. and Gebhardt, A. (2000) Implementing functions for spatial statistical analysis using the R language. *Journal of Geographical Systems*, 2(3): 307–317.

Bivand, R.S., Pebesma, E.J. and Gómez-Rubio, V. (2013) *Applied Spatial Data Analysis with R*. New York: Springer.

Bleisch, S., Duckham, M., Galton, A., Laube, P. and Lyon, J. (2014) Mining candidate casual relationships in movement patterns. *International Journal of Geographical Information Science*, 28: 363–382.

Bonabeau, E. (2002) Agent-based modelling: Methods and techniques for simulating human systems. *Proceedings of the National Academy of Sciences of the USA*, 99(3): 7280-7287.

Boulic, R., Noser, H. and Thalmann, D. (1994) Automatic derivation of curved human walking trajectories from synthetic vision. Paper read at Computer Animation '94, Geneva, 25–28 May.

Bouvier, E., Cohen, E. and Najman, L. (1997) From crowd simulation to airbag deployment: Particle systems, a new paradigm of simulation. *Journal of Electronic Imaging*, 6(1): 94-107.

Bowman, A. and Azzalini, A. (1997) *Applied Smoothing Techniques for Data Analysis: The Kernel Approach with S-Plus Illustrations*. Oxford: Oxford University Press.

Brantingham, P., Glasser, U., Kinney, B., Singh, K. and Vajihollahi, M. (2005) A computational model for simulating spatial aspects of crime in urban environments. *2005 IEEE International Conference on Systems, Man and Cybernetics*, 4: 3667–3674.

Bratman, M.E., Israel, D.J. and Pollack, M.E. (1988) Plans and resource-bounded practical reasoning. *Computational Intelligence*, 4(3): 349–355.

Brindley, T.S. and Raine, J.W. (1979) Social area analysis and planning research. *Urban Studies*, 16: 273–289.

Brooks, R.A. (1986) A robust layered control system for a mobile robot. *IEEE Journal of Robotics and Automation*, 2(1): 14–23.

Brown, L.A. and Holmes, J. (1971) The delimitation of functional regions, nodal regions, and hierarchies by functional distance approaches. *Journal of Regional Science*, 11(1): 57–72.

Brown, P. (2011) Weatherwatch: Phenology in the UK. *Guardian*, 11 April. http://www.guardian.co.uk/news/2011/apr/11/weatherwatch-phenology.

Brown, P.J.B., Hirschfield, A.F.G. and Batey, P.W.J. (2000) Adding value to census data: Public sector applications of the Super Profiles geodemographic typology. *Journal of Cities and Regions*, 10: 19–32.

Brunsdon C. (2011) pycno: Pycnophylactic Interpolation. R package version 1.1. Available at: http://cran.r-project.org/web/packages/pycno/ (accessed 13 April 2012).

Brunsdon, C. (2014) Spatial science – Looking outward. *Dialogues in Human Geography*, 4: 45–49.

Brunsdon, C. and Charlton, M. (2011) An assessment of the effectiveness of multiple hypothesis testing for geographical anomaly detection. *Environment and Planning Part B*, 38(2): 216.

Brunsdon, C. and Corcoran, J. (2006) Using circular statistics to analyse time patterns in crime incidence. *Computers, Environment and Urban Systems*, 30: 300–319.

Brunsdon, C., Fotheringham, A.S. and Charlton, M. (1996) Geographically weighted regression: A method for exploring spatial nonstationarity. *Geographical Analysis*, 28: 281–289.

Buckheit, J. and Donoho, D. L. (1995) Wavelab and reproducible research. Technical Report 474, Dept. of Statistics, Stanford University.

Budic, Z.D. and Godshalk, D.R. (1994) Implementation and management effectiveness in adoption of GIS technology in local governments. *Computers, Environment and Urban Systems*, 18(5): 285–304.

Burgess, E. (1964) Introduction. In E. Burgess and D. J. Bogue (eds), *Contributions to Urban Sociology* (pp. 7–13). Chicago: University of Chicago Press.

Burgess, W. (1925) The growth of the city: An introduction to a research project. In R.E. Park, W. Burgess and R.D. McKenzie (eds), *The City* (pp. 47–62). Chicago: University of Chicago Press, .

Burkitt, M., Walker, D., Romano, D.M. and Fazeli, A. (2011) Modelling sperm behaviour in a 3D environment. In *Proceedings of the 9th International Conference on Computational Methods in Systems Biology* (pp. 141–149). New York: Association for Computing Machinery.

Burnham, K. and Anderson, D.R. (2002): *Model Selection and Multimodal Inference: A Practical Information-Theoretic Approach*, 2nd edn. New York: Springer.

Burt W. (1943) Territoriality and home range concepts as applied to mammals. *Journal of Mammalogy*, 24(3): 346–352.

Butler, D. (2006) Virtual globes: The web-wide world. *Nature*, 439(7078): 776–778.

Butler, P. (2010) Visualizing friendships. https://www.facebook.com/notes/facebook-engineering/visualizing-friendships/469716398919 (accessed August 2014).

CACI (2013) *The Acorn User Guide: The Consumer Classification*. London: CACI Limited. http://acorn.caci.co.uk/downloads/Acorn-User-guide.pdf (accessed April 2004).

Calenge, C. (2013) adehabitat: Analysis of habitat selection by animals. Contributions from Mathieu Basille, Stephane Dray and Scott Fortmann-Roe. R project: http://cran.r-project.org/web/packages/adehabitat/ (accessed 24 November 2013).

Caprio, J.M. (1957) Phenology of lilac bloom in Montana. *Science*, 126: 1344–1345.

CartoDB (2014) CartoDB: Create beautiful dynamic data driven maps. http://cartodb.com/.

Cass, S. (2002) Mind games: To beat the competition, video games are getting smarter. *IEEE Spectrum* 39(12): 40–44.

Castle, C.J.E. (2007) Guidelines for assessing pedestrian evacuation software applications. Working Paper 115, Centre for Advanced Spatial Analysis, University College London.

Cavan, G. and Kingston, R. (2012) Development of a climate change vulnerability and risk assessment tool for urban areas. *International Journal of Disaster Resilience in the Built Environment*, 3(3): 253–269.

Cayan, D.R., Kammerdiener, S.A., Dettinger, M.D. Caprio, J.M. and Peterson, D.H. (2001) Changes in the onset of spring in the western United States. *Bulletin of the American Meteorological Society*, 82(3): 399–415.

Chapa, J., Bourgo, R.J., Greene, G.L., Kulkarni, S. and An, G. (2013) Examining the pathogenesis of breast cancer using a novel agent-based model of mammary ductal epithelium dynamics. *PLoS ONE*, 8(5): e64091.

Cheng, S.-F., Lin, L., Du, J., Lau, H.C. and Varakantham, P. (2013) An agent-based simulation approach to experience management in theme parks. In R. Pasupathy, S.-H. Kim, A. Tolk, R. Hill and M.E. Kuhl (eds), *Proceedings of the 2013 Winter Simulation Conference*, Washington, DC (pp. 1527–1538).

Cheng, T., Haworth, J. and Wang, J. (2012) Spatio-temporal autocorrelation of road network data. *Journal of Geographical Systems*, 14(4): 389–413.

Cheng, T., Tanaksaranond, G., Brunsdon, C. and Haworth, J. (2013) Exploratory visualisation of congestion evolutions on urban transport networks. *Transportation Research Part C: Emerging Technologies*, 36: 296–306.

Church, R.L. and Marston, J.R. (2003) Measuring accessibility for people with a disability. *Geographical Analysis*, 35(1): 83–96.

Church, R.L. and Murray, A.T. (2009) *Business Site Selection, Location Analysis, and GIS*. Hoboken, NJ: John Wiley.

Church, R.L. and Revelle, C. (1974) The maximal covering location problem. *Papers of the Regional Science Association*, 32: 101–118.

Church, R.L. and Sorensen, P. (1996) Integrating normative location–allocation models into GIS: Problems and prospects with the *p*-median model. In P. Longley and M. Batty (eds), *Spatial Analysis: Modelling in a GIS Environment* (pp. 167–184). Hoboken, NJ: Wiley.

Ciolek, T.M. (1978) Spatial behaviour in pedestrian areas. *Ekistics*, 45(268): 120–122.

Clærbout, J. (1992) Electronic documents give reproducible research a new meaning. In *Proceedings of the 62nd Annual International Meeting of the Society of Exploration Geophysics* (pp. 601–604).

Clark, P.J. and Evans, F.C. (1954) Distance to nearest neighbor as a measure of spatial relationships in populations. *Ecology*, 35: 445–453.

Clarke, G.P. and Wilson, A.G. (1994) A new geography of performance indicators for urban planning. In C.S. Bertuglia (ed.), *Modelling the City: Performance, Policy and Planning*. London: Routledge.

Clarke G.P., Eyre H. and Guy C. (2002) Deriving indicators of access to food retail provision in British cities: Studies of Cardiff, Leeds and Bradford. *Urban Studies*, 39(11): 2041–2060.

Clarke, K. (2013) Why simulate cities. *GeoJournal*, 79: 129–136.

Clarke, K.C., Hoppen, S. and Gaydos, L.J. (1997) A self-modifying cellular automaton model of historical urbanization in the San Francisco Bay area.
Environment and Planning B, 24(2): 247–261.

Clauset, A., Shalizi, C.R. and Newman, M.E.J. (2009) Power-law distributions in
empirical data. *SIAM Review*, 51: 661–703.

Cliff, A.D. and Ord, J.K. (1981) *Spatial Processes: Models & Applications*. London: Pion.

Cockings S., Martin, D. and Leung, S. (2010) Population 24/7: Building space–
time specific population surface models. In M. Hakley, J. Morley and H.
Rahemtulla (eds), *Proceedings of the GIS Research UK 18th Annual Conference
GISRUK 2010* (pp. 41–48). London: University College London.

Coffey, D.S. (1998) Self-organization, complexity and chaos: The new biology for
medicine. *Nature Medicine*, 4(8): 882–885.

Cohn, J.P. (2008) Citizen science: Can volunteers do real research? *BioScience*,
58(3): 192–197.

Comber, A., Fisher, P., Brunsdon, C. and Khmag, A. (2012) Spatial analysis of
remote sensing image classification accuracy. *Remote Sensing of Environment*,
127: 237–246.

Cooper, C.B., Dickinson, J., Phillips, T. and Bonney, R. (2007) Citizen science as a
tool for conservation in residential ecosystems. *Ecology and Society*, 12(2): 11.

Cooper, L. (1963) Location-allocation problems. *Operations Research*, 11(3): 331–343.

Cooper, L. (1964) Heuristic methods for location-allocation problems. *SIAM
Review*, 6(1): 37–53.

Corti, P., Mather, S.V., Kraft, T.J. and Park, B. (2014) *PostGIS Cookbook*.
Birmingham: Packt Publishing.

Cristelli, M., Batty, M. and Pietronero, L. (2012) There is more than a power law
in Zipf. *Scientific Reports*, 2: 812.

Cromley, E. and McLafferty, S. (2012) *GIS and Public Health*, 2nd edn. New York:
Guilford Press.

Crooks, A.T. (2007) Experimenting with cities: Utilizing agent-based models and
GIS to explore urban dynamics. PhD thesis, University College London.

Crooks, A.T. (2010) Constructing and implementing an agent-based model of
residential segregation through vector GIS. *International Journal of GIS*, 24(5):
661–675.

Crooks, A.T. and Castle, C. (2012) The integration of agent-based modelling and
geographical information for geospatial simulation. In A.J. Heppenstall, A.T.
Crooks, L.M. See and M. Batty (eds), *Agent-Based Models of Geographical
Systems* (pp. 219–252). Dordrecht: Springer.

Crooks, A.T. and Hailegiorgis, A. (2013) Disease modeling within refugee camps:
A multi-agent systems approach. In R. Pasupathy, S.-H. Kim, A. Tolk, R. Hill
and M.E. Kuhl (eds), *Proceedings of the 2013 Winter Simulation Conference*,
Washington, DC (pp. 1697–1706).

Crooks, A.T. and Heppenstall, A. (2012) Introduction to agent-based modelling. In A.J. Heppenstall, A.T. Crooks, L.M. See and M. Batty (eds), *Agent-Based Models of Geographical Systems* (pp. 85–108). Dordrecht: Springer.

Crooks, A.T. and Wise, S. (2013) GIS and agent-based models for humanitarian assistance. *Computers, Environment and Urban Systems*, 41: 100–111.

Crooks, A.T., Castle, C.J.E. and Batty, M. (2008) Key challenges in agent-based modelling for geo-spatial simulation. *Computers, Environment and Urban Systems*, 32(6): 417–430.

Crooks, A., Hudson-Smith, A. and Dearden, J. (2009) Agent Street: An environment for exploring agent-based models in Second Life. *Journal of Artificial Societies and Social Simulation*, 12(4): 10 (http://jasss.soc.surrey.ac.uk/12/4/10.html).

Daamen, W. and Hoogendoorn, S.P. (2003) Experimental research of pedestrian walking behavior. *Transportation Research Record*, 1828: 20–30.

Davies T., Hazelton M. and Marshall J. (2011) sparr: Analyzing spatial relative risk using fixed and adaptive kernel density estimation in R. *Journal of Statistical Software*, 39(1): 1–14.

Davison, A.C. and Hinkley, D.V. (1997) *Bootstrap Methods and Their Application*. Cambridge: Cambridge University Press.

De Smith M., Goodchild M., and Longley P. (2009) *Geospatial Analysis*, 3rd edn. Winchelsea: Winchelsea Press.

Delaney, B. (2000) Visualization in urban planning: They didn't build LA in a day. *IEEE Computer Graphics and Applications*, 20(3): 10–16.

Department of the Environment, Transport and the Regions (2000) National Strategy for Neighbourhood Renewal. Report of Policy Action Team 17: Joining it up Locally. London: DETR. http://spaa.info/wp-content/uploads/2012/09/Report-of-Policy-Action-Team-17.pdf.

Dibble, C. and Feldman, P.G. (2004) The GeoGraph 3D computational laboratory: Network and terrain landscapes for RePast. *Journal of Artificial Societies and Social Simulation*, 7(1).

Dijkstra, E.W. (1959) A note on two problems in connexion with graphs. *Numerische Mathematik*, 1: 269–271.

Dijkstra, J. and Timmermans, H. (2002) Towards a multi-agent model for visualizing simulated user behavior to support the assessment of design performance. *Automation in Construction*, 11(2): 135–145.

Dobson, A. (1990) *An Introduction to Generalized Linear Models*. London: Chapman & Hall.

Dodge, M. and Kitchin, R. (2005) Code and the transduction of space. *Annals of the Association of American Geographers*, 95(1): 162–180

Dodge, S., Laube, P. and Weibel, R. (2012) Movement similarity assessment using symbolic representation of trajectories. *International Journal of Geographic Information Science*, 26: 1563–1588.

Donnelly, F. (2010) Evaluating open source GIS for libraries. *Library Hi Tech*, 28(1): 131–151.

Doxiadis, C. A. (1968) *Ekistics: An Introduction to the Science of Human Settlements*. London: Hutchinson.

Drukker, D.M., Egger, P. and Prucha, I. R. (2013) On two-step estimation of a spatial autoregressive model with autoregressive disturbances and endogenous regressors. *Econometric Reviews*, 32(5–6): 686–733.

Duque, J., Anselin, L. and Rey, S. (2012) The max-*p*-regions problem. *Journal of Regional Science*, 53(3): 397–419.

Economist (2006) Special Report: Living a second life – Virtual online worlds. *The Economist*, 380(8497): 98.

Efron, B. and Tibshirani, R. (1993) *An Introduction to the Bootstrap*. New York: Chapman & Hall.

Egenhofer, M.J. and Franzosa, R.D. (1991) Point-set topological spatial relations. *International Journal of Geographical Information Systems*, 5: 161–174.

El-Shakhs, S. (1972) Development, primacy, and systems of cities. *Journal of Developing Areas*, 7(1): 11–36.

Epstein, J.M. and Axtell, R. (1996) *Growing Artificial Societies: Social Science from the Bottom Up*. Washington, DC: Brookings Institution Press.

Epstein, J.M., Parker, J., Cummings, D. and Hammond, R.A. (2008) Coupled contagion dynamics of fear and disease: Mathematical and computational explorations. *PLoS One*, 3(12): e3955.

European Commission (1996) *First Report on Social and Economic Cohesion*. Luxembourg: Office for Official Publications of the European Communities. http://bit.ly/1n1o6Ud (accessed 29 August 2014).

European Commission (1999) *ESDP – European Spatial Development Perspective*. Luxembourg: Office for Official Publications of the European Communities. http://bit.ly/1ljV5Yx (accessed 29 August 2014).

European Environment Agency (2008) *Impacts of Europe's Changing Climate – 2008 Indicator-Based Assessment*. Brussels: EAA. http://www.eea.europa.eu/publications/eea_report_2008_4.

Everitt, B.S. (2011) *Cluster Analysis*, 5th edn. Chichester: Wiley.

Everitt, B.S., Landau, S. and Leese, M. (2001) *Cluster Analysis*, 4th edn. London: Arnold.

Farhan, B. and Murray, A.T. (2006) Distance decay and coverage in facility location planning. *Annals of Regional Science*, 40(2): 279–295.

Farkas, I., Helbing, D. and Vicsek, T. (2002) Crowd behaves as excitable media during Mexican wave. *Nature*, 419: 131.

Fayyad, U.M., Piatetsky-Shapiro, G., Smyth, P. and Ulthurusamy, R. (eds) (1996) *Advances in Knowledge Discovery and Data Mining*. Cambridge, MA: MIT Press.

Feiner, S., MacIntyre, B., Höllerer, T. and Webster, A. (1997) A touring machine: Prototyping 3D mobile augmented reality systems for exploring the urban environment. *Personal Technologies*, 1(4): 208–217.

Feng, Z. and Flowerdew, R. (1998) Fuzzy geodemographics: A contribution from fuzzy clustering methods. In S. Carver (ed.), *Innovations in GIS 5* (pp. 119–127). London: Taylor & Francis.

Ferber, J. (1999) *Multi-Agent Systems: An Introduction to Distributed Artificial Intelligence*. Harlow: Addison-Wesley.

Ferguson, D. and Stentz, A. (2007) Field D*: An interpolation-based path planner and replanner. In S. Thrun, R. Brooks and H. Durrant-Whyte (eds), *Robotics Research*, Springer Tracts in Advanced Robotics Volume 28 (pp. 239–253). Berlin: Springer.

Ferrari, C. (2009) *The Wrapping Approach for Circular Data Bayesian Modeling*. Bolgogna: Alma Mater Studiorum – Università di Bologna.

Fisher, N.I. (1995) *Statistical Analysis of Circular Data*. Cambridge: Cambridge University Press.

Fisher, N.I. and Lee, A. (1994) Time series analysis of circular data. *Journal of the Royal Statistical Society, Series B*, 56(2): 327–339.

Flowerdew, R. and Goldstein, W. (1989) Geodemographics in practice: Developments in North America. *Environment and Planning A*, 21: 605–616.

Foltête, J.-C. and Piombini, A. (2007) Urban layout, landscape features and pedestrian usage. *Landscape and Urban Planning*, 81(3): 225–234.

Fotheringham, A.S. (1997) Trends in quantitative methods I: Stressing the local. *Progress in Human Geography*, 21: 88–96.

Fotheringham, A.S., Charlton, M. and Brunsdon, C. (1996) The geography of parameter space: An investigation into spatial nonstationarity. *International Journal of Geographic Information Systems*, 10: 605–627.

Fotheringham, A.S., Nakaya, T., Yano, K., Openshaw, S. and Ishikawa, Y. (2001) Hierarchical destination choice and spatial interaction modelling: A simulation experiment, *Environment and Planning A*, 33: 901–920.

Fotheringham, A.S., Brunsdon, C. and Charlton, M. (2002) *Geographically Weighted Regression*. Chichester: Wiley.

Friedkin, N.E. and Johnsen, E.C. (1999) Social influence networks and opinion change. *Advances in Group Processes*, 16: 1–29.

Fujiwara, O., Makjamroen, T. and Gupta, K. (1987) Ambulance deployment analysis: A case study of Bangkok. *European Journal of Operational Research*, 31(1): 9–18.

Gabaix, X. (1999) Zipf's law for cities: An explanation. *Quarterly Journal of Economics*, 114: 739–767.

Gahegan, M. (2009) Visual exploration in geography: Analysis with light. In H.J. Miller and J. Han (eds), *Geographic Data Mining and Knowledge Discovery*, 2nd edn (pp. 291–324). Boca Raton, FL: CRC Press.

Gahegan, M., Wachowicz, M., Harrower, M. and Rhyne, T.-M. (2001) The integration of geographic visualization with knowledge discovery in databases and geocomputation. *Cartography and Geographic Information Systems*, 28: 29–44.

Galler, H.P. (1997) Discrete-time and continuous-time approaches to dynamic microsimulation reconsidered. Technical Paper 13. National Centre for Social and Economic Modelling (NATSEM), University of Canberra.

Gamba, P. and Houshmand, B. (2002) Digital surface models and building extraction: Q comparison of IF-SAR and LIDAR data. *IEEE Transactions on Geoscience and Remote Sensing*, 38(4): 1959–1968.

Gayle, R., Sud, A., Andersen, E., Guy, S.J., Lin, M.C. and Manocha, D. (2009) Interactive navigation of heterogeneous agents using adaptive roadmaps. *IEEE Transactions on Visualization and Computer Graphics*, 15(1): 34–48.

Geertman, S. and Stillwell, J. (eds) (2009) *Planning Support Systems Best Practice and New Methods*. Dordrecht: Springer.

Gentleman, R. and Temple Lang, D. (2004) Statistical analyses and reproducible research. Bioconductor Project Working Papers, Working Paper 2. http://biostats.bepress.com/bioconductor/paper2/ (accessed 1 September 2014).

Gérin-Lajoie, M., Richards, C.L., Fung, J. and McFadyen, B.J. (2008) Characteristics of personal space during obstacle circumvention in physical and virtual environments. *Gait & Posture*, 27(2): 239–247.

Getis, A. and Ord, J.K. (1992) The analysis of spatial association by use of distance statistics. *Geographical Analysis*, 24(3): 189–206.

Ghosh, A. and Craig, C.S. (1984) A location–allocation model for facility planning in a competitive environment. *Geographical Analysis*, 16: 39–56.

Ghosh, A. and Rushton, G. (1987) *Spatial Analysis and Location-Allocation Models*. New York: Van Nostrand Reinhold.

Gibin M., Longley P. and Atkinson P. (2007) Kernel density estimation and percent volume contours in general practice catchment area analysis in urban areas. In A. Winstanley (ed.), *Proceedings of GIS Research UK*, National University of Ireland, Maynooth 11-13 April 2007 (pp. 270–277).

Gillies, S., Butler, H., Lautaportti, K., Junod, F., Lemoine, E., Bronn, J. and Pavlenko, A. (2008) Shapely: Geospatial geometries, predicates, and operations. https://pypi.python.org/pypi/Shapely/1.0.

Gimblett, H.R. (ed.) (2002) *Integrating Geographic Information Systems and Agent-Based Modelling Techniques for Simulating Social and Ecological Processes*. Oxford: Oxford University Press.

Gode, D.K. and Sunder, S. (1993) Allocative efficiency of markets with zero-intelligence traders: Market as a partial substitute for individual rationality. *Journal of Political Economy*, 101: 119–137.

Goffmann, E. (1971) *Relations in Public: Microstudies in the Public Order*. New York: Basic Books.

Goldberg, J., Dietrich, R., Ming Chen, J., Mitwasi, M.G., Valenzuela, T. and Criss, E. (1990) Validating and applying a model for locating emergency medical vehicles in Tuczon, AZ. *European Journal of Operational Research*, 49(3): 308–324.

Goldstein, H. (1986) Multilevel mixed linear model analysis using iterative generalised least squares. *Biometrika*, 73: 43–56.

Goldstein, H. (1987a) Multilevel covariance component models. *Biometrika*, 74: 430–431.

Goldstein, H. (1987b) *Multilevel Models in Educational and Social Research*. London: Griffin.

Golledge, R. and Stimson, R.J. (1997) *Spatial Behavior: A Geographic Perspective*. New York: Guilford Press.

Goodchild M. (1984) ILACS: A location-allocation model for retail site selection. *Journal of Retailing*, 60(1): 84–100.

Goodchild, M.F. (2007) Citizens as sensors: The world of volunteered geography. *GeoJournal*, 69(4): 211–221.

Goodchild, M.F. (2009) Geographic information systems and science: Today and tomorrow. *Annals of GIS*, 15(1): 3–9.

Goodchild, M. (2013) Prospects for a space–time GIS. *Annals of the Association of American Geographers,* 103(5): 1072–1077.

Goodchild, M.F., Guo, H., Annoni, A., Bian, L., de Bie, K., Campbell, F., Craglia, M., Ehlers, M., van Genderen, J. and Jackson, D. (2012) Next-generation Digital Earth. *Proceedings of the National Academy of Sciences of the USA*, 109(28): 11088–11094.

Goovaerts, P. (2006) Geostatistical analysis of disease data: Accounting for spatial support and population density in the isopleth mapping of cancer mortality risk using area-to-point Poisson kriging. *International Journal of Health Geographics*, 5(52): 1–31.

Gould, P.R. and Leinbach, T.R. (1966) Approach to the geographic assignment of hospital services. *Tijdschrift voor Economische en Sociale Geografie*, 57: 203–206.

Graham, M., and Shelton, T. (2013) Geography and the future of big data, big data and the future of geography. *Dialogues in Human Geography*, 3: 255–261.

Graser, A. (2013) *Learning QGIS 2.0*. Birmingham: Packt Publishing.

Gratier, T., Hazzard, E. and Spencer, P. (2015) *OpenLayers 3 Beginner's Guide*. Birmingham: Packt Publishing.

Gray, J., Chaudhuri, S., Bosworth, A., Layman, A., Reichart, D., Venkatrao, M., Pellow, F. and Pirahesh, H. (1997) Data cube: A relational aggregation operator generalizing group-by, cross-tab and sub-totals. *Data Mining and Knowledge Discovery*, 1: 29–53.

Grekousis, G. and Hatzichristos, T. (2012) Comparison of two fuzzy algorithms in geodemographic segmentation analysis: The fuzzy C-means and Gustafson–Kessel methods. *Applied Geography*, 34: 125–136.

Grimm, V. and Railsback, S.F. (2005) *Individual-Based Modeling and Ecology.* Princeton, NJ: Princeton University Press.

Gudmundsson, J. and van Kreveld, M. (2006) Computing longest duration flocks in trajectory data. In *Proceedings, ACM-GIS'06*, 10–11 November, Arlington, VA, USA. New York: Association for Computing Machinery.

Gudmundsson, J., van Kreveld, M. and Speckmann, B. (2007) Efficient detection of patterns in 2D trajectories of moving points. *Geoinformatica*, 11: 195–215.

Gulden, T., Harrison, J.F. and Crooks, A.T. (2011) Modeling cities and displacement through an agent-based spatial interaction model. Paper presented to the Computational Social Science Society of America Conference, Santa Fe, NM, 9–12 October, 2011.

Guo, D. (2003) Coordinating computational and visual approaches for interactive feature selection and multivariate clustering. *Information Visualization*, 2: 232–246.

Guo, D. (2009) Flow mapping and multivariate visualization of large spatial interaction data. *IEEE Transactions on Visualization and Computer Graphics*, 15: 1041–1048.

Guo, D. and Mennis, J. (2009) Spatial data mining and geographic knowledge discovery: An introduction. *Computers, Environment and Urban Systems*, 33: 403–408.

Guo, D., Chen, J., MacEachren, A.M. and Liao, K. (2006) A visualization system for space-time and multivariate patterns. *IEEE Transactions on Visualization and Computer Graphics*, 12: 1461–1474.

Guy, S.J., Chhugani, J., Curtis, S., Dubey, P., Lin, M. and Manocha, D. (2010) PLEdestrians: A least-effort approach to crowd simulation. In Z. Popovic and M. Otaduy (eds), *Eurographics/ACM SIGGRAPH Symposium on Computer Animation*, Madrid. New York: Association for Computer Machinery.

Habicht, A.T. and Braaksma, J.P. (1984) Effective width of pedestrian corridors. *Transportation Engineering*, 110(1): 80–93.

Hakimi, S.L. (1964) Optimal locations of switching centers and the absolute centers and medians of a graph. *Operations Research*, 12(3): 450–459.

Haklay, M., O'Sullivan, D., Thurstain-Goodwin, M. and Schelhorn, T. (2001) 'So go downtown': Simulating pedestrian movement in town centres. *Environment and Planning B*, 28(3): 343–359.

Halden, D., Jones, P. and Wixey, S. (2005) Accessibility Analysis Literature Review. Working Paper 3, Derek Halden Corporation, London. http://home.wmin. ac.uk/transport/download/SAMP_WP3_Accessibility_Modelling.pdf.

Hall, J., Henriques, J., Hickford, A., Nicholls, R., Baruah, P., Birkin, M., Chaudry, D., Curtis, T., Eyre, N., Jones, C., Kilsby, C., Leathard, A., Lorenz, A., Malleson, N., McLeod, F., Powrie, W., Preston, J., Rai, N., Street, R., Stringfellow, A., Thoung, C., Tyler, P., Velykiene, R., Watson, G. and Watson, J. (2014) Methodology for rapid assessment of long-term, cross-sectoral strategies for infrastructure systems. *Journal of Infrastructure Systems*, 20(3).

Han, J. and Kamber, J. (2012) *Data Mining: Concepts and Techniques*, 3rd edn. Amsterdam: Elsevier.

Han, J., Lee, J.-G. and Kamber, M. (2009) An overview of clustering methods in geographic data analysis. In H.J. Miller and J. Han (eds), *Geographic Data Mining and Knowledge Discovery*, 2nd edn (pp. 149–187). Boca Raton, FL: CRC Press.

Hand, E. (2010) Citizen science: People power. *Nature*, 466(7307): 685–687.

Harding, P.J., Amos, M. and Gwynne, S. (2010) Prediction and mitigation of crush conditions in emergency evacuations. In W.W.F. Klingsch, C. Rogsch, A. Schadschneider and M. Schreckenberg (eds), *Pedestrian and Evacuation Dynamics 2008* (pp. 233–246). Berlin: Springer.

Harland K. (2013) Microsimulation Model User Guide (Flexible Modelling Framework). Working Paper, ESRC National Centre for Research Methods, Southampton. http://eprints.ncrm.ac.uk.

Harland, K. and Birkin, M. (2013a) Simulating retail demand at the individual level: Stage 1 demand synthesis. Paper presented to the 4th General Conference of the International Microsimulation Association, Canberra, December.

Harland, K. and Birkin, M. (2013b) Using synthetically generated populations in agent-based models. Paper presented to the Association of American Geographers Conference: Agent-based Models and Geographical System, Los Angeles, April.

Harland, K., Heppenstall, A.J., Smith, D. and Birkin, M. (2012) Creating realistic synthetic populations at varying spatial scales: A comparative critique of population synthesis techniques. *Journal of Artificial Societies and Social Simulation*, 15(1): 1–15.

Harris, B. (1989) Beyond geographic information systems: Computers and the planning professional. *Journal of the American Planning Association*, 55: 85–90.

Harris, B. and Batty, M. (1993) Locational models, geographic information and planning support systems. *Journal of Planning Education and Research*, 12: 184–198.

Harris R., and Johnston, R. (2008) Primary schools, markets and choice: Studying polarization and the core catchment areas of schools. *Applied Spatial Analysis*, 1: 59–84.

Harris, R., Sleight, P. and Webber, R. (2005) *Geodemographics, GIS, and Neighbourhood Targeting*. Chichester: Wiley.

Harris, R., Singleton, A., Grose, D., Brunsdon, C., and Longley, P. (2010) Grid-enabling geographically weighted regression: A case study of participation in higher education in England. *Transactions in GIS*, 14(1): 43–61.

Hastie, T.J. and Tibshirani, R.J. (1990) *Generalized Additive Models*. London: Chapman & Hall.

Hastie, T., Tibshirani, R., Friedman, J. and Franklin, J. (2005) The elements of statistical learning: Data mining, inference and prediction. *Mathematical Intelligencer*, 27: 83–85.

Hastie, T., Tibshirani, R. and Friedman, J. (2009) *The Elements of Statistical Learning*, 2nd edn. New York: Springer.

Healey, P. (2005) *Collaborative Planning: Shaping Places in Fragmented Societies*, 2nd edn. Basingstoke: Palgrave Macmillan.

Helbing, D. and Balietti, S. (2011) How to do agent-based simulations in the future: From modeling social mechanisms to emergent phenomena and interactive systems design. Santa Fe Institute, Working Paper 11-06-024.

Helbing, D. and Molnár, P. (1995) Social force model for pedestrian dynamics. *Physical Review E*, 51: 4282–4286.

Helbing, D. and Molnár, P. (1997) Self-organization phenomena in pedestrian crowds. In F. Schweitzer (ed.), *Self-organization of Complex Structures: From Individual to Collective Dynamics* (pp. 569–577). London: Gordon & Breach.

Helbing, D., Farkas, I.J. and Vicsek, T. (2000) Freezing by heating in a driven mesoscopic system. *Physical Review Letters*, 84(6): 1240–1243.

Helbing, D., Buzna, L., Johansson, A. and Werner, T. (2005) Self-organized pedestrian crowd dynamics: Experiments, simulations, and design solutions. *Transportation Science*, 39(1): 1–24.

Henderson, L. F. (1971) The statistics of crowd fluids. *Nature*, 229 (5284): 381–383.

Henein, C.M. and White, T. (2007) Macroscopic effects of microscopic forces between agents in crowd models. *Physica A*, 373: 694–712.

Heppenstall, A.J., Crooks, A.T., See, L.M. and Batty, M. (eds) (2012) *Agent-Based Models of Geographical Systems*. Dordrecht: Springer.

Hey, A.J.G., Tansley, S. and Tolle, K.M. (2009) *The Fourth Paradigm: Data-Intensive Scientific Discovery*. Redmond. WA: Microsoft Research.

Hirschfield, A., Birkin, M., Brunsdon, C., Malleson, N. and Newton, A. (2013) How places influence crime: The impact of surrounding areas on neighbourhood burglary rates in a British city. *Urban Studies*. doi: 10.1177/0042098013492232.

Hodgson, M.J. (1978) Toward more realistic allocation in location–allocation models: An interaction approach. *Environment and Planning A*, 10(11): 1273–1285.

Hodgson M.J. (1981) A location–allocation model maximising consumers' welfare. *Regional Studies*, 15: 493–506.

Hodgson, M.J. (1988) An hierarchical location–allocation model for primary health care delivery in a developing area. *Social Science & Medicine*, 26(1): 153–161.

Holland, J.H. (1995) *Hidden Order: How Adaptation Builds Complexity*. Reading, MA: Addison-Wesley.

Hoogendoorn, S.P. and Bovy, P.H.L. (2000) Gas-kinetic modeling and simulation of pedestrian flows. *Transportation Research Record*, 1710: 28–36.

Hornik, K. and Grün, B. (2014) movMF: An R package for fitting mixtures of von Mises–Fisher distributions. http://cran.r-project.org/web/packages/movMF/vignettes/movMF.pdf (accessed 1 September 2014).

Huff, D. (1964) Defining and estimating a trading area. *Journal of Marketing*, 28(3): 34–38.

Huff, D. and Batsell, R. (1977) Delimiting the areal extent of a market area. *Journal of Marketing Research*, 14(4): 581–585.

Huff D. and Rust R. (1980) Measuring the congruence of market areas. *Journal of Marketing*, 48(1): 68–74.

Hunter, J. D. (2007) Matplotlib: A 2D graphics environment. *Computing in Science & Engineering*, 9(3): 90–95.

Iacovella, S. and Younglood, B. (2013) *GeoServer Beginner's Guide*. Birmingham: Packt Publishing.

Iovine, G., Di Gregorio, S. and Lupiano, V. (2003) Debris-flow susceptibility assessment through cellular automata modeling: An example from 15–16 December 1999 disaster at Cervinara and San Martino Valle Caudina (Campania, southern Italy). *Natural Hazards and Earth System Science*, 3(5): 457–468.

Jackson, C., Best, N. and Richardson, S. (2006) Improving ecological inference using individual-level data. *Statistics in Medicine*, 25: 2136–2159.

Jacobson, J. and Hwang, Z. (2002) Unreal tournament for immersive interactive theater. *Communications of the Association of Computing Machinery*, 45(1): 39–42.

Jacquez, G.M. (1996) A *k* nearest neighbour test for space–time interaction. *Statistics in Medicine*, 15(18): 1935–1949.

Jammalamadaka, S. R. and Sengupta, A. (2001) *Topics in Circular Statistics*. River Edge, NJ: World Scientific.

Janelle, D.G., Klinkenberg, B. and Goodchild, M. (1998) The temporal ordering of urban space and daily activity patterns for population role groups. *Geographical Systems*, 5(1): 117–138.

Janson, C.G. (1980) Factorial social ecology: An attempt at summary and evaluation. *Annual Review of Sociology*, 6: 433–456.

Jona-Lasinio, G., Gelfand, A. and Jona-Lasinio, M. (2012) Spatial analysis of wave direction data using wrapped Gaussian processes. *Annals of Applied Statistics*, 6: 1478–1498.

Jones, S. (1981) *Accessibility Measures: A Literature Review*. Crowthorne, Berkshire: Transport and Road Research Laboratory.

Jordan R. (2012) A social simulation of housing choice and housing policy in the EASEL Regeneration District Leeds UK. Paper presented to the Association of American Geographers Conference, Agent-based Models and Geographical Systems: Applications, New York, February.

Jordan R., Birkin M. and Evans A. (2012) Agent-based modelling of residential mobility, housing choice and regeneration. In A.J. Heppenstall, A.T. Crooks, L.M. See and M. Batty (eds), *Agent-Based Models of Geographical Systems*. Dordrecht: Springer.

Joseph, A. and Phillips, D. (1984) *Accessibility and Utilization: Geographical Perspectives on Health Care Delivery*. London: Harper & Row.

Kahle, D. and Wickham, H. (2013) ggmap: Spatial Visualization with ggplot2. *The R Journal*, 5:144–161. Retrieved from http://vita.had.co.nz/papers/ggmap.html.

Kao, S.-C., Kim, H.K., Liu,C., Cui, X. and Bhaduri, B.L. (2012) Dependence-preserving approach to synthesizing household characteristics. *Transportation Research Record: Journal of the Transportation Research Board*, 2302: 192–200.

Kaufman, L. and Rousseeuw, P.J. (1990) *Finding Groups in Data: An Introduction to Cluster Analysis*. New York: Wiley.

Kavroudakis, D., Ballas, D. and Birkin, M. (2013) Using spatial microsimulation to model social and spatial inequalities in educational attainment. *Applied Spatial Analysis and Policy*, 6: 1–23.

Keim, D., Andrienko, G., Fekete, J.-D., Görg, C., Kohlhammer, J. and Melançon, G. (2008) Visual analytics: Definition, process, and challenges. In A. Kerren, J.T. Stasko, J.-D. Fekete and C. North (eds), *Information Visualization: Human-Centered Issues and Perspectives*, Lecture Notes in Computer Science 4950 (pp. 154–175). Berlin: Springer.

Kelejian, H.H. and Prucha, I.R. (1998a) A generalized moments estimator for the autoregressive parameter in a spatial model. *Journal of Real Estate Finance and Economics*, 17, 99–121.

Kelejian, H.H. and Prucha, I.R. (1998b) A generalized spatial two-stage least squares procedure for estimating a spatial autoregressive model with autoregressive disturbances. *Journal of Real Estate Finance and Economics*, 17(1): 99–121.

Kelejian, H.H. and Prucha, I.R. (2010) Specification and estimation of spatial autoregressive models with autoregressive and heteroskedastic disturbances. *Journal of Econometrics*, 157(1): 53–67.

Kennedy, W. (2012) Modelling human behaviour in agent-based models. In A.J. Heppenstall, A.T. Crooks, L.M. See and M. Batty (eds), *Agent-Based Models of Geographical Systems* (pp. 167–180). Dordrecht: Springer.

Kevrekidis, I.G., Gear, C.W., Hyman, J.M., Kevrekidid, P.G., Runborg, O. and Theodoropoulos, C. (2003) Equation-free, coarse-grained multiscale computation: Enabling microscopic simulators to perform system-level analysis. *Communications in Mathematical Sciences*, 1(4): 715–762.

Kingston, R. (2007) Public participation in local policy decision-making: The role of web-based mapping. *Cartographic Journal*, 44(2): 138–144.

Kingston, R., Carver, S., Evans, A. and Turton, I. (2000) Web-based public participation geographical information systems: An aid to local environmental decision-making. *Computers, Environment and Urban Systems*, 24(2): 109–125.

Kitchin, R. (2013) Big data and human geography: Opportunities, challenges and risks. *Dialogues in Human Geography*, 3: 262–267.

Klosterman, R. (2008) Comment on Drummond and French: Another view of the future of GIS. *Journal of the American Planning Association*, 74(2): 174–176.

Knox, G. (1964) The detection of space–time interactions. *Applied Statistics*, 13: 25–29.

Knuth, D. (1984) Literate programming. *Computer Journal*, 27(2): 97–111.

Koenker, R. (1996) Reproducible econometric research. Technical report, Department of Econometrics, University of Illinois.

Kohonen, T. (2001) *Self-Organizing Maps*. Berlin: Springer.

Kreft, J.U., Booth, G. and Wimpenny, W.T. (1998) BacSim, a simulator for individual based modelling of bacterial colony growth. *Microbiology*, 144(12): 3275–3287.

Krishnamurthy, S. (2002) Cave or community? An empirical examination of 100 mature open source projects. *First Monday*, 7(6) http://firstmonday.org/ojs/index.php/fm/article/view/960881.

Kronholm, K. and Birkeland, K.W. (2005) Integrating spatial patterns into a snow avalanche cellular automata model. *Geophysical Research Letters*, 32(19): L19504.1–L19504.4.

Krugman, P. (1996a) Confronting the mystery of urban hierarchy. *Journal of the Japanese and International Economies*, 10: 399–418.

Krugman, P. (1996b) *The Self-Organizing Economy*. Oxford: Blackwell.

Krugman, P. (2005) Increasing returns and economic geography. *International Library of Critical Writings in Economics*, 188(2): 3–19.

Krygier, J. and Wood, D. (2011) *Making Maps: A Visual Guide to Map Design for GIS*, 2nd edn. New York: Guilford Press.

Kuijpers, B. and Othman, W. (2009) Modeling uncertainty of moving objects on road networks via space–time prisms. *International Journal of Geographical Information Science*, 23(9): 1095–1117.

Kwan, M. and Weber, J. (2003) Individual accessibility revisited: Implications for geographical analysis in the twenty-first century. *Geographical Analysis*, 35(4): 341–353.

Kyriakidis, P. (2004) A geostatistical framework for area-to-point spatial interpolation. *Geographical Analysis*, 36(3): 259–289.

Laird, J.E. (2012) *The Soar Cognitive Architecture*. Cambridge, MA: The MIT Press.

Łatek, M.M., Mussavi Rizi, S.M., Crooks, A.T. and Fraser, M. (2012) Social simulations for border security. In N. Memon and D. Zang (eds), *Proceedings of 2012 European Intelligence and Security Informatics Conference*, Odense, Denmark (pp. 340–345). Los Alamitos, CA: IEEE Computer Society.

Laube, P., Imfeld, S., and Weibel, R. (2005) Discovering relative motion patterns in groups of moving point objects. *International Journal of Geographical Information Science,* 19: 639–668.

Laube, P., Dennis, T., Forer, P. and Walker, M. (2007) Movement beyond the snapshot: Dynamic analysis of geospatial lifelines. *Computers, Environment and Urban Systems*, 31: 481–501.

Lawler, E.L., Lenstra, J.K., Rinnooy Kan, A.H.G. and Shmoys, D.B. (1985) *The Traveling Salesman Problem: A Guided Tour of Combinatorial Optimization.* Chichester: Wiley.

Lee, A. (2010) Circular data. *Wiley Interdisciplinary Reviews: Computational Statistics,* 2(4): 477–486.

Lee, K.H., Choi, M.G., Hong, Q. and Lee, J. (2007) Group behavior from video: A data-driven approach to crowd simulation. In M. Gleicher and D. Thalmann (eds), *2007 ACM SIGGRAPH/Eurographics Symposium on Computer Animation* (pp. 109–118). San Diego, CA: Eurographics Association.

Leisch, F. (2002) Dynamic generation of statistical reports using literate data analysis. In W. Härdle and B. Rönz (eds), *Compstat 2002 – Proceedings in Computational Statistics* (pp. 575–580). Heidelberg: Physica Verlag.

Lenth, R.V. (2012) StatWeave users' manual. http://homepage.stat.uiowa.edu/~rlenth/StatWeave/StatWeave-manual.pdf (accessed 6 October 2014).

Lewicka, M. (2011) Place attachment: How far have we come in the last 40 years? *Journal of Environmental Psychology,* 31: 207–230.

Lewis D. and Longley P. (2012) Patterns of patient registration with primary health care in the UK National Health Service. *Annals of the Association of American Geographers,* 102(5): 1135–1145.

Li, Z., Han, J., Ding, B. and Kays, R. (2012) Mining periodic behaviors of object movements for animal and biological sustainability studies. *Data Mining and Knowledge Discovery,* 24: 355–386.

Lindley, S.J. (2009) *Review of Climate Change Risk and Adaptation Assessment Tools.* Expert paper for the GRaBS project. http://bit.ly/1uqlZh6 (accessed 2 September 2014).

Lindley, S.J., Handley, J.F., McEvoy, D., Peet, E. and Theuray, N. (2007) The role of spatial risk assessment in the context of planning for adaptation in UK urban areas. *Built Environment,* 33(1): 46–69.

Lister, A.M. and the Climate Change Research Group (2011) Natural history collections as sources of long-term datasets. *Trends in Ecology and Evolution,* 26(4): 153–154.

Liu, S. and Zhu, X. (2004) An integrated GIS approach to accessibility analysis. *Transactions in GIS,* 8(1): 45–62.

Lloyd, C.D. (2011) *Local Models for Spatial Analysis,* 2nd edn. Boca Raton, FL: CRC Press.

Loader, C. (1999) *Local Regression and Likelihood.* New York: Springer.

Loewenstein, G.F. and Lerner, J.S. (2003) The role of affect in decision making. In R.J. Davidson, K.R. Scherer and H.H. Goldsmith (eds), *Handbook of Affective Science* (pp. 619–642). Oxford: Oxford University Press.

Logan, B.I. (1985) Evaluating public policy costs in rural development planning: The example of health care in Sierra Leone. *Economic Geography,* 61(2): 144–157.

Long, J.A. and Nelson, T.A. (2013) A review of quantitative methods for movement data. *International Journal of Geographical Information Science*, 27: 292–318.

Longford, N.T. (1993) *Random Coefficient Models*. Oxford: Clarendon Press.

Longley, P.A. (1998) Foundations. In P.A. Longley, S.M. Brooks, R. McDonnell and B. Macmillan (eds), *Geocomputation: A Primer* (pp. 3–15). Chichester: Wiley.

Longley, P.A. (2005) Geographical information systems: A renaissance of geodemographics for public service delivery. *Progress in Human Geography*, 29(1): 57–63.

Longley, P.A. (2007) Some challenges to geodemographic analysis and their wider implications for the practice of GIScience. *Computers, Environment and Urban Systems*, 31(6): 617–622.

Longley, P.A., Brooks, S.M., McDonnell R. and Macmillan B. (eds) (1998) *Geocomputation: A Primer*. Chichester: Wiley.

Longley, P.A., Goodchild, M.F., Maguire, D.J. and Rhind, D.W. (2005) *Geographical Information Systems and Science*, 2nd edn. Chichester: Wiley.

Longley, P.A., Goodchild M.F., Maguire D.J. and Rhind D.W. (2011) *Geographic Information Systems and Science*, 3rd edn. Hoboken, NJ: Wiley.

Lovelace, R. and Cheshire, J. (2014) Introduction to visualising spatial data in R. National Centre for Research Methods Working Papers, 14(03). Retrieved from https://github.com/Robinlovelace/Creating-maps-in-R

Lu, C.-T., Boedihardjo, A.P. and Shekhar, S. (2009) Analysis of spatial data with map cubes: Highway traffic data. In H.J. Miller and J. Han (eds), *Geographic Data Mining and Knowledge Discovery*, 2nd edn (pp. 69–97). Boca Raton, FL: CRC Press.

Luke, S., C. Cioffi-Revilla, L. Panait, K. Sullivan and G. Balan. (2005) Mason: A multiagent simulation environment. *Simulation*, 81(7): 517–527.

Luo, W. and Wang, F. (2003) Measures of spatial accessibility to health care in a GIS environment: Synthesis and a case study in the Chicago region. *Environment and Planning B: Planning and Design*, 30(6): 865–884.

Magliocca, N. (2012) Exploring coupled housing and land market interactions through an economic agent-based model (CHALMS). In A.J. Heppenstall, A.T. Crooks, L.M. See and M. Batty (eds), *Agent-Based Models of Geographical Systems* (pp. 543–569). Dordrecht: Springer.

Maheshwari, A., Sack, J.-R., Shahbaz, K. and Zarrabi-Zadeh, H. (2011) Fréchet distance with speed limits. *Computational Geometry*, 44: 110–120.

Malamud, B.D. and Turcotte, D.L. (2000) Cellular-automata models applied to natural hazards. *Computing in Science & Engineering*, 2(3): 42–51.

Malik, A.A., Crooks, A.T. and Root, H.L. (2013) Can Pakistan have creative cities? An agent based modeling approach with preliminary application to Karachi. Pakistan Strategy Support Program Working Paper 13, International Food Policy Research Institute, Washington, DC.

Malleson N. and Birkin M. H. (2012) Analysis of crime patterns through the integration of an agent-based model and a population microsimulation. *Computers, Environment and Urban Systems*, 36: 551–561.

Malleson, N., Heppenstall, A.J. and See, L.M. (2010) Crime reduction through simulation: An agent-based model of burglary. *Computers, Environment and Urban Systems*, 31(3): 236–250.

Malleta, D.G. and De Pillis, L.G. (2006) A cellular automata model of tumor-immune system interactions. *Journal of Theoretical Biology*, 239(3): 334–350.

Manley, E., Cheng, T., Penn, A. and Emmonds, A. (2014) A framework for simulating large-scale complex urban traffic dynamics through hybrid agent-based modelling. *Computers, Environment and Urban Systems*, 44: 27–36.

Manson, S.M., Sun, S. and Bonsal, D. (2012) Agent-based modeling and complexity. In A.J. Heppenstall, A.T. Crooks, L.M. See and M. Batty (eds), *Agent-Based Models of Geographical Systems* (pp. 125–140). Dordrecht: Springer.

Mardia, K.V. (1975) Statistics of directional data. *Journal of the Royal Statistical Society, Series B*, 37: 349–393.

Mardia, K.V. and Jupp, P.E. (2000) *Directional Statistics*. Chichester: John Wiley.

Masinick, J.P. and Teng, H. (2004) An analysis on the impact of rubbernecking on urban freeway traffic. Center for Transportation Studies at the University of Virginia Research Report No. UVACTS-15-0-62, Charlottesville, VA.

Maslow, A.H. (1943) A theory of human motivation. *Psychological Review*, 50(4): 370–396.

McCaffrey, R.E (2005) Using citizen science in urban bird studies. *Urban Habitats*, 3(1): 70–86.

McCullagh, P. and Nelder, J.A. (1989) *Generalized Linear Models*, 2nd edn. London: Chapman & Hall.

Mei, C.-L., Wang, N. and Zhang, W.-X. (2006): Testing the importance of the explanatory variables in a mixed geographically weighted regression model. *Environment and Planning A*, 38: 587–598.

Mennis, J. and Liu, J (2003) Mining association rules in spatio-temporal data. In *Proceedings of the 7th International Conference on GeoComputation*. http://www.geocomputation.org/2003/Papers/Mennis_Paper.pdf (accessed 2 September 2014).

Messner, S.F., Anselin, L., Baller, R.D., Hawkins, D.F., Deane, G. and Tolnay, S.E. (1999) The spatial patterning of county homicide rates: An application of exploratory spatial data analysis. *Journal of Quantitative Criminology*, 15(4): 423–450.

Miller, H.J. (2010) The data avalanche is here. Shouldn't we be digging? *Journal of Regional Science*, 50, 181–201.

Miller, H.J. and Goodchild, M.F. (2014) Data-driven geography. *GeoJournal,* DOI 10.1007/s/0708-014-9602-6.

Miller, H.J. and Han, J. (eds) (2009) *Geographic Data Mining and Knowledge Discovery*, 2nd edn. Boca Raton, FL: CRC Press.

Miller, J.H. and Page, S.E. (2007) *Complex Adaptive Systems*. Princeton, NJ: Princeton University Press.

Miller-Rushing, A.J., Primack, R.B., Primack, D. and Mukunda, S. (2006) Photographs and herbarium specimens as tools to document phenological changes in response to global warming. *American Journal of Botany*, 93: 1667–1674.

Milligan, G.W. (1996) Clustering validation: Results and implications for applied analyses. In P. Arabie, L.J. Hubert and G. De Soete (eds), *Clustering and Classification* (pp. 341–375). Singapore: World Scientific.

Minsky, M.L. (1967) *Computation: Finite and Infinite Machines*. Englewood Cliffs, NJ: Prentice Hall.

Møller-Jensen, L. and Kofie, R.Y. (2001) Exploiting available data sources: Location/allocation modeling for health service planning in rural Ghana. *Danish Journal of Geography*, 101: 145–153.

Monkhouse, F.J. and Wilkinson, H.R. (1971) *Maps and Diagrams Their Compilation and Construction*, 3rd edn, reprinted with revisions. London: Methuen.

Monmonier, M. (1996) *How to Lie with Maps*, 2nd edn. Chicago: University of Chicago Press.

Morrissey, K., Clarke, G., Ballas, D., Hynes, S. and O'Donoghue, C. (2008) Examining access to GP services in rural Ireland using microsimulation analysis. *Area*, 40: 354–364.

Moseley, M. (1979) *Rural Accessibility*. London: Methuen.

Moussaïd, M., Helbing, D. and Theraulaz, G. (2011) How simple rules determine pedestrian behavior and crowd disasters. *Proceedings of the National Academy of Sciences*, 108(17): 6884–6888.

Myers, J.L., Well, A. and Lorch, R.F. (2010) *Research Design and Statistical Analysis*, 3rd edn. New York: Routledge.

Nagel, K. and Schreckenberg, M. (1992) A cellular automaton model for freeway traffic. *Journal de Physique*, 1(2): 2221–2229.

Nagel, K., Rasmussen, S. and Barrett, C.L. (1997) Network traffic as a self-organized critical phenomenon. In F. Schweitzer (ed.), *Self-organization of Complex Structures: From Individual to Collective Dynamics* (pp. 579–592). Amsterdam: Gordon and Breach Science Publishers.

Nakaya, T., Fotheringham, A.S., Brunsdon, C. and Charlton, M. (2005) Geographically weighted Poisson regression for disease association mapping. *Statistics in Medicine*, 24: 2695–2717.

Nakaya, T., Fotheringham, A.S., Hanaoka, K., Clarke, G., Ballas, D. and Yano, K. (2007) Combining microsimulation and spatial interaction models for retail location analysis. *Journal of Geographical Systems*, 9(4): 345–369.

Nakaya, T., Fotheringham, A.S., Charlton, M. and Brunsdon, C. (2009) Semiparametric geographically weighted generalised linear modelling in GWR4.0. In B.G. Lees and S.W. Laffan (eds), *Proceedings of the 10th International Conference on GeoComputation*. http://www.geocomputation.org/2009/PDF/Nakaya_et_al.pdf (accessed 2 September 2014).

Nanni, M., Kuijpers, B., Körner, C., May, M. and Pedreschi, D. (2008) Spatio-temporal data mining. In F. Giannotti and D. Pedreschi (eds), *Mobility, Data Mining and Privacy* (pp. 267–296). Berlin: Springer.

Nareyek, A. (2002) Intelligent agents for computer games. In T.A. Marsland and I. Frank (eds), *Computers and Games: Second International Conference*, Lecture Notes in Computer Science 2063 (pp. 414–422). Berlin: Springer.

Nemet, G.F. and Bailey, A.J. (2000) Distance and health care utilization among the rural elderly. *Social Science & Medicine*, 50(9): 1197–1208.

Nightingale, F. (1858) *Notes on Matters Affecting the Health, Efficiency, and Hospital Administration of the British Army: Founded Chiefly on the Experience of the Late War*. London: Harrison & Sons.

O'Brien, O. (2014) Rank Clock Visualiser. http://casa.oobrien.com/rankclocks/ (accessed 5 January 2014).

O'Kelly, M.E. (1986) Activity levels at hub facilities in interaction networks. *Geographical Analysis*, 18: 343–356.

O'Sullivan, D. (2004) Complexity science and human geography. *Transactions of the Institute of British Geographers*, 29(3): 282–295.

O'Sullivan, D. (2014) Spatial network analysis. In M.M. Fischer and P. Nijkamp (eds), *Handbook of Regional Science* (pp. 1253–1273).

O'Sullivan, D. and Perry, G.L. (2013) *Spatial Simulation: Exploring Pattern and Process*. Chichester: Wiley.

Office for National Statistics (2000) *National Strategy for Neighbourhood Renewal. Report of Policy Action Team 18: Better Information*. London: The Stationery Office. http://www.neighbourhood.statistics.gov.uk/HTMLDocs/downloads/better_information.pdf.

Office for National Statistics (2013) Population: Overview. http://www.ons.gov.uk/ons/taxonomy/index.html?nscl=Population (accessed 11 January 2014).

Okabe A., and Satoh, T. (2009) Spatial analysis on a network. In A.S. Fotheringham and P. Rogers (eds), *The SAGE Handbook of Spatial Analysis* (pp. 443–464). London: Sage.

Oliphant, T.E. (2007) Python for scientific computing. *Computing in Science & Engineering*, 9(3): 10–20.

Openshaw, S. and Abrahart, R. J. (1996). Geocomputation. In R.J. Abrahart (ed.), *Proceedings of the 1st International Conference on Geocomputation* (pp. 665–666). Leeds: University of Leeds.

Openshaw, S. and Wymer, C. (1995) Classifying and regionalizing census tracts. In S. Openshaw (ed.), *Census User's Handbook* (pp. 239–270). Cambridge: GeoInformation International.

Openshaw, S., Charlton, M.E., Wymer, C. and Craft, A. (1987) A Mark 1 Geographical Analysis Machine for the automated analysis of point data sets. *International Journal of Geographical Information Systems*, 1(4): 335–358.

Orcutt, G. (1957) A new type of socio-economic system. *Review of Economics and Statistics*, 39(2): 116–123.

Ord, J.K. (1975) Estimation methods for models of spatial interaction. *Journal of the American Statistical Association*, 70(349): 120–126.

Papadias, D., Tao, Y., Kalnis, P. and Zhang, J. (2002) Indexing spatio-temporal data warehouses. In R. Agrawal et al. (eds), *Proceedings of the 18th International Conference on Data Engineering* (pp. 166–175). Los Alamitos, CA: IEEE Computer Society Press.

Parker, E.G., and O'Brien, J.F. (2009) Real-time deformation and fracture in a game environment. In R. Tamstorf, D. Fellner and S. Spencer (eds), *Proceedings of the ACM SIGGRAPH/Eurographics Symposium on Computer Animation* (pp. 156–166). New York: Association for Computing Machinery.

Parrish, J.K. and Hamner, W.M. (eds) (1997) *Animal Groups in Three Dimensions: How Species Aggregate*. New York: Cambridge University Press.

Partridge, B. L. (1982) The structure and function of fish schools. *Scientific American*, 246(6): 114–123.

Patel, A., Crooks, A.T. and Koizumi, N. (2012) Slumulation: An agent-based modeling approach to slum formations. *Journal of Artificial Societies and Social Simulation*, 15(4). http://jasss.soc.surrey.ac.uk/15/4/2.html.

Patel, A., Koizumi, N. and Crooks, A.T. (2014) Measuring slum severity in Mumbai and Kolkata: A household-based approach. *Habitat International*, 41: 300–306.

Pelechano, N., Allbeck, J. and Badler, N.I. (2008) *Virtual Crowds: Methods, Simulation, and Control*. San Rafael, CA: Morgan & Claypool.

Pelekis, N., Raffaetà, A., Damiani, M.-L., Vanegot, C., Marketos, G., Frentzos, E., Ntoutsi, I. and Thedoridis, Y. (2008) Towards trajectory data warehouses. In F. Giannotti and D. Pedreschi (eds), *Mobility, Data Mining and Privacy* (pp. 189–211). Berlin: Springer.

Penn, R. (2011) Spring's here: Skylarks overhead, moles in the garden, moths in the bathroom. *Guardian*, 27 March. http://www.guardian.co.uk/environ ment/2011/mar/27/spring-wildlife-black-mountains-wales.

Pérez, F. and Granger, B.E. (2007) IPython: A system for interactive scientific computing. *Computing in Science & Engineering*, 9(3): 21–29.

Petres, C., Pailhas, Y., Patron, P., Petillot, Y., Evans, J. and Lane, D. (2007) Path planning for autonomous underwater vehicles. *IEEE Transactions on Robotics*, 23(2): 331–341.

Pfoser, D. and Jensen, C. S. (1999) Capturing the uncertainty of moving-object representations. In R.H. Güting, D. Papadias and F. Lochovsky (eds), *Advances in Spatial Databases: 6th International Symposium (SSD'99)*, Lecture Notes in Computer Science 1651 (pp. 111–131). Berlin: Springer.

Pheasant, S. and Haslegrave, C.M. (2006) *Bodyspace: Anthropometry, Ergonomics and the Design of Work*, 3rd edn. London: Taylor & Francis.

Pickles, J. (1995) *Ground Truth: The Social Implications of Geographical Information Systems*. New York: Guilford Press.

Pint, B., Crooks, A.T. and Geller, A. (2010) An agent-based model of organized crime: Favelas and the drug trade. Paper presented to 2nd Brazilian Workshop on Social Simulation, São Bernardo do Campo, Brazil 23–28 October, 2010.

Pumain, D. (2012) Multi-agent system modelling for urban systems: The series of SIMPOP models. In A.J. Heppenstall, A.T. Crooks, L.M. See and M. Batty (eds), *Agent-Based Models of Geographical Systems* (pp. 721–738). Dordrecht: Springer.

R Core Team (2013) *R: A Language and Environment for Statistical Computing*. Vienna: R Foundation for Statistical Computing. http://www.R-project.org/.

Raafat, R.M., Chater, N. and Frith, C. (2009) Herding in humans. *Trends in Cognitive Sciences*, 13(10): 420–428.

Rahman, S. and Smith, D.K. (2000) Use of location-allocation models in health service development planning in developing nations. *European Journal of Operational Research*, 123: 437–452.

Ramasubramanian, L. (2008) *Geographic Information Science and Public Participation*. Berlin: Springer-Verlag.

Ramsey, N. (1994) Literate programming simplified. *IEEE Software*, 11(5): 97–105.

Ramsey, P. and Dubovsky, D. (2013) Geospatial software's open future. *GeoInformatics*, 16(4) http://www.geoinformatics.com/blog/in-the-spotlight/geospatial-softwares-open-future.

Raney, B., Çetin, N., Völlmy, A., Vrtic, M., Axhausen, K.W. and Nagel, K. (2003) An agent-based microsimulation model of Swiss travel: First results. *Networks and Spatial Economics*, 3(1): 23–42.

Rao, A.S. and Georgeff, M.P. (1991) Modeling rational agents within a BDI-architecture. In J. Allen, R. Fikes and E. Sandewall (eds), *Principles of Knowledge Representation and Reasoning: Proceedings of the Second International Conference*. San Mateo, CA: Morgan Kaufmann.

Ratti, C., Frenchman, D., Pulselli, R.M. and Williams, S. (2006) Mobile landscapes: Using location data from cell-phones for urban analysis. *Environment and Planning B: Planning and Design*, 33(5): 727–748.

Ravindran, P. (2003) Bayesian analysis of circular data using wrapped distributions. PhD thesis, North Carolina State University.

Reaney, S. (2008) The use of agent based modelling techniques in hydrology: Determining the spatial and temporal origin of channel flow in semi-arid catchments. *Earth Surface Processes and Landforms*, 33(2): 317–327.

Rees, P. (1972) Problems of classifying subareas within cities. In B.J.L. Berry and K.B. Smith (eds), *City Classification Handbook: Methods and Applications* (pp. 265–330). New York: Wiley-Interscience.

Reilly, W. (1931) *The Law of Retail Gravitation*. New York: Pilsbury.

Rey, S. (2001) Spatial empirics for economic growth and convergence. *Geographical Analysis*, 33(3): 195–214.

Rey, S.J. (2004) Spatial analysis of regional income inequality. In M. Goodchild and D. Janelle (eds), *Spatially Integrated Social Science: Examples in Best Practice* (pp. 280–299). Oxford: Oxford University Press.

Rey, S.J. (2014) Fast algorithms for a space-time concordance measure. *Computational Statistics*, 29: 799–811.

Rey, S.J. and Anselin, L. (2007) PySAL: A Python library of spatial analytical methods. *Review of Regional Studies*, 37: 5–27.

Rey, S.J. and Janikas, M.V. (2006) STARS: Space–time analysis of regional systems. *Geographical Analysis*, 38(1): 67–86.

Rey, S.J. and Smith, R.J. (2013) A spatial decomposition of the Gini coefficient. *Letters in Spatial and Resource Sciences*, 6: 55–70.

Rey, S.J., Murray, A.T. and Anselin, L. (2011) Visualizing regional income distribution dynamics. *Letters in Spatial and Resource Sciences*, 4(1): 81–90.

Rey, S.J., Mack, E. and Koschinsky, J. (2012) Exploratory space–time analysis of burglary patterns. *Journal of Quantitative Criminology*, 28: 509–531.

Reynolds, C.W. (1999) Steering behaviors for autonomous characters. In *Proceedings of the Game Developers Conference, 1999* (pp. 763–782). San Jose, CA: Miller Freeman Game Group.

Rivest, S., Bédard, Y., Proulx, M.-J., Nadeau, M., Hubert, F. and Pastor, J. (2005) SOLAP technology: Merging business intelligence with geospatial technology for interactive spatio-temporal exploration and analysis of data. *ISPRS Journal of Photogrammetry and Remote Sensing*, 60: 17–33.

Robbirt, K.M., Davy, A.J, Hutchings, M.J. and Roberts, D.L. (2010) Validation of biological collections as a source of phenological data for use in climate change studies: A case study with the orchid *Ophrys sphegodes*. *Journal of Ecology*, 99(1): 235–241.

Robinson, D.T. and Brown, D. (2009) Evaluating the effects of land-use development policies on ex-urban forest cover: An integrated agent-based GIS approach. *International Journal of Geographical Information Science*, 23(9): 1211–1232.

Robinson, D.T., Brown, D., Parker, D.C., Schreinemachers, P., Janssen, M.A., Huigen, M., Wittmer, H., Gotts, N., Promburom, P., Irwin, E., Berger, T., Gatzweiler, F. and Barnaud, C. (2007) Comparison of empirical methods for building agent-based models in land use science. *Journal of Land Use Science*, 2(1): 31–55.

Rosero-Bixby, L. (2004) Spatial access to health care in Costa Rica and its equity: A GIS-based study. *Social Science & Medicine*, 58(7): 1271–1284.

Ross, N.A., Rosenberg, M.W. and Pross, D.C. (1994) Siting a women's health facility: A location-allocation study of breast screening services in Eastern Ontario. *Canadian Geographer*, 38(2): 150–161.

Rothman, D.H. and Keller, J.M. (1988) Immiscible cellular-automaton fluids. *Journal of Statistical Physics*, 52(3–4): 1119–1127.

Roy, J.R. and Thill, J.-C. (2004) Spatial interaction modelling. *Papers in Regional Science*, 83(1): 339–361.

Schelling, T.C. (1971) Dynamic models of segregation. *Journal of Mathematical Sociology*, 1(1): 143–186.

Schildt, H. (2002) *Java 2: The Complete Reference*, 5th edn. Berkeley, CA: McGraw-Hill/Osborne.

Schmidt, B. (2000) *The Modelling of Human Behaviour*. San Diego, CA: SCS Publications.

Schmidt, B. (2002) The modelling of human behaviour: The PECS reference model. In *Simulation in Industry: 14th European Simulation Symposium*. Society for Modeling and Simulation International.

Schmidt, C.R., Rey, S.J. and Skupin, A. (2011) Effects of irregular topology in spherical self-organizing maps. *International Regional Science Review*, 34(2): 215–229.

Schmidt, S.K. and Griffin, W.A. (2007) The signals of play: An ABM of affective signatures in children's playgroups. In T. Terano, S. Takahashi, D. Sallach and J. Rouchier (eds), *Advancing Social Simulation: The First World Congress* (pp. 283–294). New York: Springer.

Schroeder, M. (1991) *Fractals, Chaos, Power Laws: Minutes from an Infinite Paradise*. New York: Dover.

Schwartz, M.D. (1994) Monitoring global change with phenology – the case of the spring green wave. *International Journal of Biometeorology*, 38(1): 18–22.

Schwartz, M.D. (1997) Phenology of seasonal climates. In H. Lieth and M.D. Schwartz (eds), *Spring Index Models: An Approach to Connection Satellite and Surface Phenology* (pp. 23–38). Leiden: Backhuys.

Schwartz, M.D. (1998) Green-wave phenology. *Nature*, 394(6696): 839–840.

Schwartz, M.D. and Caprio, J.M. (2003) North American first leaf and first bloom lilac phenology data. IGBP PAGES/World Data Center for Paleoclimatology Data; Contribution Series # 2003-078; NOAA/NGDC Paleoclimatology Program, Boulder, CO.

Schwartz, M.D. and Reiter, B.E. (2000) Changes in North American spring. *International Journal of Climatology*, 20(8): 929–932.

Schweitzer, F. (1997) Active Brownian particles: Artificial agents in physics. In T. Pöschel and L. Schimansky-Geier (eds), *Stochastic Dynamics* (pp. 358–371). Berlin: Springer.

Sembolini, F., Assfalg, J., Armeni, S., Gianassi, R. and Marsoni, F. (2004) CityDev, an interactive multi-agents urban model on the web. *Computers, Environment and Urban Systems*, 28(1/2): 45–65.

Senn, S. (2001) Two cheers for P-values? *Journal of Epidemiology and Biostatistics*, 6: 193–204.

Shaw, D.L., McCombs, M., Weaver, D.H. and Hamm, B.J. (1999) Individuals, groups, and agenda melding: A theory of social dissonance. *International Journal of Public Opinion Research*, 11(1): 2–24.

Shekhar, S., Lu, C.T., Tan, X., Chawla, S. and Vatsavai, R. (2001) Map cube: A visualization tool for spatial data warehouses. In H.J. Miller and J. Han (eds), *Geographic Data Mining and Knowledge Discovery*, 2nd edn (pp. 74–109). Boca Raton, FL: CRC Press.

Sherman, G. (2008) *Desktop GIS: Mapping the Planet with Open Source Tools.* Raleigh, NC: Pragmatic Bookshelf.

Shevky, E. and Bell, W. (1955) *Social Area Analysis.* Stanford: Stanford University Press.

Silverman, B. (1986) *Density Estimation for Statistics and Data Analysis.* London: Chapman & Hall.

Simon, H.A. (1955) On a class of skew distribution functions. *Biometrika*, 42: 425–440.

Simon, H.A. (1956) Rational choice and the structure of the environment. *Psychological Review*, 63:129–138.

Simpson, E.H. (1951) The interpretation of interaction in contingency tables. *Journal of the Royal Statistical Society, Series B*, 13(2): 238–241.

Singleton, A.D. (2010) The geodemographics of educational progression and their implications for widening participation in higher education. *Environment and Planning A*, 42(11): 2560–2580.

Singleton, A. (2014a) A GIS approach to modelling CO_2 emissions associated with the pupil-school commute. *International Journal of Geographical Information Science*, 28(2): 256–273.

Singleton, A. (2014b) Why geographers should learn to code. *Geographical Magazine*, 4 January: 77.

Singleton, A.D. and Longley, P.A. (2009) Geodemographics, visualisation, and social networks in applied geography. *Applied Geography*, 29(3): 289–298.

Singleton, A.D. and Spielman, S.E. (2014) The past, present and future of geodemographic research in the United States and United Kingdom. *Professional Geographer*, 66(4): 558–567.

Singleton, A.D., Wilson, A.G. and O'Brien, O. (2010) Geodemographics and spatial interaction: An integrated model for higher education. *Journal of Geographical Systems*, 14(2): 223–241.

Singleton, A., Longley, P., Allen, R. and O'Brien, O. (2011) Estimating secondary school catchment areas and the spatial equity of access. *Computers, Environment and Urban Systems*, 35(3): 241–249.

Skupin, A. and Hagelman, R. (2005) Visualizing demographic trajectories with self organizing maps. *GeoInformatica*, 9(2): 159–179.

Sleight, P. (1997) *Targeting Customers: How to Use Geodemographic and Lifestyle Data in Your Business.* Henley-on-Thames: NTC Publications.

Smirnov, O. and Anselin, L. (2001) Fast maximum likelihood estimation of very large spatial autoregressive models: A characteristic polynomial approach. *Computational Statistics & Data Analysis*, 35(3): 301–319.

Smith, D., Clarke, G.P., Ransley, J. and Cade, J. (2006) Food access and health: A microsimulation framework for analysis. *Studies in Regional Science*, 35(4): 909–927.

Smith, D.M., Clarke, G.P. and Harland K. (2009) Improving the synthetic data generation process in spatial microsimulation models. *Environment and Planning A*, 41: 1251–1268.

Sobel, R.S. and Lillith, N. (1975) Determinants of nonstationary personal space invasion. *Journal of Social Psychology*, 97(1): 39–45.

Sofianopoulou, E., Rushton, S., Rubin, G. and Pless-Mulloli, T. (2012) Defining GP practice areas based on true service utilisation. *Health & Place*, 18(6): 1248–1254.

Song, Y. and Miller, H.J. (2012) Exploring traffic flow databases using spacetime plots and data cubes. *Transportation*, 39: 215–234.

Spielauer, M. (2009a) What is dynamic social science microsimulation? Statistics Canada. http://www.statcan.gc.ca/microsimulation/modgen/new-nouveau-eng.htm (accessed 23 October 2013).

Spielauer, M. (2009b) Microsimulation approaches. Statistics Canada. http://www.statcan.gc.ca/microsimulation/modgen/new-nouveau-eng.htm (accessed 23 October 2013).

Spielman, S.E. and Thill, J.C. (2008) Social area analysis, data mining, and GIS. *Computers, Environment and Urban Systems*, 32: 110–122.

Stanilov, K. (2012) Space in agent-based models. In A.J. Heppenstall, A.T. Crooks, L.M. See and M. Batty (eds), *Agent-Based Models of Geographical Systems* (pp. 253–271). Dordrecht: Springer.

Stanley, H.E. (2000) Non-equilibrium physics: Freezing by heating. *Nature*, 404(6779): 718–719.

Stefanovic, N., Han, J. and Kopersky, K. (2000) Object-based selective materialization for efficient implementation of spatial data cubes. *IEEE Transactions on Knowledge and Data Engineering*, 12: 938–958.

Steiniger, S. and Bocher, E. (2009) An overview on current free and open source desktop GIS developments. *International Journal of Geographical Information Science*, 23(10): 1345–1370.

Steiniger, S. and Hunter, A. (2013) The 2012 Free and Open Source GIS software map – a guide to facilitate research, development and adoption. *Computers, Environment and Urban Systems*, 39: 136–150.

Stephenson, N. (1993) *Snow Crash*. New York: Bantam Books.

Strogatz, S. (2004) The physics of crowds. *Nature*, 428(6981): 367–368.

Sugiyama, Y., Fukui, M., Kikuchi, M., Hasebe, K., Nakayama, A., Nishinari, K., Tadaki, S. and Yukawa, S. (2008) Traffic jams without bottlenecks – experimental evidence for the physical mechanism of the formation of a jam. *New Journal of Physics*, 10: 033001–033008.

Sui, D. (2014) Opportunities and impediments for open GIS. *Transactions in GIS*, 18(1): 1–24.

Sui, D. and Goodchild, M. (2011) The convergence of GIS and social media: Challenges for GIScience. *International Journal of Geographical Information Science*, 25(11): 1737–1748.

Svennerberg, G. (2010) *Beginning Google Maps API 3*. New York: Apress.

Takama, T. and Preston, J. (2008) Forecasting the effects of road user charge by stochastic agent-based modelling. *Transportation Research Part A: Policy and Practice*, 42(4): 738–749.

Tanser, F., Gething, P. and Atkinson, P. (2010) Location–allocation planning. In T. Brown, S. McLafferty and G. Moon (eds), *A Companion to Health and Medical Geography* (pp. 540–566). Oxford: Blackwell.

Tanton, R. and Edwards, K. (2013) Introduction to spatial microsimulation: History, method and applications. In R. Tanton and K. Edwards (eds), *Spatial Microsimulation: A Reference Guide for Users* (pp. 3–8). Dordrecht: Springer.

Teitz, M.B. and Bart, P. (1968) Heuristic methods for estimating the generalized vertex median of a weighted graph. *Operations Research*, 16(5): 955–961.

Tesfatsion, L. (1997) How economists can get a life. In W.B. Arthur, S. Durlaf and D. Lane (eds), *The Economy as an Evolving Complex System II* (pp. 533–564). Reading, MA: Addison-Wesley.

Thomas, J.J. and Cook, K.A. (2004) *Illuminating the Path: The R&D Agenda for Visual Analytics*. Richland, WA: National Visualization and Analytics Center.

Tobler, W. (1970) A computer movie simulating urban growth in the Detroit region. *Economic Geography*, 46: 234–240.

Tobler W. (1979) Smooth pycnophylactic interpolation for geographical regions. *Journal of the American Statistical Association*, 74(367): 519–530.

Tomintz, M.N., Clarke, G.P. and Rigby, J.E. (2008) The geography of smoking in Leeds: Estimating individual smoking rates and the implications for the location for stop smoking services. *Area*, 40(3): 341–353.

Tomintz, M.N., Clarke, G.P. and Rigby, J.E. (2009) Planning the location of stop smoking services at the local level: A geographic analysis. *Journal of Smoking Cessation*, 4(2): 61–73.

Tomintz, M.N., Clarke, G.P., Rigby, J.E. and Green J.M. (2013) Optimizing the location of antenatal classes. *Midwifery*, 29(1): 33–43.

Tong, D. and Murray, A.T. (2009) Maximising coverage of spatial demand for service. *Papers in Regional Science*, 88(1): 85–97.

Toregas, C., Swain, R., Revelle, C. and Bergman, L. (1971) The location of emergency service facilities, *Operations Research*, 19(6): 1363–1373.

Torfs, P. and Brauer, C. (2014) *A (Very) Short Introduction to R*. http://cran.r-project.org/doc/contrib/Torfs+Brauer-Short-R-Intro.pdf.

Torrens, P.M. (2004) Simulating sprawl: A dynamic entity-based approach to modelling North American suburban sprawl using cellular automata and multi-agent systems. PhD thesis, University College London.

Torrens, P.M. (2006) Simulating sprawl. *Annals of the Association of American Geographers*, 96(2): 248–275.

Torrens, P.M. (2009) Process models and next-generation geographic information technology. In *GIS Best Practices: Essays on Geography and GIS* (pp. 63–75). Redlands, CA: ESRI Press.

Torrens, P.M. (2012) Moving agent pedestrians through space and time. *Annals of the Association of American Geographers*, 102(1): 35–66.

Torrens, P.M. (2014) High-resolution space–time processes for agents at the built–human interface of urban earthquakes. *International Journal of Geographical Information Science*, 28(5): 964–986.

Torrens, P.M. and Benenson, I. (2005) Geographic automata systems. *International Journal of Geographical Information Science*, 19(4): 385–412.

Torrens, P.M. and McDaniel, A. (2013) Modeling geographic behavior in riotous crowds. *Annals of the Association of American Geographers*, 103(1): 20–46.

Torrens, P.M. and Nara, A. (2013) Polyspatial agents for multi-scale urban simulation and regional policy analysis. *Regional Science Policy and Practice*, 44(4): 419–445.

Torrens, P.M., Li, X. and Griffin, W.A. (2011) Building agent-based walking models by machine-learning on diverse databases of space–time trajectory samples. *Transactions in Geographic Information Science*, 15(s1): 67–94.

Torrens, P.M., Nara, A., Li, X., Zhu, H., Griffin, W.A. and Brown, S.B. (2012) An extensible simulation environment and movement metrics for testing walking behavior in agent-based models. *Computers, Environment and Urban Systems*, 36(1): 1–17.

Torrens, P.M., Kevrekidis, I., Ghanem, R. and Zou, Y. (2013) Simple urban simulation atop complicated models: Multi-scale equation-free computing of sprawl using geographic automata. *Entropy*, 15(7): 2606–2634.

Tran, M., Hall, J., Hickford, A., Nicholls, R., Alderson, D., Barr, S., Baruah, P., Beavan, R., Birkin, M., Blainey, S., Byers, E., Chaudry, M., Curtis, T., Ebrahimy, R., Eyre, N., Hiteva, R., Jenkins, N., Jones, C., Kilsby, C., Leathard, A., Manning, L., Otto, A., Oughton, E., Powrie, W., Preston, J., Qadrdan, M., Thoung, C., Tyler, P., Watson, J., Watson, G. and Zuo, C. (2014) *National Infrastructure Assessment: Analysis of Options for Infrastructure Provision in Great Britain, Interim Results*. Oxford: Environmental Change Institute.

Treuille, A., Cooper, S. and Popović, Z. (2006) Continuum crowds. *ACM Transactions on Graphics*, 25(3): 1160–1168.

Tryon, R.C. (1955) *Identification of Social Areas by Cluster Analysis: A General Method with an Application to the San Francisco Bay Area*. Berkeley: University of California Press.

Tsou M. and Smith J. (2011) Free and open source software for GIS education. GeoTech Center white paper. http://www.grossmont.edu/judd.curran/OpenSource_GIS.pdf.

Tuan, Y.-F. (1974) *Topophilia: A Study of Environmental Perception, Attitudes, and Values*. New York: Columbia University Press.

Turing, A.M. (1936) On computable numbers, with an application to the Entscheidungsproblem. *Proceedings of the London Mathematical Society, Series 2*, 42: 230–265.

Turing, A.M. (1950) Computing machinery and intelligence. *Mind*, 49: 433–460.

Turton, I. and Openshaw, S. (1998) High-performance computing and geography: Developments, issues, and case studies. *Environment and Planning A*, 30(10): 1839–1856.

Twigg, L., Moon, G. and Jones, K. (2000) Predicting small-area health-related behaviour: A comparison of smoking and drinking indicators. *Social Science & Medicine*, 50(7–8): 1109–1120.

UK H2 Mobility (2014) *Hydrogen: Fuelling Cleaner Motoring*. http://www.ukh2mobility.co.uk/.

United Nations Human Settlements Programme (2003) *The Challenge of Slums: Global Report on Human Settlements*. London: Earthscan.

Urban, C. (2000) PECS: A reference model for the simulation of multi-agent systems. In R. Suleiman, K.G. Troitzsch and N. Gilbert (eds), *Tools and Techniques for Social Science Simulation* (pp. 83–114). Heidelberg: Physica-Verlag.

Uricchio, W. (2011) The algorithmic turn: Photosynth, augmented reality and the changing implications of the image. *Visual Studies*, 26(1): 25–35.

USA National Phenology Network (2011) History of lilac and honeysuckle phenological observations in the USA. http://www.usanpn.org/?q=node/36.

van den Berg, J., Patil, S., Sewall, J., Manocha, D. and Lin, M. (2008) Interactive navigation of multiple agents in crowded environments. In E. Haines and M. McGuire (eds), *Proceedings of the 2008 Symposium on Interactive 3D Graphics and Games* (pp. 139–147). New York: Association for Computing Machinery.

Van Imhoff, E. and Post, W. (1998) Microsimulation methods for population projection. *Population: An English Selection*, 10(1): 97–138.

van Kempen, R. (2002) The academic formulations: explanations for the partitioned city. In P. Marcuse and R. van Kempen (eds), *Of States and Cities: The Partitioning of Urban Space* (pp. 35–56). Oxford: Oxford University Press.

van Oort, P.A.J., Zhang, T., de Vries, M.E., Heinemann, A.B. and Meinke, H. (2011) Correlation between temperature and phenology prediction error in rice (*Oryza sativa* L.). *Agricultural and Forest Meteorology*, 151(12): 1545–1555.

Venables, W.N., Smith, D.M. and R Development Core Team (2013) *An Introduction to R*. http://cran.ma.imperial.ac.uk/doc/manuals/r-devel/R-intro.pdf.

Verhein, F. and Chawla, S. (2008) Mining spatio-temporal patterns in object mobility databases. *Data Mining and Knowledge Discovery*, 16: 5–38.

Vickers, D. and Rees, P. (2007) Creating the UK national statistics 2001 output area classification. *Journal of the Royal Statistical Society, Series A*, 170(2): 379–403.

Villaverde, R. (2007) Methods to assess the seismic collapse capacity of building structures: State of the art. *Journal of Structural Engineering*, 133(1): 57–66.

Vincent, O.O. (2009) Exploring spatial growth pattern of informal settlements through agent-based simulation. MS thesis in Geographical Information Management & Applications (GIMA). Utrecht University, Delft University of Technology, Wageningen University and the International Institute for Geo-Information Science and Earth Observation. Wageningen, Netherlands.

Voas, D. and Williamson, P. (2001) The diversity of diversity: A critique of geodemographic classification. *Area*, 33(1): 63–76.

von Neumann, J. (1951) The general and logical theory of automata. In L.A. Jeffress (ed.), *Cerebral Mechanisms in Behavior* (pp. 1–41). New York: Wiley.

Vonk, G., Geertman, S. and Schot, P. (2005) Bottlenecks blocking widespread usage of planning support systems. *Environment and Planning A*, 37: 909–924.

Wagner, C.H. (1982) Simpson's paradox in real life. *American Statistician*, 36(1): 46–48.

Wan, N., Zhan, B., Zou, B. and Chow, E. (2012) A relative spatial access assessment approach for analyzing potential spatial access to colorectal cancer services in Texas. *Applied Geography*, 32(2): 291–299.

Wang, S. (2010) A cyberGIS framework for the synthesis of cyberinfrastructure, GIS, and spatial analysis. *Annals of the Association of American Geographers*, 100(3): 535–557.

Ware, C., Arsenault, R., Plumlee, M. and Wiley, D. (2006) Visualizing the underwater behavior of humpback whales. *Computer Graphics*, 26(4):14–18.

Webber, R.J. (1975) Liverpool social area study. 1971 Data. PRAG Technical Paper 14. London: Centre for Environmental Studies.

Webber, R.J. (1977) *An Introduction to the National Classification of Wards and Parishes*. London: Centre for Environmental Studies.

Webber, R.J. (1978) Making the most of the census for strategic analysis. *Town Planning Review*, 49(3): 274–284.

Wickham, H. (2009) *ggplot2: Elegant Graphics for Data Analysis.* New York: Springer.

Wickham, H. (2010) A layered grammar of graphics. *Journal of Computational and Graphical Statistics*, 19(1): 3–28.

Wilensky, U. (1997) NetLogo Traffic Basic model. Center for Connected Learning and Computer-Based Modeling, Northwestern University, Evanston, IL. http://ccl.northwestern.edu/netlogo/models/TrafficBasic.

Wilensky, U. (1999) NetLogo. Center for Connected Learning and Computer-Based Modeling, Northwestern University, Evanston, IL. http://ccl.northwestern.edu/netlogo/.

Wilkinson, L. (2005) *The Grammar of Graphics*, 2nd edn. New York: Springer.

Willekens, F. (1983) Specification and calibration of spatial interaction model: A contingency-table perspective and an application to intra-urban migration in Rotterdam. *Tijdschrift voor economische en sociale geografie*, 74(4): 239–252.

Wilson, A.G. (1974) *Urban and Regional Models in Geography and Planning*. London: Wiley.

Wilson, A.G. (2000) *Complex Spatial Systems: The Modelling Foundations of Urban and Regional Analysis*. New York: Pearson Education.

Wilson, M. and Graham, M. (2013) Neogeography and volunteered geographic information: A conversation with Michael Goodchild and Andrew Turner. *Environment and Planning A*, 45: 3–9.

Winter, D.A. (2009) *Biomechanics and Motor Control of Human Movement*. Hoboken, NJ: Wiley.

Wise, S. and Crooks, A.T. (2012) Agent based modelling and GIS for community resource management: Acequia-based agriculture. *Computers, Environment and Urban Systems*, 36(6): 562–572.

Worboys, M.F. and Duckham, M. (2004) *GIS: A Computing Perspective*, 2nd edn. Boca Raton, FL: CRC Press.

Worton, B. (1989) Kernel methods for estimating the utilization distribution in home-range studies. *Ecology*, 70(1): 164–168.

Wu, B.M., Birkin, M.H. and Rees, P.H. (2008) A spatial microsimulation model with student agents. *Computers, Environment and Urban Systems*, 32: 440–453.

Wu, B.M., Birkin, M.H. and Rees, P.H. (2011) A dynamic microsimulation model with agent elements for spatial demographic forecasting. *Social Science Computing Review*, 29(1): 145–160.

Xie, Y., Batty, M. and Zhao, K. (2007) Simulating emergent urban form: Desakota in China. *Annals of the Association of American Geographers*, 97(3): 477–495.

Yano, K., Nakaya, T. and Ishikawa, Y. (2000) An analysis of inter-municipal migration flows in Japan using GIS and spatial interaction models. *Geographical Reports of Japan*, 73B: 165–177.

Yu, H. and Shaw, S. (2008) Exploring potential human activities in physical and virtual spaces: A spatio-temporal GIS approach. *International Journal of Geographical Information Science*, 22(4): 409–430.

Yuan, Y. and Raubal, M. (2014) Measuring similarity of mobile phone user trajectories: A spatio-temporal edit distance method. *International Journal of Geographical Information Science*, 28: 496–520.

Zhang, Q. (2009) Simulated annealing and crowd dynamics: Approaches for intelligent control. In H. Wang, Y. Shen, T. Huang and Z. Zeng (eds), *The Sixth*

International Symposium on Neural Networks, Advances in Intelligent and Soft Computing 56 (pp. 501–506). Berlin: Springer.

Zipf, G.K. (1965) *Human Behavior and the Principle of Least Effort*. New York: Hafner. First published by Addison-Wesley in 1949.

Zlatanova, S., Rahman, A.A. and Pilouk, M. (2002) Trends in 3D GIS development. *Journal of Geospatial Engineering*, 4(2): 71–80.

Zou, Y., Torrens, P.M., Ghanem, R. and Kevrekidis, I.G. (2012) Accelerating agent-based computation of complex urban systems. *International Journal of Geographical Information Science*, 26(10): 1917–1937.

Zuo C., Birkin M., Clarke G., McEvoy F. and Bloodworth A. (2013) Modelling the transportation of primary aggregates in England and Wales: Exploring initiatives to reduce CO_2 emissions. *Land Use Policy*, 34: 112–124.

Zyda, M. (2005) From visual simulation to virtual reality to games. *Computer*, 38(9): 25–33.

INDEX

Note: Page numbers in **bold** indicate a more comprehensive coverage of the topic.
Page numbers in *italic* refer to figures and tables.

accessibility, spatial flow and 221–2
agent-based modelling (ABM) **63–77**
  background and key aspects 63–5, 77
    complexity and emergent phenomena 65–7
    integration of GIS with ABM 67–70
  comparison/links with microsimulation 82–4,
      89, 95–6
  review of applications 70–1
    border control 73–5
    pedestrian evacuation/movement 71–3
    traffic congestion 66–7, 73
    urban growth and movement 68–70
  see also three-dimensional agent-based models
association rules 105–6
attribute space 152–3
automata-based models 25
  see also polyspatial automata

bandwidth
  kernel density estimation 173–6
  selection for GWGLM models 208–9
best matching unit 159
Big Data 322–3
border control model 73–5
Brisbane commuting study 125–32
  bus commuter modelling 125–32
buildings
  exits and egress 35
  fracture and collapse 36–8
  skyscrapers 42, 54–6
  streetscapes 29–30, *31, 38*

Cauchy distribution, 116, *118*
census data
  geodemographics 139–40
  Self-Organizing Maps 162–7
choropleth mapping 238, 243–6, 258, *259*

circular statistics **110–34**
  advantages and uses 110–11, 132–3
  descriptive statistics 111–12
  kernel smoothing/non-parametric method 121
  mixture models/semi-parametric method
      118–21
  p-values and testing 121–5
  bus commuter application 125–32
cities
  size distribution of 42, 43–7
  study of dynamics 52–6
  urban growth models 68–9, 76
  urban streetscapes 29–30, *31, 38*
citizen science 267–8
climate change 304
cloud computing 323
cluster analysis 104, 142
    clustering algorithms 153, 236
    for geodemographic classification 143–6
    primer on 154–7
    and PySAL 236
    Self-Organizing Maps 167
cognitive architectures 65
collective motion patterns 107
computational geometry 236
crossed-effects models 277
crowd-sourced information **267–80**
    citizen science 267–9
    phenology 269–79
crowds, modelling 30, 31–6, *38*
    pedestrian evacuation models 71–3

daily mobility models 88–9
data cubes 101–3
data mining 98, 322
    spatio-temporal data 99–100
    see also geodemographic analysis, SOM

data warehousing 100–1
decentralised energy production 313–17
desktop GIS 284–5
directional relations 99
distributions
  circular 113–18
  normal 40, 41, 43, 113
  Poisson 202, 204–5, 207, 209
dynamic microsimulation 79
DYNAMOD model 92

earthquakes 36–8
ecological data
  phenology 269, 279–80
  see also environmental data
electrical power networks 313–17
emergency services 188–9
energy
  consumption 314
  demand projection 86–7
environmental change
  climate change within EU 304–12
  lilac blossoming study 269–78
evacuation models 71–3
  room-egress 35
  see also crowds
extraction, transformation and load (ETL)
    functions 100–2

Fisher–von Mises distribution 113
flow, spatial see spatial interaction
Fréchet distance 104

geodemographic analysis **137–51**
  background and history 137–9
  classification systems 139–41
    cluster analysis 143–6
    scale, variables and evaluation 141–2
geographically weighted generalised linear
    modelling (GWGLM) **201–20**
  background
    regression analysis (GWR) 201–2
    S-GWGLM semi-parametric models 202–3
  GWGLM and S-GWGLM methodology 203–4
    automated model building 210–11
    bandwidth selection 208–9
    estimating coefficients 206–7
    geographical variability test coefficients
      209–10
    specification/regression model 204–6

geographically weighted generalised linear
    modelling (GWGLM) cont.
  tutorial/GWR4 application 203, 211–13
    automated variable selection 210–11
    considerations 211–13
    using GWR4 213–20
GeoServer 287
Geospatial Data Abstraction Library (GDAL) 5
Geoweb 287–90
ggplot2 package 4–5, 6–7
  see also R software
GIS software 3, 19, **281–300**, 320–1
  described 281–4, 298–300
  general aspects of
    linking with agent-based models 6–70
    location-allocation models 187
    reproducible research 262–3
    role in planning support systems 302–3, 304,
      307, 317
  open source packages 282–4, 290
    desktop GIS 284–5
    Geoweb client libraries 287–90
    spatial databases 286–7
    web map servers 287
GISTools 258
  see also R software
global properties 99
Google Maps API 14, 287, 307, 309, *310–11*, 313
GP surgeries
  location-allocation analysis 188
  service area analysis 182–3
  spatial flows 224, 225–31
GRaBs project see under online geocomputational
    mapping
graphical user interface 3, 91
GRASS 281, 284–8
gvSIG projects 285
GWGLM models see geographically weighted
    generalised linear modelling
GWR4 application 201, 205, 210–11
  demonstration tutorial 213–19
  place attachment analysis 211–13

Hausdorff distance 104, *105*
health-care services
  location-allocation 188–9
  service area analysis 180–2
  spatial flow 224, 225–31
home range estimation 171
hospital trust service areas 180–2

housing decisions 84–6
human agents 24
    polyspatial modelling 26–9
    three-dimensional 31–6
human behaviour, modelling 64–5
    see also human agents

income, distribution of *88*
interactions, three-dimensional aspects of 24
interval-based properties 99

kernel density estimation (KDE) 121, **172–8**
    adaptive KDE 176–7
    bandwidth 173–6
    cell size 173
    density surfaces 172–3
    Parzen's window 121
    percent volume contour 177, 179
    zonal data and network KDE 177–9
Kohonen Self-Organizing Map algorithm 154
    see also self-organizing maps (SOM)
kriging technique 178

LaTeX 257, *258*, 262
Leaflet 288, 289–90
Local Indices of Spatial Autocorrelation (LISA)
    237, 249–52
literate programming 257–8
location-allocation models **185–96**
    calculating distance travelled 187
    combination with GIS 187
    model types/formulation 185–7
    required input data 188
    studies and models 188–9
        police stations in Kuwait 194–6
        stop-smoking services 189–93

map cube 103
map projection 7, 14–15
Mapnik 286
mapping tools 281, 318
    Google Maps API 307, 309, *310–11*, 313
    public participation in *see* online
        geocomputational mapping
MapServer 287
Markov chains 246–52
MASON ABM toolkit 93
MATLAB 111, 112, 113, 321
medical services *see* GP surgeries; health care
meta-modelling 25

microsimulation modelling **78–96**
    basic principles 79–80
        spatial *versus* non-spatial models 81–2
        static *versus* dynamic microsimulation
            79–80
        transition and event-based approaches 80–1,
            92–3
    design questions 95
    dynamic microsimulation models 84
        daily mobility 88–9
        decision support/social housing 84–5
        strategic planning/infrastructure 86–8
    links to agent-based approaches 82–4, 89, 95–6
    technical architecture considerations
        abstraction level 89–91
        actions and event handling 91–2
        time handling 92–3
moving objects, properties of 99
multi-modelling 25
MySQL database 286

NASA 'Blue Marble' image 18
Natural Earth website 5
neogeography 324
network KDE 177–9
normal distribution 40, 41, 43, 113
    wrapped normal 116–17

OGC organisation 283–4
online mapping tools **301–17**
    public participation 301–2
        and governance strategies 303
        planning support systems 302–3
        regulatory *vs* discretionary planning 304
        web-based technologies 317–18
    study of climate change/GRaBS project 304–7
        assessment tool development 307–8
        data types 308
        user needs and requirements analysis 307–10
        workshops and user feedback 311–13
    study of decentralised energy production
        313–17
open data 321
open practices 320–1
open-source software 381–2
    supporting organisations 383–4
OpenLayers 287, 288
OpenStreetMap (OSM) 14, 288
Opticks project 285
Ordnance Survey 290–1
OSGeo organisation 283

p-values 123–4
Parzen's window 121
pedestrian evacuation/movement models 71–3
people modelling *see* human agents
percent volume contour (PVC) 177–9
periodic patterns 106–7
phenological data 269, 279–80
  lilac study 269–78
place
  conceptualizations of 323–4
  place attachment 211–13
  sense of space and place 23–4
police stations, location-allocation of 194–6
policy analysis 78, 84–8
  polyspatial modelling 22
  building fracture and collapse 36–8
  polyspatial automata 26–31
  three-dimensional crowds and people 31–6
population models 69–70, 93–4
PostgreSQL database 286
power laws 41, 42, 43–5
power networks, electrical 313–17
probability density function
  and kernel density estimation 121
  von Mises distribution 114, *115*, *116*
  wrapped distributions 117, *118*
probability distributions
  and microsimulation 80
  in spatial analysis 40–3
public participation approaches *see* crowd-sourced
    information; online geocomputational
    mapping
pycnophylactic surface model 178, 258, *260*
python spatial analysis library (PySAL) **233–53**
  background and development 233–4
  key component modules 234, *235*
    clustering 236
    computational geometry 236
    exploratory spatial data analysis 236–7
    spatial dynamics 237
    spatial econometrics 237
    spatial weights 234–6
    supplementary modules 238
  support for end-users 239
    GIS toolkits 241
    graphical user interfaces 239–41
    interactive computing 239
    web services 242–3
  tutorial/spatial dynamics 243
    choropleth maps of homicide 238, 243–6
    Markov chains 246–52

QGIS 281
  tutorial 290–8

R software packages
  circular statistics 111, 117
  ggplot2 4–5, 6–7
  GISTools 258
  pycnophylactic surface 258, *260*
  self-organizing map 154, 159–62
R software, visualisation with **3–20**
  background 3–4, 19–20
    base graphics and ggplot2 package 4–5, 6–7
    command line data input 3, 19
    illustrative case study of shipping map 16–19
    tutorial/world map 5–19
      colour palette and breaks 8–11
      dataset and slot names 5–6
      line width and colour 11–12
      map adornments and annotations 12–14
      map reprojection 7, 14–15
      plotting over a base map 14–15
railway
  hubs 55–7
  transport accessibility map 290–6
rank clocks 42, 46–7
  visualizer 57–9
  city dynamics 52–6
  trajectories and morphologies of 47–52
rank-size distributions *see* rank clocks
raster data 5, 69
regression models 201–3
Reilly's law 171
reproducible research **254–63**, 321–2
  factors affecting 254–6
    software barriers 256–7
    literate programming and Sweave package
        257–8
      example geocomputation 258–60
    practicalities and implications 261–3
reprojection 7, 14–15
rose diagrams 114, *115*, *128*
Routino library 290

S-GWGLM models *see* geographically weighted
    generalised
SANET toolkit 179
school service areas 179–80, 182
segregation model 69–70
self-organizing map (SOM) 104, **154–68**
  described 154, 157–8, 167
  SOM creation in R 159–62

self-organizing map (SOM) *cont.*
  training a SOM 158–9
  using SOMs (case study) 162–7
service area analysis **169–84**
  assessing service provision 169–70
  service area analysis methodology 170–2
    service area creation 172–9
  studies/examples of service area analysis 179–83
services
  accessibility and spatial flow 221–2
  location *see* location-allocation models
set-oriented relations 98
SEXTANTE 284, 285
shipping map study 16–19
Simpson's paradox 272, 274
size distributions, city 42, 43–7
skyscrapers 42, 54–6
SLEUTH model 67–8
slots 6
slums 76
smoking services, location-allocation of 189–93
smuggling 73, 75
social area analysis *see* self-organizing map and
    geodemographics
social housing 84–6
social media 324
social policy analysis 78, 84–8
socio-economic profiles 139–40
space-time association rules (STARs) 106
space-time attribute cube 103–4
spatial analysis **40–59**
  power laws 41, 42, 43–5
  rank clocks 42, 46–7
    the Rank Clock Visualiser 57–9
    and studies of city dynamics 52–6
    trajectories and morphologies of 47–52
  spatial probability distributions 40–3
  studies of cities 42–7
    dynamics of US cities 52–6
    London rail hubs 56–7
    skyscrapers 42, 54–6
  tutorial QGIS application 290–300
  see also python spatial analysis; spatio-temporal
    data
spatial data warehouses 101
spatial databases, GIS 285–7
spatial dynamics 237
spatial econometrics 237–8
spatial flow studies *see* spatial interaction
spatial interaction models (SIMs) **221–32**
  accessibility 221–2
  modelling steps and framework 222–4

spatial interaction models (SIMs) *cont.*
    travel impedance 224–5
    types of SIMs 225–6
  study of flows to medical centres 224, 225–31
spatial models
  microsimulation 81–2
  see also three-dimensional agent-based models
spatial relations, types of 98
spatio-temporal data **97–109**
  data and knowledge discovery 97–8, 108–9
  data processing
    approaches and challenges 98–9
    data warehousing 100–1
  spatio-temporal data exploration 101
    collective motion patterns 1–7
    sequence mining 106–7
    space-time cubes 101–4
      associations 105–6
      clustering 104–5
      visualization/visual analytics 107–8
  see also circular statistics
spring, modelling onset of 269–79
stamen design 15
STARs 106
StatWeave package 262–3
streetscape 29–30, *31*
surgeries *see* GP surgeries
Sweave package 257–8, 322

temporal dynamics 323–4
temporal relations, types of 98–9
three-dimensional agent-based models **21–39**
  attributes of 3-D space
    characterization 22–3
    enhancing geographic information 23
    interactions/human agents 24
    meta and multi-modelling 25
    sense of space and place/virtual worlds 23–4
  polyspatial automata
    basic principles 25–9
    building fracture and collapse 36–8
    human agents 27–31
    three-dimensional crowds and people 31–6
Tobler's law 138, 276
topological relations 99
traffic models 67–9, 73
  direction of travel of bus commuters 125–32
transition-based microsimulation 80–1
travel impedance 224–5
travel and transport
  bus commuters 125–32
  daily mobility models 88–9

travel and transport *cont.*
 direction of travel 111
 pedestrian movement 71–3
 rail transport accessibility study 290–6
 spatial flow models *see* spatial interaction
 see also allocation-location models; traffic
  models

urban growth models 68–9, 76
urban streetscape 29–30, *31, 38*

vector data 5, 69
virtual worlds 23–4

visual analytics 107–8
visualisation, spatial data *see* R software
volunteered geographical information 267–8
von Mises distribution 113–17, *119, 120*
voxel-based strategies 30–1

web map servers, GIS 287
web-based *see* online
world map, R software 5–6, 7, *9, 12*
world shipping study 16–19
wrapped distributions 116–17, *118*

Zipf plot 45–6